NITROSOUREAS

CURRENT STATUS AND NEW DEVELOPMENTS

Edited by

Archie W. Prestayko
Bristol Laboratories
Syracuse, New York

Laurence H. Baker
Department of Oncology
Wayne State University
Detroit, Michigan

Stanley T. Crooke
SmithKline and French
Philadelphia, Pennsylvania

Stephen K. Carter
Northern California Cancer Program
Palo Alto, California

Philip S. Schein
Departments of Medicine and Pharmacology
Georgetown University Medical Center
Washington, D.C.

Assisted by

Nancy A. Alder
Philadelphia, Pennsylvania

D1457757

ACADEMIC PRESS

A Subsidiary of Harcourt Brace Jovanovich, Publishers

New York   London   Toronto   Sydney   San Francisco   1981

ACADEMIC PRESS, INC.
111 Fifth Avenue, New York, New York 10003

United Kingdom Edition published by
ACADEMIC PRESS, INC. (LONDON) LTD.
24/28 Oval Road, London NW1 7DX

Library of Congress Cataloging in Publication Data
Main entry under title:

Nitrosoureas: current status and new developments.

   Includes bibliographical references.
   1. Nitrosoureas--Therapeutic use--Congresses.
2. Cancer--Chemotherapy--Congresses. 3. Nitrosoureas
--Testing--Congresses. 4. Nitrosoureas--Metabolism--
Congresses. I. Prestayko, Archie W. [DNLM: 1. Drug
therapy, Combination--Congresses. 2. Neoplasma--Drug
therapy--Congresses. 3. Nitrosoureas compounds--
Therapeutic use--Congresses. 4. Nitrosoureas compounds--
Administration and dosage--Congresses. QZ 267 N731 1980]
RC271.N56N57   616.99'4061   81-413
ISBN 0-12-565060-4

PRINTED IN THE UNITED STATES OF AMERICA

81 82 83 84   9 8 7 6 5 4 3 2 1

CONTENTS

CONTRIBUTORS

Numbers in parentheses indicate the pages on which the authors' contributions begin.

MUHYI AL-SARRAF (301), Department of Oncology, Wayne State University, Grace Hospital, Detroit, Michigan 48201

MICHAEL ALEXANDER (221), Department of Medicine, Stanford University Medical Center, Stanford, California 94305

M. BERGER (27), Institute of Toxicology and Chemotherapy, German Cancer Research Center, Heidelberg, Federal Republic of Germany.

CLARA D. BLOOMFIELD (181), School of Medicine, University of Minnesota, Minneapolis, Minnesota 55455

ROBERT A. BRUNDRETT, JR. (43), Pharmacology Laboratory, Johns Hopkins Oncology Center, Baltimore, Maryland 21205

PAUL A. BUNN, JR. (233), National Cancer Institute, Veterans Administration Medical Oncology Branch, Division of Cancer Treatment, Washington, D.C. 20422

TAUSEEF BUTT (123), Department of Biochemistry and The Vincent T. Lombardi Cancer Center, Georgetown University Medical Center, Washington, D.C. 20007

PATRICK J. BYRNE (367), Department of Medicine, Division of Medical Oncology, Georgetown University Hospital, Washington, D.C. 20007

GEORGE CANELLOS (155), Division of Medicine, Sidney Farber Cancer Institute, Boston, Massachusetts 02130

STEPHEN K. CARTER (221, 411), Northern California Cancer Program, Palo Alto, California 94304

LINDA CHAK (221), Department of Radiology, Stanford University Medical Center, Stanford, California 94305

MARTIN H. COHEN (245), National Cancer Institute, Veterans Administration Medical Oncology Branch, Veterans Administration Hospital, Washington, D.C. 20422

MICHAEL J. COLVIN (43), Pharmacology Laboratory, Johns Hopkins Oncology Center, Baltimore, Maryland 21205

ROBERT L. COMIS (209), Division of Hematology/Oncology, State University of New York, Upstate Medical Center, Syracuse, New York 13210

M. ROBERT COOPER (181), Bowman Gray School of Medicine, Winston-Salem, North Carolina 27103

THOMAS H. CORBETT (9), Chemotherapy Research Department, Southern Research Institute, Birmingham, Alabama 35255

GIBBONS G. CORWELL (313), Norris Cotton Cancer Center and Dartmouth-Hitchcock Medical Center, Hanover, New Hampshire 03755

E. CSÁNYI (27), Institut Drug Res., Semmelweis University, Budapest, Hungary

M. ELAINE CURRY (361), Tissue Cloning, Bristol Laboratories, Syracuse, New York 13201

JOAN C. D'AOUST (293), Department of Clinical Cancer Research, Bristol Laboratories, Syracuse, New York 13201

JOHN R. DANIELS (221), University of Southern California, Division of Medical Oncology, LAC–USC Medical Center, Stanford University Medical Center, Stanford, California 94305

JONATHAN DUCORE (69), Laboratory of Molecular Pharmacology, Developmental Therapeutics Program, Division of Cancer Treatment, National Cancer Institute, National Institutes of Health, Bethesda, Maryland 20205

LUZ DUQUE-HAMMERSHAIMB (145, 387), Cancer Therapy Evaluation Program, Division of Cancer Treatment, National Cancer Institute, Bethesda, Maryland 20205

JOHN R. DURANT (199), Comprehensive Cancer Center, University of Alabama, University Station, Birmingham, Alabama 35294

G. EISENBRAND (27), Institute of Toxicology and Chemotherapy, German Cancer Research Center, Heidelberg, Federal Republic of Germany

LEONARD C. ERICKSON (69), Laboratory of Molecular Pharmacology, Developmental Therapeutics Program, Division of Cancer Treatment, National Cancer Institute, National Institutes of Health, Bethesda, Maryland 20205

REGINA A. EWIG (69), Laboratory of Molecular Pharmacology, Developmental Therapeutics Program, Division of Cancer Treatment, National Cancer Institute, National Institutes of Health, Bethesda, Maryland 20205

P. FALAUTANO (27), Mario Negri Institute for Pharmacological Research, Milan, Italy

JOSEPH W. FAY (337), Department of Internal Medicine, Washington University School of Medicine, St. Louis, Missouri 63110

H. H. FIEBIG (27), Medizinische Universitatklinik, Freiburg, Federal Republic of Germany

S. FILIPPESCHI (27), Mario Negri Institute for Pharmacological Research, Milan, Italy

EMIL FREI (155), Division of Medicine, Sidney Farber Cancer Institute, Boston, Massachusetts 02130

MICHAEL A. FRIEDMAN (379), Cancer Research Institute, Department of Medicine, University of California, San Francisco, California 94143

RICHARD A. GAMS (199), Comprehensive Cancer Center, University of Alabama, University Station, Birmingham, Alabama 35294

SANDRA J. GINSBERG (209), Department of Hematology/Oncology, State University of New York, Upstate Medical Center, Syracuse, New York 13210

ARVIN S. GLICKSMAN (181), Rhode Island Hospital, Providence, Rhode Island 02908

LINDA P. GLOWIENKA (313), Cancer and Leukemia Group B, Scarsdale, New York 10583

ARLAN J. GOTTLIEB (181), Department of Hematology, State University of New York, Upstate Medical Center, Syracuse, New York 13210

DANIEL P. GRISWOLD, JR. (9), Chemotherapy Research Department, Southern Research Institute, Birmingham, Alabama 35255

M. HABS (27), Institute of Toxicology and Chemotherapy, German Cancer Research Center, Heidelberg, Federal Republic of Germany

JOHN HARLEY (313), Memorial Hospital, Johnstown, Pennsylvania 15907

GEOFFREY P. HERZIG (337), Department of Internal Medicine, Washington University School of Medicine, St. Louis, Missouri 63110

ROGER H. HERZIG (337), Department of Internal Medicine, Washington University School of Medicine, St. Louis, Missouri 63110

FRED HOCHBERG (155), Division of Medicine, Sidney Farber Cancer Institute, Boston, Massachusetts 02130

JAMES F. HOLLAND (181), Department of Neoplastic Disease, The Mount Sinai Hospital School of Medicine, New York, New York 10029

DANIEL F. HOTH (387), Investigational Drug Branch, Division of Cancer Treatment, National Cancer Institute, National Institutes of Health, Bethesda, Maryland 20205

BRIAN F. ISSELL (293, 361), Department of Clinical Cancer Research, Bristol Laboratories, Syracuse, New York 13201

DON JUMP (123), Department of Biochemistry and The Vincent T. Lombardi Cancer Center, Georgetown University Medical Center, Washington, D.C. 20007

HERBERT E. KANN, JR. (95), Department of Medicine, Emory University School of Medicine, Atlanta, Georgia 30322

SHAUL KOCHWA (313), The Mount Sinai Hospital, New York, New York 10029

MARSHA KOHLER (221), Stanford University Medical Center, Northern California Cancer Program, Stanford, California 94305

KURT W. KOHN (69), Laboratory of Molecular Pharmacology, Developmental Therapeutic Program, Division of Cancer Treatment, National Cancer Institute, National Institutes of Health, Bethesda, Maryland 20205

ROBERT KYLE (313), Department of Hematology, Mayo Clinic, Rochester, Minnesota 55905

W. RUSSELL LASTER, JR. (9), Chemotherapy Research Department, Southern Research Institute, Birmingham, Alabama 35255

GUY LAURENT (69), Laboratory of Molecular Pharmacology, Developmental

Therapeutics Program, Division of Cancer Treatment, National Cancer Institute, National Institutes of Health, Bethesda, Maryland 20205

HILLARD M. LAZARUS (337), Washington University School of Medicine, St. Louis, Missouri 63110

LOUIS LEONE (313), Rhode Island Hospital, Providence, Rhode Island 02908

VICTOR A. LEVIN (171, 259), Brain Tumor Chemotherapy Service, Brain Tumor Research Center, Department of Neurological Surgery, School of Medicine, and Department of Pharmaceutical Chemistry, School of Pharmacy, University of California, San Francisco, California 94143

DAVID B. LUDLUM (85), Division of Oncology, Department of Medicine, Albany Medical College, Albany, New York 12208

JOHN S. MACDONALD (145), Cancer Therapy Evaluation Program, Division of Cancer Treatment, National Cancer Institute, National Institutes of Health, Bethesda, Maryland 20205

L. HERBERT MAURER (209), Section of Hematology/Oncology, Dartmouth-Hitchcock Medical Center, Hanover, New Hampshire 03755

O. ROSS McINTYRE (313), Norris Cotton Cancer Center and Dartmouth-Hitchcock Medical Center, Hanover, New Hampshire 03755

GERALD METTER (343), Department of Biostatistics, City of Hope Medical Center, Duarte, California 91010

JOHN A. MONTGOMERY (3), Southern Research Institute, Birmingham, Alabama 35205

NIS I. NISSEN (181), Finsen Institute, Copenhagen, Denmark

NANCY NOLAN (123), Department of Biochemistry and The Vincent T. Lombardi Cancer Center, Georgetown University Medical Center, Washington, D.C. 20007

MAKOTO OGAWA (399), Division of Clinical Chemotherapy, Cancer Chemotherapy Center, Japanese Foundation for Cancer Research, Kami-Ikebukuro, Toshima-ku, Tokyo 170 Japan

THOMAS F. PAJAK (181, 209, 313), Cancer and Leukemia Group B, Scarsdale, New York 10583

LEROY PARKER (155), Division of Medicine, Sidney Farber Cancer Institute, Boston, Massachusetts 02130

GORDON L. PHILLIPS (337), Washington University School of Medicine, St. Louis, Missouri 63110

DONALD POSTER (145), Cancer Therapy Evaluation Program, Division of Cancer Treatment, National Cancer Institute, Bethesda, Maryland 20205

CARY A. PRESANT (343), Department of Medical Oncology, City of Hope Medical Center, Duarte, California 91010

ARCHIE W. PRESTAYKO (293), Research and Development, Bristol Laboratories, Syracuse, New York 13201 and Department of Pharmacology, Baylor College of Medicine, Houston, Texas 77030

RICHARD PROFFITT (343), Department of Medical Oncology, City of Hope Medical Center, Duarte, California 91010

DONALD J. REED (51), Biochemistry and Biophysics, Oregon State University, Corvallis, Oregon 97331

SAUL E. RIVKIN (325), Tumor Institute of the Swedish Hospital Medical Center, Seattle, Washington 98104

JEFFREY G. ROSENSTOCK (269), Division of Oncology, University of Pennsylvania, Children's Hospital of Philadelphia, Philadelphia, Pennsylvania 19104

FRANK M. SCHABEL, JR. (9), Chemotherapy Research Department, Southern Research Institute, Birmingham, Alabama 35225

PHILIP S. SCHEIN (367), Departments of Medicine and Pharmacology, Division of Medical Oncology, Georgetown University Medical Center, Washington, D.C. 20007

D. SCHMÄHL (27), Institute of Toxicology and Chemotherapy, German Cancer Research Center, Heidelberg, Federal Republic of Germany

NANCY SHARKEY (69), Laboratory of Molecular Pharmacology, Developmental Therapeutics Program, Division of Cancer Treatment, National Cancer Institute, National Institutes of Health, Bethesda, Maryland 20205

BRANIMIR I. SIKIC (221), Department of Medicine (Oncology and Clinical Pharmacology), Stanford University Medical Center, Stanford, California 94305

T. SMINK (27), Radiobiological Institute TNO, Rijswijk, The Netherlands

MARK SMULSON (123), Department of Biochemistry and The Vincent T. Lombardi Cancer Center, Georgetown University Medical Center, Washington, D.C. 20007

S. SOMFAI-RELLE (27), Research Institute of Oncopathology, Budapest, Hungary

FEDERICO SPREAFICO (27), Mario Negri Institute for Pharmacological Research, Milan, Italy

SWAROOP SUDHAKAR (123), Department of Biochemistry and The Vincent T. Lombardi Cancer Center, Georgetown University Medical Center, Washington, D.C. 20007

TAK TAKVORIAN (155), Division of Medicine, Sidney Farber Cancer Institute, Boston, Massachusetts 02130

KENNETH D. TEW (107), Division of Medical Oncology, Department of Biochemistry, Georgetown University, Washington, D.C. 20007

CLAUDE TIHON (361), Tissue Cloning, Bristol Laboratories, Syracuse, New York 13201

WILLIAM P. TONG (85), Division of Oncology, Department of Medicine, Albany Medical College, Albany, New York 12208

MARY W. TRADER (9), Chemotherapy Research Department, Southern Research Institute, Birmingham, Alabama 35255

FREDERICK VALERIOTE (343), MaillincKrogt Institute of Radiology, Washington University School of Medicine, St. Louis, Missouri 63110

PREFACE

The nitrosoureas constitute one of the most extensively studied classes of anticancer agents. Beginning with BCNU in the early 1960s, a total of six nitrosourea compounds have been actively studied in the United States. Two of these drugs, BCNU and CCNU, have reached commercial application. Additional nitrosourea compounds have been placed into clinical study throughout the world in places such as Japan, France, and the Soviet Union. As a class, the nitrosoureas are actively used in a range of solid tumors and the malignant lymphomas. Their continuity in central nervous system neoplasms is particularly noteworthy.

Both the preclinical and clinical status of these agents are reviewed. The literature and data on these compounds are so vast that this volume can touch only on some areas in broad overview. It is hoped that both practicing oncologists and scientists working in drug development will find the book valuable.

Section I
PRECLINICAL STUDIES

Chapter 1

THE DEVELOPMENT OF THE NITROSOUREAS: A STUDY IN CONGENER SYNTHESIS

John A. Montgomery

I. N-NITROSO FUNCTION

In 1955 the United States government launched a national cooperative cancer drug research program under the auspices of the National Cancer Institute (NCI). It began with the screening in animal tumor systems of chemical compounds that might show activity against human cancer (Endicott, 1965). Four years later N-methyl-N'-nitro-N-nitrosoguanidine (MNNG), a compound marketed by the Aldrich Chemical Company for the preparation of diazomethane used by organic chemists for synthetic work, was tested in the laboratories of the Wisconsin Alumni Research Foundation, one of the NCI screening contractors, and found to have significant activity against leukemia L1210, although it was inactive against sarcoma 180 and adenocarcinoma 755, the other two tumors used in the primary screen at that time. In later, confirmatory tests at Hazleton Laboratories and Southern Research Institute, MNNG barely qualified as an L1210 active (Leiter and Schneiderman, 1959).

Nevertheless, these findings stimulated the author to submit N-methyl-N-nitrosourea, another, less stable precursor of diazomethane, to the primary screen, with negative results in all the preliminary tests. Since the dose selected for the initial L1210 test—based on the prior results in the S 180 and Ca 755 system—proved to be toxic, no further tests were scheduled. Lower doses, tested at the author's request, produced increases in life span of leukemic mice and established MNU as a new type of anticancer agent, the mechanism of action of which was not known. MNU was also active by oral administration and, more importantly, was the only one of seven agents—including methotrexate, 6-mercaptopurine, and cyclophosphamide—active against intraperitoneally (ip) or intravenously (iv) implanted leukemia L1210 to show activity against this leukemia when cells were implanted intracerebrally (ic) to model human leukemia in the brain (Skipper *et al.*, 1961).

The discovery of the ability of MNU to cross the blood–brain barrier at sufficiently high levels to show therapeutic effectiveness led to the development at Southern Research Institute of a synthesis program to produce more active nitrosoureas and related compounds, as judged by the leukemia L1210 system, with the aim of producing a clinically effective agent.

The inactivity of methylurea, and subsequently other ureas, showed the necessity of the N-nitroso function. For systematic modification, the MNU molecule was then divided into three parts, labeled A, B, and C in the accompanying structure (Fig. 1), and alterations in the amino group (A), the carbonyl group (B) and the N-methyl-N-nitroso group (C) were undertaken. Although the nitrosoguanidines and two nitrosoamides showed some activity in the L1210 system, the nitrosoureas, for reasons that remain unclear, proved to be superior to all other N-nitroso compounds evaluated (Montgomery, 1976).

II. CHLOROETHYL MOIETY

The next phase of the synthesis program was the optimization of the C portion of the molecule. A number of N-alkyl-N-nitrosoureas, N-(substituted alkyl)-N-nitrosoureas, and N,N'-disubstituted nitrosoureas were prepared and evaluated against ip leukemia L1210. (Fig. 2). Of these only one specific structural type, N-(2-haloethyl)-N-nitrosoureas, proved superior to MNU itself with respect to activity on a weight basis, or to maximum activity (% increase in lifespan or cures) at the LD_{10} or maximum tolerated dose (Johnston *et al.*, 1963). At this point, N,N'-bis(2-chloroethyl)-N'-nitrosourea (BCNU) was found to be curative against both ip and ic L1210 leukemia (Johnston *et al.*, 1966). These results were the basis for clinical trials of BCNU and focused the synthetic program on the 2-haloethyl compounds. The 2-iodoethyl compounds proved to be inactive, whereas the 2-bromoethyls were inferior to the 2-chloroethyl compounds

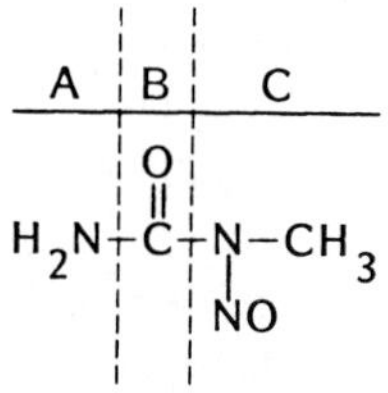

Fig. 1. MNU molecule.

(Johnston *et al.*, 1966). The 2-fluoroethyl compounds, on the other hand, showed the same activity as the 2-chloroethyl compounds, a somewhat surprising result in view of the lack of activity of the fluoro analog of cyclophosphamide (Papanastassiou *et al.*, 1966). The fluoroethyl compounds have one liability not shared by the chloroethyl compounds that is associated with the chemistry of these compounds. Decomposition of the nitrosoureas at pH 7.4 *in vitro* or *in vivo* occurs to give a diazohydroxide and an isocyanate (Montgomery *et al.*, 1967; 1975). In the case of the chloroethyl compounds an anomalous decomposition occurs giving rise to roughly equal amounts of 2-chloroethanol and ionic chloride. The other product of the anomalous reaction, acetaldehyde, is difficult to quantitate but should be formed in an amount equivalent to the chloride ion release. In any event, the fluoroethyl compounds decompose normally to give only 2-fluoroethanol which is metabolized *in vivo* to fluoroacetate, a compound well known for its toxicity to rodents. Although this toxicity can be reversed by administration of sodium acetate in ethanol (Montgomery *et al.*, 1967), enhanced therapeutic activity has yet to be demonstrated to result from this reversal. The fluoroethyl compounds are of particular interest because Wodinsky (1970; personal communication) has shown that they are much less marrow toxic than the corresponding chloroethyl compounds, using the Bruce spleen colony assay technique (Bruce *et al.*, 1966).

The next phase of the synthesis program was the study of alterations in the 2-chloroethyl group—again to optimize activity. Although some of these alterations resulted in active compounds, none gave compounds equal to those containing the 2-chloroethyl group itself. At this point, work focused on the A portion of the molecule from which the isocyanate derives and which can be varied widely without loss of activity. Disubstitution of the nitrogen of A does result in inactivity (Johnson *et al.*, 1963)—unless oxidative removal of one substituent can occur in the liver (Brundrett *et al.*, 1976)—since the normal chemical decomposition cannot occur. Within the rather broad log P range of -2 to $+4.5$ a large number of highly active compounds were found (Montgomery, 1976), as judged by the L1210 system, although a number of them fail to cross the blood–brain barrier in sufficient quantity to affect the ic disease. Among the most active compounds against both ip and ic L1210 were N-(2-chloroethyl)-N'-cyclohexyl-N-nitrosourea (CCNU) and N-(2-chloroethyl)-N'-(2, 6-dioxo-3-piperidyl)-N-nitrosourea (PCNU) (Fig. 2).

III. ANTITUMOR ACTIVITY IN ANIMAL TUMORS

Attention now turned to the development of nitrosoureas with optimal activity against solid tumors in rodents, and the Lewis lung carcinoma was selected as the model system (Schabel, 1976). After subcutaneous implantation this tumor rapidly metastasizes to the lungs of mice causing death. Initially, 39 nitrosoureas selected chiefly on the basis of their activity against leukemia L1210

(Montgomery, 1976; Johnston *et al.*, 1963; 1971) were evaluated against established Lewis lung tumors; that is, treatment was initiated 7 days after implantation of the tumor. Seventeen showed moderate regression followed by regrowth with no cures at nontoxic doses, and one, N-(2-chloroethyl)-N'-(*trans*-4-methyl-cyclohexyl)-N-nitrosourea (MeCCNU), showed complete temporary regressions with some cures at nontoxic doses (Montgomery *et al.*, 1977). The 1,4-*trans* configuration of MeCCNU prompted a search for appropriate *cis-trans* pairs for comparison since the less toxic *cis* form of MeCCNU was virtually inactive against the established Lewis lung tumors, even though it was on an equitoxic basis as active against L1210 as was the *trans* form. The more toxic *cis* form of 1,1'-(*trans*-1,4-cyclohexylene)bis-3-(2-chloroethyl)-3-nitrosourea, although highly active against leukemia L1210, was also definitely less active against Lewis lung than was the *trans* form (Johnston *et al.*, 1971; Montgomery *et al.*, 1977). Other *cis–trans* pairs such as the 4-carboxycyclohexyl and the 4-methoxycarbonylcyclohexyl compounds, however, failed to show this clear superiority of the *trans* isomer, although with the exception of the hydroxy compound the N-[*trans*-4-(substituted cyclohexyl)]-N'-(2-chloroethyl)-N'-nitrosoureas that have been evaluated are all active against the established Lewis lung carcinoma, and this group contains all of the compounds highly active against this tumor (Montgomery *et al.*, 1977).

The high level of activity of MeCCNU prompted its evaluation against other

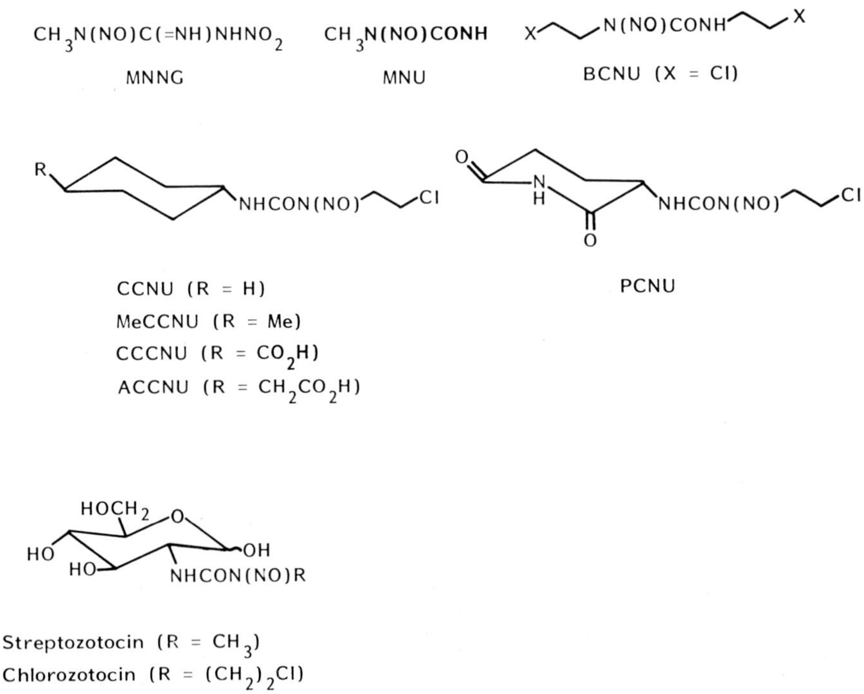

CH$_3$N(NO)C(=NH)NHNO$_2$ CH$_3$N(NO)CONH   X⌢⌢N(NO)CONH⌢⌢X

MNNG   MNU   BCNU (X = Cl)

R⌢⌢⌢NHCON(NO)⌢⌢Cl   O=⌢⌢N—⌢⌢NHCON(NO)⌢⌢Cl

CCNU (R = H)   PCNU
MeCCNU (R = Me)
CCCNU (R = CO$_2$H)
ACCNU (R = CH$_2$CO$_2$H)

HOCH$_2$—O
HO⌢⌢⌢OH
HO— NHCON(NO)R

Streptozotocin (R = CH$_3$)
Chlorozotocin (R = (CH$_2$)$_2$Cl)

Fig. 2.

solid tumors and the synthesis of congeners. In particular the high lipid solubility of MeCCNU, which caused problems in the formulation of the drug for injection, led to the synthesis of the carboxy (CCCNU) and acetic acid (ACCNU) analogs. (Fig. 2). The sodium salts of these compounds are readily soluble in water and are equal to, if not better than, MeCCNU against leukemias and solid tumors (Johnston et al., 1977). These three compounds are the most active compounds tested in these laboratories against a variety of solid tumors (as well as against leukemia L1210).

The antibiotic streptozotocin, a methylnitrosourea derived from 2-amino-2-deoxyglucose (Fig. 2), is marginally active against leukemia, diabetogenic, and reported to have some activity against malignant insulinomas in man. Because of these observations, the tetraacetate of the 2-chloroethyl analog of streptozotocin was prepared (Johnston et al., 1971) and shown to be highly active against leukemia L1210. This tetraacetate, like streptozotocin, showed reduced bone marrow toxicity (reported to be the limiting toxicity of other nitrosoureas in man), but unlike streptozotocin was nondiabetogenic (Schein et al., 1973). These results led to the synthesis of the 2-chloroethyl analog itself (Johnston et al., 1975), now called chlorozotocin, (Fig. 2) which also showed reduced bone marrow toxicity (Anderson et al., 1975) and activity against human cancers (Hoth et al., 1979; Woolley et al., 1980). The basis of the reduced marrow toxicity, once thought to result from reduced carbamoylating activity, is not known but is under study. It should be noted that despite its reduced marrow toxicity, chlorozotocin is more toxic (to mice) on a molar basis than CCNU, but the limiting toxicity has not been established. Furthermore, chlorozotocin is not active against ic leukemia L1210 nor the Lewis lung carcinoma. Efforts to overcome these limitations have been at least partially successful by the development of structural hybrids of MeCCNU and chlorozotocin (deoxychlorozotocin triacetate and dihydroxyMeCCNU) that are active against Lewis lung (Montgomery, 1979).

The nitrosoureas have shown a broader spectrum of activity against experimental neoplasms than any other class of agents reported in the scientific literature and are one of the few classes that cross the so-called blood–brain barrier at therapeutically effective levels. Certain trans-4-substituted derivatives of CCNU are the most active compounds studied against a number of solid tumors in rodents—for example, Lewis lung carcinoma, B16 melanoma, and colon tumor 26 (Schabel, 1976).

Therapeutic synergism against one or more rodent tumors with several nitrosoureas in two-drug combinations with representatives of most of the major classes of anticancer agents has been reported. With a number of advanced-stage mouse tumors, generally considered to be refractory to treatment with most anticancer agents, long-term cures have been obtained with combination drug or combined-modality (surgery plus chemotherapy) treatment (Schabel, 1976). The lack of cross-resistance of several leukemias and solid tumors of mice selected for resistance to the nitrosoureas or to other alkylating agents such as cyclophosphamide and melphalan has been demonstrated, suggesting the use of alkylating agent–drug combinations alone or in combination with other modalities in the treatment of human cancer.

IV. CONCLUSION

Study of nitrosourea synthesis and clinical development has focused on solubilities and activities and/or toxicities in a variety of animal model tumors. Many of these animal tumors, including human xenograft tumors, have not predicted for clinical activity of the nitrosoureas. The future of a rational approach to new nitrosourea drug design lies in the better understanding of specific biological targets within tumors and of how this class of drugs interacts with these targets.

REFERENCES

Anderson, T., McMenamin, M. G., and Schein, P. S. (1975). *Cancer Res. 35,* 761–765.
Bruce, W. R., Meeker, B. E., and Valeriote (1966). *J. Nat. Cancer Inst. 37,* 233–245.
Brundrett, R. B., Cowens, J. W., and Colvin, M. (1976). *Proc. Am. Assoc. Cancer Res. and ASCO 17,* 102.
Endicott, K. M. (1965). *New Scientist,* 856–858.
Hoth, D., Robichaud, K., Woolley, P. V., Macdonald, J. S., Price, N., Gullo, J., and Schein, P. S. (1979). *Proc. Am. Assoc. Cancer Res. and ASCO 20,* 413.
Johnston, T. P., McCaleb, G. S., and Montgomery, J. A. (1963). *J. Med. Chem. 6,* 669–681.
Johnston, T. P., McCaleb, G. S., Opliger, P. S., and Montgomery, J. A. (1966). *J. Med. Chem. 9,* 892–911.
Johnston, T. P., McCaleb, G. S., Opliger, P. S., Laster, W. R. Jr., and Montgomery, J. A. (1971). *J. Med. Chem. 14,* 600–614.
Johnston, T. P., McCaleb, G. S., and Montgomery, J. A. (1975). *J. Med. Chem. 18,* 104–106.
Johnston, T. P., McCaleb, G. S., Clayton, S. D., Frye, J. L., Krauth, C. A., and Montgomery, J. A. (1977). *J. Med. Chem. 20,* 279–290.
Leiter, J., and Schneiderman, M. A. (1959). *Cancer Res. 19,* 31–189.
Montgomery, J. A. (1976). *Cancer Treat. Rep. 60,* 651–664.
Montgomery, J. A. (1979). *In* "Medicinal Chemistry VI" (Proceedings of the 6th International Symposium on Medicinal Chemistry) (M. A. Simkins, ed.), pp. 313–321. Cotswold Press Ltd., Oxford, U.K.
Montgomery, J. A., James, R., McCaleb, G. S., and Johnston, T. P. (1967). *J. Med. Chem. 10,* 668–674.
Montgomery, J. A., James, R., McCaleb, G. S., Kirk, M. C., and Johnston, T. P. (1975). *J. Med. Chem. 18,* 568–571.
Montgomery, J. A., McCaleb, G. S., Johnston, T. P., Mayo, J. G., and Laster, W. R., Jr. (1977). *J. Med. Chem. 20,* 291–295.
Papanastassiou, Z. B., Bruni, R. J., Fernandes, F. P., and Levins, P. L. (1966). *J. Med. Chem. 9,* 357–359.
Schabel, F. M., Jr. (1976). *Cancer Treat. Rep. 60,* 665–698.
Schein, P. S., McMenamin, M. G., and Anderson, T. (1973). *Cancer Res. 33,* 2005–2009.
Skipper, H. E., Schabel, F. M., Jr., Trader, M. W., and Thomson, J. R. (1961). *Cancer Res. 21,* 1154–1164.
Woolley, P. V., Pavlovsky, S., Schein, P. S., and Rosenoff, S. (1980). *Proc. Am. Assoc. Cancer Res. and ASCO 21,* 156.

Chapter 2

COMBINATION CHEMOTHERAPY WITH NITROSOUREAS PLUS OTHER ANTICANCER DRUGS AGAINST ANIMAL TUMORS[1]

Frank M. Schabel, Jr.
W. Russell Laster, Jr.
Mary W. Trader
Thomas H. Corbett
Daniel P. Griswold, Jr.

I. INTRODUCTION

While the search for new and better anticancer drugs continues, a clearly indicated investigative activity for laboratory and clinical oncologists is to improve the therapeutic effectiveness of available drugs. One of the most promising and rewarding research activities of this kind is the search for combinations of two or more anticancer drugs with greater therapeutic effectiveness when used together than can be obtained by using any single drug or groups of drugs in the combination. Combination chemotherapy is the common practice of medical oncologists treating patients with malignant neoplasms in whom evident or likely metastasis has occurred when the tumor is first detected. For an excellent discussion of clinical combination chemotherapy of cancer in man, see Blum and Frei

[1]Work from the Southern Research Institute reported herein was supported by Contracts NO1-CM-43756 and NO1-CM-97309 from the Division of Cancer Treatment and Grant CA17303 from the National Large Bowel Cancer Project, National Cancer Institute, National Institutes of Health, Department of Health and Human Services.

(1979), and for a listing of documented effective drug combinations in clinical use, see Dorr and Fritz (1980).

Therapeutic synergism[2] between the drugs in a combination can be anticipated if (*a*) they are less than additive in toxicity for vital normal cells; (*b*) they kill tumor cells by different biochemical mechanisms of action; or (*c*) tumor cells resistant to one or more drugs in the combination retain cytotoxic sensitivity to one or more other drugs in the combination. Drug combinations with some or all of the above assets would be expected to be therapeutically synergistic, and those with greatest promise for therapeutic gain over the individual drugs in the combination commonly have all three assets.

II. BIOLOGIC ACTIVITY OF NITROSOUREAS IN COMBINATION WITH OTHER ANTICANCER DRUGS

The structures of the nitrosoureas to be discussed in greatest detail are shown in Fig. 1. Many of the data included here have been previously published or reviewed (Schabel, 1968; 1975; 1976; Schabel *et al.*, 1978).

A. Toxicity of Nitrosoureas plus Other Anticancer Drugs for Vital Normal Cells of the Mouse

Skipper (1974) has described a procedure for determining whether two anti-cancer drugs when used in combination are additive, less than additive, or more than additive in toxicity for vital normal cells; he has called this the combination toxicity index (CTI). Briefly stated, the CTI is determined as follows: (*a*) The doses of individual drugs used in the combination toxicity study are tabulated as decimal fractions of the previously determined LD_{10} doses for the same treatment schedule; and (*b*) the decimal fraction of the LD_{10}'s of the individual drugs used in the drug combination are added and the sums are plotted against observed mortality on log-probit paper. The CTI is the LD_{10} (sum) of the combination read from the line of best fit. A significant and reproducible CTI of <1.0 indicates greater-than-additive toxicity, a CTI of 1.0 indicates additive toxicity, and a CTI of >1.0 indicates less-than-additive toxicity for vital normal cells. A high CTI is not essential for demonstrated therapeutic synergism, but it is an asset to be sought in planning combination chemotherapy trials, since most two-drug combinations that have been reported to be therapeutically synergistic for one or more murine tumors have CTIs of $\geqslant 1.3$ (Schabel, 1975).

The CTI in mice of a number of the most active nitrosoureas in combination

---

[2]Therapeutic synergism = greater therapeutic effect against a drug-sensitive tumor at equitoxic doses ($\leqslant LD_{10}$) of the drugs in combination that can be obtained with any component(s) of the drug combination.

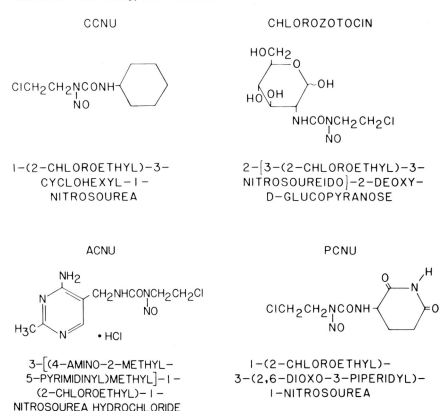

CCNU

$CICH_2CH_2NCONH$⟨cyclohexyl⟩
        NO

1-(2-CHLOROETHYL)-3-
CYCLOHEXYL-1-
NITROSOUREA

CHLOROZOTOCIN

$HOCH_2$

HO OH OH

$NHCONCH_2CH_2CI$
      NO

2-[3-(2-CHLOROETHYL)-3-
NITROSOUREIDO]-2-DEOXY-
D-GLUCOPYRANOSE

ACNU

NH_2

$CH_2NHCONCH_2CH_2CI$
          NO

H_3C N • HCI

3-[(4-AMINO-2-METHYL-
5-PYRIMIDINYL)METHYL]-1-
(2-CHLOROETHYL)-1-
NITROSOUREA HYDROCHLORIDE

PCNU

$CICH_2CH_2NCONH$⟨2,6-dioxopiperidyl⟩
        NO

1-(2-CHLOROETHYL)-
3-(2,6-DIOXO-3-PIPERIDYL)-
1-NITROSOUREA

Fig. 1. Chemical structures of CCNU, chlorozotocin, ACNU, and PCNU.

with members of other chemical and functional classes of anticancer drugs have been determined (Skipper, 1974; Schabel, 1975; 1976; Schabel *et al.*, 1978; 1979). These are listed in Table I. It is apparent from the data in Table I that the nitrosoureas with greatest antitumor activity against a variety of animal tumors (1,3-bis(2-chloroethyl)-1-nitrosourea (BCNU); CCNU; N-(2-chloroethyl)-N'-(*trans*-4-methylcyclohexyl)-*N*-nitrosourea (MeCCNU); PCNU) are either additive or less than additive in toxicity against vital normal cells in non-tumor-bearing mice. The only exceptions are the combinations of 5-(3,3-dimethyl-1-triazeno)-imidazole-4-carboxamide (DTIC) plus BCNU or CCNU, which are greater than additive in toxicity (Schabel, 1974).

Less-than-additive toxicity of sufficient magnitude likely to be of real therapeutic value (e.g., CTI of ≥1.3) is commonly, although not exclusively, seen with combinations of two anticancer drugs that show marked and readily reproducible therapeutic synergism against drug-sensitive animal tumors (Schabel, 1975; 1976; Schabel *et al.*, 1978; 1979).

TABLE I. Combination Toxicity Indexes of Selected Nitrosoureas Plus Other Anticancer Drugs in BDF1 Mice

Drug combination	Treatment schedule	Combination toxicity index
BCNU + vincristine	Simult., single doses	~1.9
BCNU + 6-MeMPR	Offset, qd × 15; single dose	~1.8
BCNU + *cis*-DDPt II	Simult., single doses	1.7
MeCCNU + bleomycin	Overlap, qd × 10 + days 7 & 11	1.6
BCNU + hydroxyurea	Simult., q3hr × 8; q4d × 4	1.6
MeCCNU + adriamycin	Offset, qd × 8 + single dose	1.5
MeCCNU + cyclophosphamide	Simult., single doses	1.5
MeCCNU + 5-FU	Simult., q4d × 3	1.5
CCNU + bleomycin	Overlap, qd × 10 + days 3 & 10	~1.5
MeCCNU + BIC	Simult., single doses	1.4
BCNU + prednisone	Simult., single doses	~1.4
BCNU + cyclophosphamide	Simult., single doses	1.2–1.4
BCNU + ara-C	Simult., q3hr × 8; q4d × 4	1.3
MeCCNU + 5-FU	Simult., q4d × 3	1.3–1.5
CCNU + ara-C	Simult., q3hr × 8; q4d × 4	1.3
PCNU + 6-TG	Simult., q4d × 4[a]	1.25
BCNU + 5-FU	Simult., single doses	~1.2
CCNU + BIC	Simult., single doses	1.1
CCNU + L-PAM	Simult., single doses	1.1
CCNU + cyclophosphamide	Simult., single doses	~1.2
BCNU + DTIC	Simult., single doses	0.75
CCNU + DTIC	Simult., single doses	0.45

[a] CDF1 mice.

B. Therapeutic Synergism of Nitrosoureas in Combination with Other Anticancer Drugs

A review of the reports of therapeutic synergism of BCNU, CCNU, and MeCCNU in combination with other drugs was published by Schabel (1976). Marked therapeutic synergism was observed with various nitrosoureas in combination with representatives of several of the chemical and functional classes of anticancer drugs including antimetabolites such as 1-β-D-arabinofuranosylcytosine (ara-C), 5-fluorouracil (5-FU), 5-FUdR; alkylating agents such as cyclophosphamide (CPA), L-phenylalanine mustard (L-PAM, melphalan), 4(5)-[3,3-bis (2-chloroethyl)-1-triazenyl] imidazole-5(4)-carboxamide (BIC); and drugs such as adriamycin that bind to or intercalate with DNA against a variety of transplantable leukemias and solid tumors of rodents.

1. *Nitrosoureas plus Other Alkylating Agents*

Therapeutic synergism of BCNU or CCNU in combination with CPA against a transplantable line of AKR lymphoma was first observed by Valeriote *et al.*

(1968). Their observations were of great interest and theoretical importance because they provided the first data from therapy trials indicating that all alkylating agents do not have a common cytotoxic mechanism of action against mammalian cells, either tumor or normal (see Table I). We have extended the original observations of Valeriote and co-workers to other combinations of alkylating agents (including nitrosoureas) and have reported additional examples of therapeutic synergism and lack of cross-resistance among other alkylating agent combinations. The therapeutically synergistic activity of these drug combinations may be explained on the basis of likely different alkylating moieties among different chemical classes of alkylating agents (Schabel *et al.*, 1978). Data at hand support the clear indication for expecting therapeutic gain from certain alkylating agent combinations, and they provide a logical basis for understanding the therapeutic synergism of combinations of alkylating agents that have been observed.

CCNU plus chlorozotocin. Wheeler and associates (in press) have recently reported that the combination of CCNU plus chlorozotocin is therapeutically synergistic against leukemia L1210. CCNU has high alkylating and carbamoylating activity and marked therapeutic activity against intraperitoneally (ip), intravenously (iv), or intracerebrally (ic) implanted L1210. It also has undesirable toxicity for bone marrow. Chlorozotocin has high alkylating activity but very little carbamoylating activity. It is markedly active against ip implanted L1210,

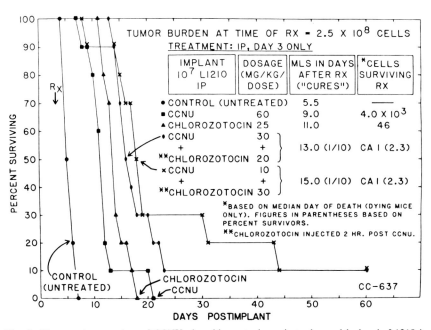

Fig. 2. Therapeutic synergism of CCNU plus chlorozotocin against advanced leukemia L1210 in CDF1 mice. Cumulative mortality plots of optimum therapeutic responses with single $\leq LD_{10}$ doses given ip 3 days after ip implant of 10^7 viable L1210 cells.

somewhat less active against iv implanted L1210, but is inactive against ic implanted L1210. It has been reported to be less toxic for the bone marrow of the mouse than CCNU (Panasci *et al.*, 1979). Available evidence indicates that the anticancer activity of the nitrosoureas is related to alkylating activity and that carbamoylating activity probably inhibits the repair of sublethal damage to DNA caused by alkylation. It was thought that the combination of CCNU plus chlorozotocin might be therapeutically synergistic against drug-sensitive tumor cells in the mouse because (*a*) both agents would contribute to the total alkylating activity but the combination would be less toxic for bone marrow than an equitoxic dose ($\leq LD_{10}$) of CCNU; (*b*) the CCNU might control the L1210 cells present in the central nervous system (CNS) in advanced disease; and (*c*) the carbamoylating activity of CCNU might inhibit repair of sublethal damage to DNA of tumor cells caused by the alkylating activity of the two drugs.

The data presented in Fig. 2 are optimal therapeutic activity at $\leq LD_{10}$ doses observed in an internally controlled experiment in which single doses of CCNU or chlorozotocin alone and the two drugs in simultaneous combination were given to mice about 2.5 days before median day of death in untreated controls. At that time, the total body burden of viable (clonogenic) tumor cells was estimated to be $>10^8$/mouse, and historical experience indicated that the L1210 cells following ip implant were widely distributed in the major lymphoid and other vital organs, including the brain, at the time of drug treatment (Skipper *et al.*, 1961). Estimates of the body burden of tumor cells present at the time of drug treatment and surviving drug treatment were made by use of published procedures (Schabel *et al.*, 1977). These consisted of objectively determining the viability (clonogenicity) of the L1210 cells implanted into mice and determining the population growth kinetics of the control and drug-treated tumor cells in each experiment. The plotted therapeutic responses were optimal for each drug used alone or in combination at equitoxic doses ($\leq LD_{10}$) derived from a series of doses ranging from frankly toxic ($>LD_{10}$) to less than optimally active in each experiment. Thus, all therapeutic activities reported were from experiments in which all biologic parameters (tumor growth kinetics, drug(s) toxicity, and therapeutic activity) were internally controlled. Therapeutic synergism of CCNU plus chlorozotocin was demonstrated, based on (*a*) greater increase in life span of drug-treated but dying animals; and (*b*) survivors consistent with the estimate of about one viable tumor cell surviving therapy in animals dying of leukemia. Confirmatory trials have essentially reproduced these results.

2. *Nitrosoureas or Other Alkylating Agents plus Antimetabolite Anticancer Agents*

L1210 implanted ip or iv. The limited solubility of the highly active nitrosoureas (BCNU,[3] CCNU, MeCCNU) makes parenteral administration difficult

---

[3]BCNU is quite water soluble, but its rate of solution is slow.

and absorption from the gastrointestinal tract following oral administration variable and unpredictable. Antitumor activity, even at identical doses on a mg/kg dosage basis, is quite variable, probably because of variation in particle size of insoluble compound in the drug preparations administered (Montgomery, 1980). A number of nitrosoureas with greater water solubility and equal or greater activity against L1210 than CCNU or MeCCNU and/or a higher rate of solution than BCNU have been prepared by medicinal chemists in several countries (Shimizu and Arakawa, 1975; Fiebig *et al.*, 1977). One of the most therapeutically active nitrosoureas against a number of transplantable murine tumors, including L1210, is ACNU (Shimizu and Arakawa, 1978).

Fujimoto *et al.* (1977) reported that ACNU and several other alkylating agents, including CPA and mitomycin C, were therapeutically synergistic in combination with 6-thioguanine (6-TG) against L1210. Sartorelli and Booth (1965) had previously reported that mitomycin C or porfiromycin were synergistic in combination with either 6-TG or 5-FU against both sarcoma 180 and L1210, and they had previously reported similar results with other antimetabolite drugs in combination with other alkylating agents. The unique aspect of the report of Fujimoto and co-workers was the marked activity of ACNU plus 6-TG against advanced L1210. They reported cure of advanced-stage L1210 when ACNU plus 6-TG was given about 3 days before median day of death in untreated control leukemic mice that had been implanted ip with 10^5 L1210 cells (Fujimoto *et al.*, 1977). Similar high cure rates of even more advanced L1210 were reported earlier by Schabel (1968) using CCNU plus ara-C. Therapeutic synergism of BCNU plus 5-FU against transplanted AKR lymphoma (Valeriote *et al.*, 1968) and of MeCCNU plus 5-FU against colon tumor 26 (Corbett *et al.*, 1977) and colon tumor 38 (Griswold and Corbett, 1978) in mice has been reported, and MeCCNU plus 5-FU has been reported to be more effective in treating colon cancer in man than either drug alone (Moertel *et al.*, 1975). However, none of these therapeutic synergisms were sufficiently marked to result in cure of advanced disease in man or animals. A recent report that 6-TG alone has some limited clinical activity against colon cancer in man (Horton *et al.*, 1975) suggests that therapeutic synergism of nitrosoureas in combination with 6-TG might hold clinical promise.

We conducted experiments to confirm and extend the observations of Fujimoto *et al.* (1977). Confirmation was reproducibly accomplished and extended. Data in Fig. 3 indicate that reduction of a body burden of $>10^8$ viable L1210 cells could be accomplished by treatment with a single dose of ACNU plus 6-TG 4 days after implant of 10^7 L1210 cells ip (2 days before median day of death in untreated controls). At this stage of disease, ACNU showed no therapeutic activity. This is to be expected since Ehmann and Wheeler (1979), working with rat brain sarcoma (9L) cells in culture, and Barlogie and Drewinko (1977; 1978), using a number of different human lymphoma cells in culture, have shown that these drug-sensitive tumor cells are not immediately killed on exposure to lethal doses of BCNU or several other alkylating agents, but that they can

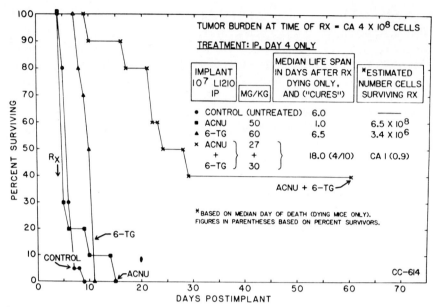

Fig. 3. Therapeutic synergism of ACNU plus 6-TG against advanced leukemia L1210 in CDF1 mice. Cumulative mortality plots of optimum therapeutic responses with single $\leq LD_{10}$ doses given ip 4 days after ip implant of 10^7 viable L1210 cells.

go through at least one and sometimes two or three additional cell divisions after drug exposure before the cell division cycle is blocked. With a body burden of 4 $\times$ 10^8 at time of treatment with BCNU (Schabel, 1968), and presumably also with ACNU (Fig. 3) or CCNU (Schabel, 1968), the lethal body burden of tumor cells (*ca.* 10^9) would be exceeded before the lethal lesion to the tumor cell DNA caused by the nitrosourea would stop all cell division. The presence of 6-TG in the combination, which halts DNA synthesis at time of lethal drug exposure and killed about 99% of the cells (Fig. 3), allowed the lethal action of the ACNU to exert its full therapeutic effectiveness. ACNU plus 6-TG caused a net log reduction of tumor cells of >8 orders of magnitude and resulted in the marked increase in life span of dying animals and the 40% cure rate observed. Very similar data were obtained when advanced L1210 leukemia was treated following iv implant L1210 (Fig. 4). Here the body burden of tumor cells at time of drug treatment was about one-half that in the experiment shown in Fig. 3. As a result, the marked activity of ACNU alone was demonstrated since the body burden of tumor cells did not exceed the lethal number prior to expression of the lethal activity of the drug. Therapeutic synergism, based on $\log_{10}$ reduction of tumor cells, was not as apparent as in Fig. 3 since ACNU alone reduced the body burden of tumor cells to about two cells on the average (reduction to less than one cell necessary for cure), but no cures were obtained. Both the increased life span of drug-treated mice dying of leukemia and a 50% cure rate document the reality

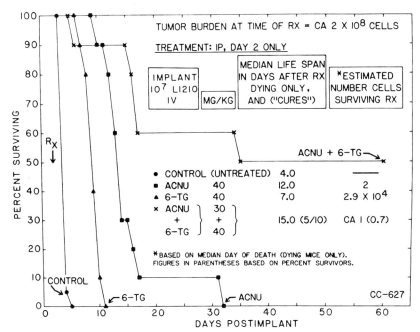

Fig. 4. Therapeutic synergism of ACNU plus 6-TG against advanced leukemia L1210 in CDF1 mice. Cumulative mortality plots of optimum therapeutic responses with single $\leq LD_{10}$ doses given ip 2 days after iv implant of 10^7 viable L1210 cells.

of the therapeutic synergism of ACNU plus 6-TG against advanced and widely systemic L1210.

The range of therapeutic synergistic activity of alkylating agents in combination with 6-TG or 6-mercaptopurine (6-MP) and its ready reproducibility are shown in summary data from a number of other similar experiments (Table II). We consider that reproducible $\log_{10}$ reduction of the body burden of tumor cells by two drugs in combination of ≥ 2 orders of magnitude more than that obtained with either drug alone, in dose-response studies in two or more separate experiments, establishes therapeutic synergism under the experimental conditions used. Data in Table II indicate that ACNU is therapeutically synergistic against advanced L1210 (body burdens $>10^8$ viable tumor cells) in simultaneous treatment with 6-TG or 6-MP. Additionally, 6-TG is therapeutically synergistic in combination with MeCCNU, chlorozotocin, CPA, L-PAM, or dianhydrogalactitol, but not in combination with *cis*-diamminedichloroplatinum(II) (*cis*-DDPt). The failure to show therapeutic synergism of 6-TG with *cis*-DDPt provides additional basis to question whether *cis*-DDPt is indeed functionally similar to other active alkylating agents.

Data in Table II provide additional examples of the observation common to a variety of structually and functionally different alkylating agents—that delaying drug treatment until the body burden reaches a point where a very few (perhaps

TABLE II. Therapeutic Synergism with Nitrosoureas or Other Alkylating Agents in Combination with 6-TG or 6-MP versus Advanced L1210[a]

	Body burden at R_x	Log_{10} reduction by R_x[b] Expt. No.			"Cures" survivors 60 days postimplant
		1	2	3	
ACNU		0	0[c]	5	0
6-TG	3 to 4 × 10⁸	2	2[c]	1	0
ACNU + 6-TG		8	8[c]	8	40–80%
ACNU		0	0	5	0
6-MP	4 × 10⁸	1	2	2	0
ACNU + 6-MP		6	6	8	0–50%
MeCCNU		0	0		0
6-TG	4 × 10⁸	3	0		0
MeCCNU + 6-TG		8	8		60–70%
Chlorozotocin		0	0		0
6-TG	4 × 10⁸	2	2		0
Chloroz + 6-TG		6	5		0
Cyclophosphamide		0	0		0
6-TG	3 to 4 × 10⁸	2	2		0
CPA + 6-TG		8	8		0
L-PAM		2	2	6[d]	0
6-TG	2[d] to 4 × 10⁸	2	2	4[d]	0
L-PAM + 6-TG		6	5	8[d]	0–10%[d]
Dianhydrogalactitol		0	0		0
6-TG	4 × 10⁸	2	2		0
DAG + 6-TG		5	5		0
cis-DDPt II		0	0	4[d]	0
6-TG	2[d] to 4 × 10⁸	2	2	4[d]	0
cis-DDPt + 6-TG		1	1	3[d]	0

[a] 10⁷ viable L1210 cells implanted ip; R_x—single dose ($\leq LD_{10}$), ip, day 4.

[b] Log_{10} reduction = order of magnitude reduction as determined by published procedures (Schabel *et al.*, 1977). With a body burden of viable tumor cells of 4 × 10⁸ at time of treatment and a median life span of 18 days (dying animals only) and 4/10 cures (see Fig. 3), there was an 8-log_{10} reduction in viable tumor cells obtained by treatment with ACNU plus 6-TG.

[c] Detailed data plotted in Fig. 3.

[d] Single-dose treatment on day 3 postimplant.

no more than one or two) additional cell divisions of all viable tumor cells will exceed the lethal body burden completely removes all evidence of therapeutic activity (based on increased life span) of these alkylating agents that show high therapeutic activity against smaller body burdens of tumor cells. These are not inactive drugs. They are only therapeutically inactive against very advanced disease, that is, disease staging close to death at time of treatment. Under these circumstances, alkylating agent plus antimetabolite drug combinations are markedly effective, at least with selected experimental tumors.

L1210 implanted ic. The need for effective drugs to treat either primary tumors of the brain or metastases to the brain from other primary sites is apparent. As with the leukemias of children, the problem will become progressively larger as our ability to effectively control metastatic disease outside the CNS improves and an increasing proportion of patients with drug-controlled systemic disease outside the CNS develops clinically evident brain metastases. Certain primary human tumors commonly metastasize to the brain (lung, breast, melanoma, and others), and it has been estimated that about one-fourth of patients dying of cancer have intracranial (including dural) metastases (Posner, 1980).

One of the major assets of nitrosoureas is their demonstrated activity against drug-sensitive tumors in the CNS in animals, either primary or metastatic from other sites (Schabel, 1976). Clinical experience in treating brain tumors in man, either primary (Levin and Wilson, 1975) or metastatic from other sites (Shapiro, 1980), with nitrosoureas has been much less effective than in animals (Schabel, 1975), but usually more advanced-staged disease has been treated in man than in animals. The prospect of marked improvement of nitrosourea therapy of brain tumors with therapeutically synergistic drug combinations suggested by the data presented above provided the indication for the experiments to be described.

ACNU, PCNU, BCNU, dianhydrogalactitol, or procarbazine was tested alone and in combination with 6-TG, 5-FU, or ara-C against ic implanted L1210. Figs. 5–8 contain detailed data indicating therapeutic synergism of 6-TG plus ACNU or PCNU and of 5-FU or ara-C plus ACNU against large body burdens (5×10^6 to 6×10^7) of viable L1210 cells in the brains of mice.

Johnston *et al.* (1966) synthesized PCNU. It was selected for extended experimental trial because it has a favorable log P (1-octanol/water partition coefficient) indicating lipid solubility believed to favor treatment of brain tumors (Levin and Kabra, 1974) and is markedly effective against ic implanted rat sarcoma 9L (Levin and Kabra, 1974) and L1210 (Fig. 6). Dianhydrogalactitol and procarbazine were also included in these studies since they are among the most active anticancer drugs against ic implanted mouse ependymoblastoma (Geran *et al.*, 1974), and procarbazine has been reported to be comparable to BCNU in effectiveness against human malignant brain tumors (Levin and Kabra, 1974). 5-FU was included because of its synergistic activity in combination with nitrosoureas in particular and alkylating agents in general against a number of tumors of mice and man. Ara-C was included because of its well-established activity against ic implanted L1210 (Skipper *et al.*, 1967; Lee *et al.*, 1977).

Table III summarizes data from these experiments. Therapeutic synergism was widely observed, although with the exception of dianhydrogalactitol plus 6-TG and ACNU plus either 5-FU or ara-C (single unconfirmed experiments), greater tumor cell kill by the drug combination of ≥ 2 orders of magnitude over

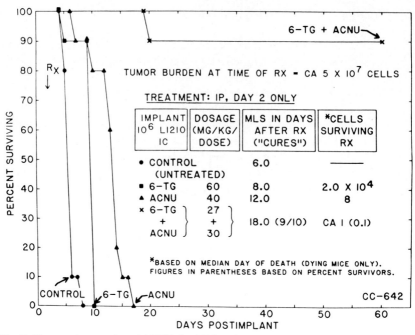

Fig. 5. Therapeutic synergism of ACNU plus 6-TG against advanced CNS leukemia L1210 in CDF1 mice. Cumulative mortality plots of optimum therapeutic responses with single $\leq LD_{10}$ doses given ip 2 days after ic implant of 10^6 viable L1210 cells.

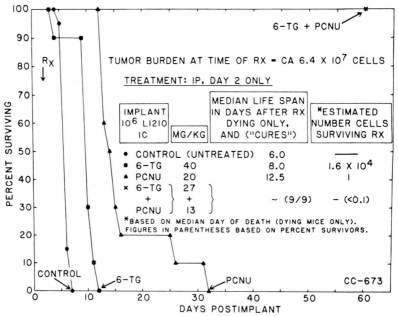

Fig. 6. Therapeutic synergism of PCNU plus 6-TG against advanced CNS leukemia L1210 in CDF1 mice. Cumulative mortality plots of optimum therapeutic responses with single $\leq LD_{10}$ doses given ip 2 days after ic implant of 10^6 viable L1210 cells.

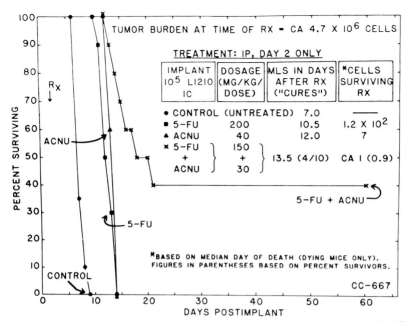

Fig. 7. Therapeutic synergism of ACNU plus 5-FU against advanced CNS leukemia L1210 in CDF1 mice. Cumulative mortality plots of optimum therapeutic responses with single $\leq LD_{10}$ doses given ip 2 days after ic implant of 10^5 viable L1210 cells.

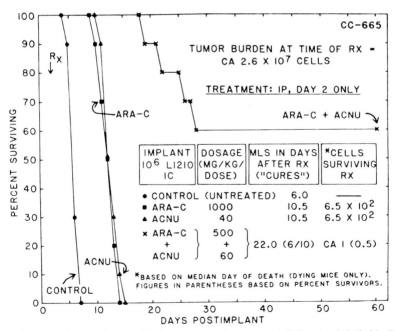

Fig. 8. Therapeutic synergism of ACNU plus ara-C against advanced CNS leukemia L1210 in CDF1 mice. Cumulative mortality plots of optimum therapeutic responses with single $\leq LD_{10}$ doses given ip 2 days after ic implant of 10^6 viable L1210 cells.

21

TABLE III. Therapeutic Synergism with Nitrosoureas or Other Alkylating Agents in Combination with 6-TG, 5-FU, or Ara-C versus Intracerebrally Implanted L1210[a]

	Body burden at R_x	Log$_{10}$ reduction by R_x[b] Expt. No. 1	2	"Cures" survivors 60 days postimplant
ACNU		7[c]	6	0
6-TG	5×10^7 to 1×10^8	3[c]	5	0
ACNU + 6-TG		8[c]	8	30–90%
PCNU		6	7	0
6-TG	5 to 6×10^7	4	3	0
PCNU + 6-TG		8	8	60–100%
BCNU		7	8	0
6-TG	4×10^7 to 1×10^8	3	5	0
BCNU + 6-TG		7	8	0
ACNU		5		0
5-FU	5×10^{6d}	4		0
ACNU + 5-FU		7		40%
ACNU		4		0
Ara-C	3×10^7	4		0
ACNU + Ara-C		8		70%
Dianhydrogalactitol		4	4	0
6-TG	7 to 8.5×10^{6d}	4	4	0
DAG + 6-TG		6	6	0
Procarbazine[e]		1	1	0
6-TG	7 to 8.5×10^{6d}	4	4	0
Procarbazine + 6-TG		4	4	0

[a] 10^6 or 10^5 viable L1210 cells implanted ic; R_x—single dose ($\leq LD_{10}$), ip, day 2 postimplant.

[b] Log$_{10}$ reduction = order of magnitude reduction as determined by published procedures (Schabel *et al.*, 1977). With a body burden of viable tumor cells of 5×10^7 at time of treatment, one mouse dying of leukemia on day 18 and 9/10 cures (see Fig. 5), there was an 8-log$_{10}$ reduction in viable tumor cells obtained by treatment with ACNU plus 6-TG.

[c] Detailed data plotted in Fig. 5.

[d] Implant was 10^5 L1210 cells ic.

[e] Probably acts as a methylating agent.

that of either drug alone was not reproducibly observed. This was probably because ACNU, PCNU, and BCNU are so effective that drug treatment approached cure with these drugs when used alone. With ACNU or PCNU in combination with 6-TG, high and reproducible cure rates were seen with the drug combinations. Reproducible therapeutic synergism, based on ≥ 2 log$_{10}$ greater cell kill, but no cures, was reproducibly obtained with dianhydrogalactitol plus 6-TG. No therapeutic synergism was seen with procarbazine plus 6-TG.

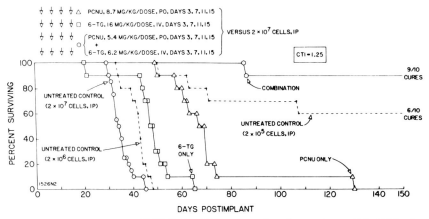

Fig. 9. Therapeutic synergism of PCNU plus 6-TG against colon adenocarcinoma 51 in CDF1 mice. Cumulative mortality plots of optimum therapeutic responses with single oral doses of PCNU or 6-TG iv alone or in combination on days 3, 7, 11, and 15 after ip implant of 2×10^7 counted cells (about 100 times the number of cells necessary to establish a fatal tumor).

Colon adenocarcinoma no. 51 implanted ip. Fig. 9 illustrates therapeutic synergism of PCNU plus 6-TG against colon adenocarcinoma 51 (Corbett *et al.*, 1977). Plotted data were optimal therapeutic responses from dose-response studies with PCNU or 6-TG alone and in combination at $\leq LD_{10}$ doses. The data show a $>1-\log_{10}$ but $<2-\log_{10}$ cell kill with either drug alone but a $>2-\log_{10}$ cell kill with the drug combination and a high cure rate against a body burden of tumor cells probably about 500 times that necessary to establish fatal disease. These results have been confirmed in another experiment. Studies are under way against the subcutaneously implanted tumor.

C. Resistance and Cross-Resistance of Nitrosourea-Resistant Tumor Cells to Other Anticancer Drugs

Extensive data have been published clearly indicating that drug-resistant tumor cells selected from populations originally very sensitive to nitrosoureas (tumor cell populations resuming growth under drug treatment that originally caused multiple $\log_{10}$ reductions of tumor cells) are markedly but not completely cross-resistant to other nitrosoureas at $\leq LD_{10}$ doses (Schabel, 1976; Schabel *et al.*, 1978; 1980). However, with the exception of BIC (Tyrer *et al.*, 1969; Schabel *et al.*, 1978), BCNU-resistant murine tumor cells retain marked sensitivity to all other alkylating agents (Schabel, 1976; Schabel *et al.*, 1978; 1980), and BIC-resistant L1210 cells are cross-resistant to BCNU and CCNU but retain sensitivity to other alkylating agents (Tyrer *et al.*, 1969). On the other hand, L1210 cells selected for resistance to DTIC, an alkylating agent similar to BIC but lacking the bis-chloroethyl groups, retain full sensitivity to BCNU, CCNU, and BIC (Kline *et al.*, 1971; and unpublished data).

With the very limited exception of cross-resistance of BCNU-resistant L1210 cells to BIC, tumor cells selected for resistance to nitrosoureas retain marked, if not comparable, sensitivity to all other chemical and functional classes of anticancer drugs, including other alkylating agents, and tumor cells selected for resistance to other anticancer drugs show no cross-resistance to nitrosoureas (Schabel, 1976; Schabel *et al.*, 1978; 1980).

III. DISCUSSION

The broad spectrum activity of anticancer nitrosoureas against animal tumors is widely recognized, and they have proved to be useful in treatment of a number of human tumors. Developmental cancer chemotherapy seeks to improve existing drugs by two methods: congener synthesis of parent structures seeking enhanced therapeutic usefulness of single drugs; and by improving application of the best of any chemical or functional class of anticancer drugs by combining it with one or more active drugs. This approach serves to (*a*) lessen drug toxicity for vital normal cells; (*b*) obtain synergistic cell kill of drug-sensitive cells by proper drug class selection; and/or (*c*) control overgrowth of drug-resistant cells in initially drug-sensitive tumor cell populations by adding other drugs with useful cytotoxic activity against these drug-resistant mutant tumor cells that cause ultimate therapeutic failure if not effectively contained or destroyed.

Data from previously published and unpublished studies establish that active nitrosoureas, several of which are available for treatment of human cancer, are less than additive in toxicity for vital normal cells in combination with other drugs, at least in the mouse; therapeutically synergistic against drug-sensitive tumors in combination with other drugs; and active against tumor cells resistant to all other anticancer drugs. The drugs, investigated in combination with the nitrosoureas, are representative of all of the chemical and functional classes of anticancer drugs with recognized usefulness for drug treatment of cancer in man.

Lacking treatment schedule dependency for optimal therapeutic effect, at least against animal tumors, the nitrosoureas possess all of the theoretical assets indicated for consideration in prospective protocols for combination chemotherapy of drug-sensitive human tumors.

IV. SUMMARY

Selected nitrosoureas (BCNU, CCNU, MeCCNU, PCNU) with anticancer activity against human and/or animal tumors are less than additive in toxicity for vital normal cells and therapeutically synergistic in combination with representatives of all of the major chemical and functional classes (including alkylating agents) of clinically useful anticancer drugs. Also they retain marked activity against tumor cells that have been selected for resistance to these other drugs.

ACNU, MeCCNU, chlorozotocin, cyclophosphamide, L-PAM, and dianhydrogalactitol but not *cis*-DDPt II show marked therapeutic synergism in combina tion with 6-TG, 6-MP, 5-FU, or ara-C against advanced L1210 following ip or ic implant into mice. PCNU plus 6-TG is therapeutically synergistic against about 500 times the number of colon adenocarcinoma 51 cells necessary to establish fatal disease in mice.

The ability of 6-TG, 5-FU, and ara-C to kill significant numbers (about 4 orders of magnitude) of L1210 cells in the brains of mice plus their therapeutic synergism in combination with both nitrosoureas and other selected alkylating agents, which are cytotoxic when used alone against L1210 cells in the brains of mice, indicates additional drug combinations worthy of consideration for prospective trial against primary brain tumors or those metastatic to the CNS from other primary sites.

REFERENCES

Barlogie, B., and Drewinko, B. (1977). *Cancer Treat. Rep. 61,* 425-436.

Barlogie, B., and Drewinko, B. (1978). *Europ. J. Cancer 14,* 741-745.

Blum, R. H., and Frei, E., III (1979). *In* "Cancer Drug Development, Part B. Methods in Cancer Research" (V. DeVita, Jr. and H. Busch, eds.), Vol. 17, pp. 215-257. Academic Press, New York.

Corbett, T. H., Griswold, D. P., Jr., Roberts, B. J., Peckham, J. C., and Schabel, F. M., Jr. (1977). *Cancer 40,* 2660-2680.

Dorr, R. T., and Fritz, W. L. (1980). *In* "Cancer Chemotherapy Handbook," pp. 50-74. Elsevier North Holland, New York.

Ehmann, U. K., and Wheeler, K. T. (1979). *Europ. J. Cancer 15,* 461-473.

Fiebig, H. H., Eisenbrand, G., Zeller, W. J., and Deutsch-Wenzel, T. (1977). *Europ. J. Cancer 13,* 937-945.

Fujimoto, S., Inagaki, J., Horikoshi, N., and Hoshino, A. (1977). *Gann 68,* 543-552.

Geran, R. I., Congleton, G. F., Dudeck, L. E., Abbott, B. J., and Gargus, J. L. (1974). *Cancer Chemother. Rep., Part 2, 4,* 53-87.

Griswold, D. P., Jr., and Corbett, T. H. (1978). *In* "Gastrointestinal Tract Cancer" (M. Lipkin and R. A. Good, eds.), pp. 399-418. Plenum Publishing, New York.

Horton, J., Mittelman, A., Taylor, S. G., III, Jurkowitz, L., Bennett, J. M., Ezdinli, E., Colsky, J., and Hanley, J. A. (1975). *Cancer Chemother. Rep., Part 1, 59,* 333-340.

Johnston, T. P., McCaleb, G. S., Opliger, P. S., and Montgomery, J. A. (1966). *J. Med. Chem. 9,* 892-911.

Kline, I., Woodman, R. J., Gang, M., and Venditti, J. M. (1971). *Cancer Chemother. Rep. 55,* 9-28.

Lee, S. H., Caron, N., and Kimball, A. P. (1977). *Cancer Res 37,* 1953-1955.

Levin, V. A., and Kabra, P. (1974). *Cancer Chemother. Rep., Part 1, 58,* 787-792.

Levin, V. A., and Wilson, C. B. (1975). *Semin. Oncol. 2,* 63-67.

Moertel, C. G., Schutt, A. J., Hahn, R. G., and Reitemeier, R. J. (1975). *J. Nat. Cancer Inst. 54,* 69-71.

Montgomery, J. A. (1980). *Cancer Treat. Rep. 64,* 987-988.

Panasci, L. C., Green, D., and Schein, P. S. (1979). *J. Clin. Invest. 64,* 1103-1111.

Posner, J. B. (1980). *In* "Brain Metastasis" (L. Weiss, H. A. Gilbert, and J. B. Posner, eds.), pp. 2-29. G. K. Hall, Boston.

Sartorelli, A. C., and Booth, B. A. (1965). *Cancer Res. 25,* 1393–1400.

Schabel, F. M., Jr. (1968). *In* "The Proliferation and Spread of Neoplastic Cells" (21st Annual Symposium on Fundamental Cancer Research, 1967, The University of Texas M.D. Anderson Hospital & Tumor Institute at Houston), pp. 379–408. Williams & Wilkins, Baltimore.

Schabel, F. M., Jr. (1974). *Biochem. Pharmacol., Suppl. 2,* pp. 163–176.

Schabel, F. M., Jr. (1975). *In* "Pharmacological Basis of Cancer Chemotherapy" (27th Annual Symposium on Fundamental Cancer Research, 1974, The University of Texas M.D. Anderson Hospital & Tumor Institute at Houston), pp. 595–621. Williams & Wilkins, Baltimore.

Schabel, F. M., Jr. (1976). *Cancer Treat. Rep. 60,* 665–698.

Schabel, F. M., Jr., Griswold, D. P., Jr., Laster, W. R., Jr., Corbett, T. H., and Lloyd, H. H. (1977). *Pharmacol. Ther. A 1,* 411–435.

Schabel, F. M., Jr., Trader, M. W., Laster, W. R., Jr., Wheeler, G. P., and Witt, M. H. (1978). *In* "Antibiotics and Chemotherapy. Fundamentals in Cancer Chemotherapy" (F. M. Schabel, Jr., ed.), Vol. 23, pp. 200–215. S. Karger, Basel.

Schabel, F. M., Jr., Trader, M. W., Laster, W. R., Jr., Corbett, T. H., and Griswold, D. P., Jr. (1979). *Cancer Treat. Rep. 63,* 1459–1473.

Schabel, F. M., Jr., Skipper, H. E., Trader, M. W., Laster, W. R., Jr., Corbett, T. H., and Griswold, D. P., Jr. (1980). *In* "Breast Cancer: Experimental and Clinical Aspects" (H. T. Mouridsen and T. Palshof, eds.), pp. 199–211. Pergamon Press, Oxford.

Shapiro, W. R. (1980). *In* "Brain Metastasis" (L. Weiss, H. A. Gilbert, and J. B. Posner, eds.), pp. 328–339. G. K. Hall, Boston.

Shimizu, F., and Arakawa, M. (1975). *Gann 66,* 149–154.

Shimizu, F., and Arakawa, M. (1978). *Gann 69,* 545–548.

Skipper, H. E. (1974). *Cancer Chemother. Rep., Part 2, 4,* 137–145.

Skipper, H. E., Schabel, F. M., Jr., Trader, M. W., and Thomson, J. R. (1961). *Cancer Res. 21,* 1154–1164.

Skipper, H. E., Schabel, F. M., Jr., and Wilcox, W. S. (1967). *Cancer Chemother. Rep. 51,* 125–165.

Tyrer, D. D., Kline, I., Gang, M., Goldin, A., and Venditti, J. M. (1969). *Cancer Chemother. Rep. 53,* 229–241.

Valeriote, F. A., Bruce, W. R., and Meeker, B. E. (1968). *J. Nat. Cancer Inst. 40,* 935–944.

Wheeler, G. P., Schabel, F. M., Jr., and Trader, M. W. (in press). *Cancer Treat. Rep.*

Chapter 3

EORTC STUDIES WITH NOVEL NITROSOUREAS

F. Spreafico
S. Filippeschi
P. Falautano
G. Eisenbrand
H. H. Fiebig
M. Habs
V. Zeller
M. Berger
D. Schmähl
L. M. Van Putten
T. Smink
E. Csányi
S. Somfai-Relle

I. INTRODUCTION

In the clear progress achieved over the past 10 to 15 years in the therapeutic effectiveness of cancer chemotherapy, a significant role has been played by the progressive development of strategies and philosophies for less empirical uses of cytotoxic antineoplastic drugs. However, it can also be easily argued that a major, if not determinant, factor in this progress has been represented by the introduction of a series of novel chemicals possessing high activity against a larger number of human neoplasms. Among the more recent classes of agents, the nitrosoureas (NUs) are undoubtedly important additions to the armamentarium of chemotherapy.

During the early development of NUs, it was recognized that certain agents in this class of compounds possessed remarkably high antitumor activity in a wide spectrum of responsive animal tumors. An analysis of the experimental antitumor activity and pharmacology of these agents has been reported (Schabel, 1976) but is beyond the scope of this article. However, a general aspect emerging from the preclinical investigations of NUs should be mentioned as it justifies the interest given to this class of compounds and the attention given to new developments in this field. In addition to their capacity to penetrate the blood–brain barrier, and thus to be uniquely effective on CNS neoplasms, NUs are synergistic with agents representative of the other major classes of cytotoxic drugs in the absence of additive toxicity for normal cells.

As a result of the first demonstrations of activity on a variety of human neoplasms, much effort has been expended by various investigators to obtain novel analogs possessing still higher antineoplastic effectiveness and/or better therapeutic indexes. This report will present a number of selected experimental results obtained by the EORTC Screening Pharmacology Group on some novel European NUs. An exhaustive review of the European efforts in this field is not intended.

II. THE CHEMOTHERAPEUTIC ACTIVITY OF HECNU AND OTHER NU DERIVATIVES

In studies investigating the effect of the insertion of potentially bifunctional carbamoylating functions into the 2-chloroethylnitrosoureido structure, Eisenbrand *et al.* (1976) described a series of 1,1'-polymethylenbis-3-(2-chloroethyl)-3-nitrisoureas possessing high activity in the rat L5222 leukemia model. Single doses of a number of these compounds (e.g., the 1,1' ethylenebis, propylenebis- and tetramethylenebis derivatives) were capable of inducing 50–70% cures in this tumor even when the tumor was in advanced stage. None of these compounds, which were designed as transport forms of polymethylene diisocyanates,

TABLE I. Comparison of Bifunctional and Water-Soluble NU Derivatives with BCNU

Compound	L5222 % cures	sc Walker tumor weight inhibition (%)
BCNU	70	83
1,1'-ethylene-bis-CNU	60	75
1,1'-propylene-bis-CNU	60	72
1,1'-tetramethylene-bis-CNU	75	77
1,1'-pentamethylene-bis-CNU	30	68
1,1-hexamethylene-bis-CNU	50	70
1-(2-hydroxyethyl)-CNU	90	85
1-(3-hydroxyethyl)-CNU	10	0
1-(4-hydroxybutyl)-CNU	5	4

was better than BCNU. These data suggest that carbamoylating activity plays a minor role in the chemotherapeutic potential of this class of drugs (Table I). On the other hand, 1-(2-hydroxyethyl)-3-(2-chloroethyl)-3-nitrosourea (HECNU) was significantly more active than BCNU in both the sc Walker carcinoma and early or preterminal ip and ic L5222 rat models (Zeller *et al.*, 1978). These findings thus prompted the synthesis of a number of both cyclic and open-chain derivatives and extensive biological testing of this water-soluble compound.

A first system in which the high activity of HECNU was confirmed was the L1210 Cr Leukemia. As shown by the representative data of Table II, very high proportions of animals surviving longer than 90 days were consistently observed employing single injections of the agent. Values of 40-60% and 80-100% cures were seen with doses of 10 and 20 mg/kg, respectively.

Higher doses were already toxic as indicated by lower incidences of cures. Under the same conditions, BCNU at the maximal nontoxic dose of 20 mg/kg produced lower (45-65%) proportions of cures. Chlorozotocin (CZT) was somewhat less effective than BCNU in terms both of cure rates and increase in life span, methyl-CCNU being the least active in this system. In confirmation of previous data in rats concerning the very high effectiveness of HECNU on tumor cells in the brain, the same table shows that this compound, which is approximately 30 times more water soluble than BCNU (11%, compared to 0.4%), was as active on intracerebral as on ip L1210, showing again higher activity than seen

TABLE II. Effect of Different NUs on Intraperitoneal and Intracerebral L1210 Leukemia

Drug	Dose (mg/kg)	L 1210 Cr ip		L 1210 Cr ic		L 1210 Ha ip	
		% T/C	% cures	% T/C	% cures	% T/C	% cures
BCNU	40	213	35	204	40	236	45
	30	205	55	232	50	292	85
	20	162	25	198	20	235	50
	10	151	0	160	0	192	10
Methyl-CCNU	40	149	0	134	0	132	10
	30	177	40	163	35	225	60
	20	208	10	152	20	216	40
	10	152	0	133	0	173	0
CZT	40	147	10	102	15	140	20
	30	161	25	184	30	201	30
	20	206	50	187	50	198	45
	10	195	0	142	5	177	0
HECNU	40	83	0	92	0	101	0
	30	186	30	219	25	247	30
	20	–	100	–	100	–	100
	10	254	55	228	40	210	55

10^5 cells were injected on day 0 in $CD2F_1$ mice; drugs were given ip on day 1.

with BCNU; CZT and methyl-CCNU were again significantly less potent. The effective penetration in the brain was confirmed by obtaining 80–90% long-term survivors after single 15 mg/kg injections in a murine ependymoblastoma model in which equivalent results were seen with BCNU at doses 3 times larger.

The comparative activity of HECNU and other clinically used NUs was also assessed in a series of solid murine neoplasms, and Table III presents some representative results. In the colon 26 carcinoma model, HECNU and methyl-CCNU at 20 and 40 mg/kg ×1, respectively, gave 50–65% long-term (> 90 days) survivors. HECNU produced somewhat better results in terms of tumor growth delays and prolongations in survival also at other doses. Both substances were more active than BCNU which at the optimal 40 mg/kg dose gave a significantly ($p < .05$) lower percentage (25–40%) of long-term survivors in this system. This agent was in turn slightly but consistently more effective than CZT. In the mammary 16 carcinoma, similar delays in tumor growth (13 and 12 days) and increases in survival (187 and 178 T/C %) were seen with BCNU and HECNU at 30 and 20 mg/kg ×1, respectively, injected on day 10, whereas 40

TABLE III. Effect of Different NUs on Colon 26 and Mammary 16 Carcinomas in Mice

Drug	Dose (mg/kg)	Colon 26[a]			Mammary 16[b]	
		TGD[c]	% T/C	LTS[d]	TGD[c]	% T/C
BCNU	40	35	213	7/20	–	93
	30	35	198	2/20	33	178
	20	28	163	0/20	29	159
	10	21	139	0/20	27	141
Methyl-CCNU	40	38	>300	12/20	32	181
	30	30	181	5/20	29	158
	20	23	158	0/20	29	137
	10	21	138	0/20	25	133
CZT	40	36	184	5/20	31	179
	30	29	161	0/20	27	150
	20	23	158	0/20	27	148
	10	21	137	0/20	23	115
HECNU	40	–	84	0/20	–	88
	30	37	186	3/20	37	136
	20	38	>300	11/20	34	187
	10	32	205	5/20	30	168

[a] 2.10^5 cells injected im in CD2F$_1$ mice; drugs given ip on day 1; median survival time of controls 22 days.

[b] 10^5 cells injected im in C3H mice; drugs given ip on day 10; median survival time of controls 35 days.

[c] Tumor growth delay: time in days to reach 1 cm diameter (18 and 21 days in controls for colon 26 and mammary 16, respectively).

[d] Over 90 days survivors.

mg/kg were required with CZT and methyl-CCNU to observe comparable results. As presented in Table IV, slightly greater reductions in primary tumor weight, percentage of animals with lung metastases, and greater increases in survival were seen with HECNU than with CZT and BCNU in the 3LL carcinoma. The optimal dose of HECNU again being lower (20, 30, and 40 mg/kg, respectively) than for the two other NUs. In this system, however, methyl-CCNU was the most effective of the four compounds. In a total of 11 mouse and rat tumor models in which a comparison was made between HECNU, BCNU, and one other NU, HECNU was more effective than the most potent NU in nine systems. Table V gives a list of further experimental systems in which HECNU was found active, indicating that, as for other NUs, this compound possesses a broad system of activity, which has recently been observed to extend to various types of human xenografts in nude mice.

It is known for a number of cancer chemotherapeutic agents (e.g., cyclophosphamide, adriamycin, methotrexate) that their therapeutic activity is related to, and at least partially dependent on, tumor immunogenicity (Spreafico, 1980). In a further attempt comparatively to evaluate HECNU with other NUs, it was of interest to compare the effect of HECNU on the nonimmunogenic CR subline of L1210 leukemia (i.e., the subline commonly employed in screening tests) with its effect on the highly immunogenic Ha subline. These lines possess equal growth rates *in vivo* and *in vitro* and have repeatedly been shown to be equally

TABLE IV. Comparative Effect of HECNU and other NUs in the Murine 3LL Carcinoma

Drug	Dose (mg/kg)	% TWI[a]	% Mice w. metast.	% T/C
BCNU	40	61	55	203
	30	51	70	172
	20	27	90	156
	10	21	100	128
Methyl-CCNU	40	–	–	67
	30	75	35	>225
	20	63	60	204
	10	50	75	172
CZT	40	–	–	59
	30	54	70	191
	20	33	80	168
	10	22	85	143
HECNU	40	–	–	62
	30	71	20	158
	20	59	50	215
	10	40	75	166

[a] Tumor weight inhibition on day 21 after im transplant of 2.10^5 cells in C57B1/6 mice. Drugs were administered on day 1; 100% controls had lung metastases.

TABLE V. Antineoplastic Activity of HECNU in Some Rodent Systems

Tumor system	Dose (mg/kg ip)	% cures	% T/C	TGD[a] (days)
Mouse				
L1210 ip	20×1	100		
L1210 ic	20×1	100		
TLX 9 ip lymphoma	20×1	70		
3LL ca	10×3	100		
3645/75 mammary	25×1			17
Mammary 16	20×1		187	
TM2 mammary	10×3		165	
Colon 26	10×3	100		
Colon 51	20×1			13
P 815 mastocyt.	10×1	50		
C22LR osteosarc.	25×1	40		
Glioma 26	10×3		182	
Ependymobl.	10×1	80		
Rat				
L5222	15×1	90		
NK leuk.	20×1		>260	
Glioma 60	10×1	90		

[a] TGD = tumor growth delay.

sensitive in culture to a large variety of antitumor, antineoplastic agents. The data of Table II show that for both BCNU and methyl-CCNU the effectiveness was clearly better on the immunogenic line as indicated by the higher percentages of cures over a wider range of doses. Conversely, in the case of HECNU and CZT, the level of antineoplastic activity was not significantly dependent on tumor immunogenicity since these drugs displayed equal activity on both sublines. This point is of possible practical relevance considering that the immunogenicity of human neoplasms is believed to be generally low.

In the last year a number of other NUs have also been examined within the EORTC group. In view of their more recent synthesis, results are still limited, and no conclusions on their comparative value can be advanced. The ensuing brief description of some of these compounds is essentially to illustrate the types of chemical modification that are being explored.

Fiebig *et al.* (1980), examining the influence of different substituents at the N'-position of the urea molecule, have observed that the water-soluble derivative 1-(2-chloroethyl)-1-nitroso-3(methylenecarboxcimado)-urea(acetamido-CNU) and the methanesulfonate derivative of HECNU (HECNUMS), while not better than HECNU in the Walker carcinoma model, possess better therapeutic indexes than does BCNU, a superiority confirmed also on a DMBA-induced mammary carcinoma in rats. The values of the LD_{50}/EC_{50} ratios in the Walker cancer model

TABLE VI. Single-Dose Therapy of Preterminal L5222 Rat Leukemia with 2-Chloroethylnitrosoureas[a]

Drug	4	7	11.1	14	17.6	22.1	27.9	35.1	44.2	55.6
BCNU	170	197(1/6)	190(1/6)	215(1/6)	675(2/6)	680(2/5)	595	370(2/6)	70	–
CCNU	180	170	280(1/5)	540(3/6)	(5/5)	(5/6)	(5/6)	625(3/6)	420	–
MeCCNU	160	220(1/6)	500(3/6)	(5/5)	(5/5)	(6/6)	(6/6)	(6/6)	(4/6)	110
HECNUMS	160	170	185(1/6)	405(1/4)	270(1/5)	420(2/5)	395(2/6)	65	–	–
Acetamido-CNU	225(1/6)	235(2/6)	(5/6)	(5/6)	800(4/6)	615(2/6)	35	–	–	–
Dihydroxypropyl-CNU	145	190	205		150(2/8)		20		–	–
Carboxyethyl-CNU	110	145	165		165		210(1/7)		200(1/8)	–
Cyanoethyl-CNU		155	155(1/6)		325(1/6)		230(2/6)		20	–
Morpholino-CNU		155(1/6)	250(2/6)		180(1/6)		(6/6)	130	40	–
Piperidino-CNU	160	155(1/6)	250(2/6)		180(1/6)		(6/6)	130	40	–
Methylene-dioxybenzyl-CNU	125	150(1/8)	160(1/8)		180(2/8)		200(1/8)		650(3/8)	–
Methylene-3 pyridyl-CNU	230(2/5)	260(2/6)	470(3/8)		235(3/8)		50(2/6)		20	–
Methylene-4 pyridyl-CNU	145	155	160		170		490(4/8)	70	–	–

[a] Column head numbers are drug doses in mg/kg. Results shown are percentage ILS. Numbers in parentheses refer to cures/total animals.

were in fact 2:1 for BCNU, 2:4 for acetamido-CNU, 2:7 for HECNUMS, and 2:9 for HECNU. HECNUS, which at variance with HECNU is as poorly water soluble as BCNU, was on the other hand inferior to HECNU and to BCNU in the preterminal L5222 rat leukemia model. The activity in the latter system of acetamido-CNU and of a series of open-chain and cyclic derivatives is shown in Table VI. Because of the different chemical approach followed, 1-[3-(2-chloro-ethyl)-3-nitrosoureido]-4-[3-(2-chloroethyl)-1-nitrosoureido]-1,4-dideoxy-D,L-threithrol is of interest to discuss. This substance, so far the most interesting of a series of similar derivatives synthesized in Hungary, has shown potent activity at initial testing. When given orally at 50 mg/kg ×2 this compound has induced 90–100% cures in the L1210 and P388 murine leukemias, a q3d treatment from day 6 to 23 with 3 mg/kg ip producing over 50% cures in the 3LL carcinoma system. In view of the fact that very significant effects have also preliminaryly been observed in a series of other rodent tumors of widely different histology, but in a limited number of human tumor xenografts, this substance appears an interesting candidate for further, more extended testing.

III. SELECTING AMONG NU ANALOGS

For many antineoplastic agents, including NUs, the synthesis of a number of novel derivatives that possess high and comparable effectiveness in animal tumors but that are frequently only marginally better than commonly used compounds raises the problem of the most appropriate criteria to be adopted in selecting for clinical testing (Carter, 1978).

On the basis of the data described in the previous section, the conclusion can be advanced that HECNU, in addition to favorable physico-chemical characteristics, possesses a greater antitumor effectiveness than commercially available NUs in the majority of tested experimental tumors. However, this superiority may not be considered sufficiently high in comparison to clinically employed compounds of this class. In addition, the profile of the dose–response curve, with a relatively narrow margin between active and toxic doses, does not appear to be more favorable than for other NUs. Current data indicate that the spectrum of antitumor activity of HECNU does not appear to be substantially different from that of other NUs. It is, however, well known that the predictive ability of animal systems for responsiveness of given types of human neoplasms is limited. In the case of NUs, the argument can be advanced that the main objective in the synthesis of novel derivatives should be the development of substances displaying lower and/or different toxicity. This section will thus present a series of results comparing new analogs to clinically employed NUs.

Prolonged bone marrow depression is considered the most severe toxic effect of NUs. The traditional first-level approach to evaluate the hematotoxicity of antitumor agents in animals consists of determining the kinetics of WBC depression. When HECNU was thus compared with other NUs in mice, it was seen that the level of WBC and time to nadir as well as the total duration of WBC depression after single doses were essentially comparable with those observed with BCNU and methyl-CCNU at maximum tolerated doses, while CZT was less depressant. Given the generally higher antitumor activity of HECNU, the therapeutic index was therefore still more favorable for this compound. An alternative approach to a rapid screening system for the evaluation of the bone marrow toxicity of NUs was an attempt to derive information on this effect from mouse mortality data, following the hypothesis that prolonged bone marrow depression could result in greater sensitivity to repeated drug administrations. Graded drug doses were thus administered ip to normal mice either in single injections or repeated at 1, 2, or 3-week intervals up to 3 doses. Table VII shows one typical experiment in which melphalan was used as a negative control and busulphan as a presumably positive control. Since no differences were found to depend on the time interval between doses, data obtained using different time intervals were pooled. It may be observed that the LD_{50} in mg/kg per dose does not significantly decrease when NUs are administered repeatedly in single injections. Although the data obtained with this approach are still limited, these results do not favor the possibility that a short-term test for identifying NUs with reduced bone marrow toxicity on the basis of split-dose mortality may be effective.

Another parameter of toxicity that was investigated as a possible basis for selection among NUs as well as for its possible correlation with bone marrow toxicity was immunodepressive activity. Infections are a frequent and important complication of chemotherapy, and prolonged immunodepression may in principle contribute to the carcinogenicity of antitumor agents. In addition to the possibility that different interactions with host resistance mechanisms may contribute to the understanding of differential *in vivo* effectiveness among analogs (Spreafico, 1980), less immunodepressive agents may be expected to show better synergism with immunotherapeutic compounds. The data in Table VIII are illustrative of the latter point, showing that RFCNU, the recent sugar-containing NU of French origin which has been shown to have a comparatively lower immune interfering capacity (Imbach *et al.*, 1975), is more synergistic with immunomodulators such as *C. parvum* (or Pyran copolymer) than methyl-CCNU, which in our studies was significantly more immunedepressive.

TABLE VII. Lethality of NUs in Mice When Administered in Single or Repeated Doses

Drug	LD_{50} (mg/kg per dose)			
	1 dose	2 doses	3 doses	p
Melphalan	23.8	29.7	23.2	n.s.
Busulphan	79.1	62.7	54.2	$p < 0.02$
CCNU	49.3	67.0	55.2	n.s.
Methyl-CCNU	57.5	55.5	53.4	n.s.

TABLE VIII. Antileukemia Activity of Methyl-CCNU and RFCNU in Combination with *C. parvum*[a]

Exp. Group	Dose (mg/kg)	T/C %	% cures
Methyl-CCNU	30	213	35
	15	197	10
	7.5	155	0
Methyl-CCNU + *C. parvum*	30	277	60
	15	268	30
	7.5	241	0
RFCNU	15	202	40
	10	185	10
	5	147	0
RFCNU + *C. parvum*	15	–	80
	10	348	50
	5	235	30

[a] 10^6 L1210 Ha cells ip on day 0 in CD2F$_1$ mice, drugs on day 3 ip and *C. parvum* (1 mg iv) on day 6; this agent alone gave a 106 T/C %.

In the first system employed to evaluate immunodepressive activity, inhibition of primary antibody response to a standard antigen such as sheep erythrocytes was assessed by counting the number of antibody-forming cells (AFC) in the spleen. The time chosen for evaluation coincided with the maximum normal immune response to antigen innoculation and with the time of maximum drug-induced immunosuppression. Under these conditions exponential dose–response curves were obtained, and it was found that single HECNU doses were significantly more immundepressive on a mg/kg basis ($p < .05$) than BCNU and CZT. Methyl-CCNU was slightly less inhibitory than the two latter drugs. However, when the ED$_{50}$ in this condition was related to the acute (14 days) LD$_{50}$ and to the optimal antineoplastic dose in the L1210 model, the results indicated that both ratios were not significantly more unfavorable for HECNU than for the other

TABLE IX. Comparative Immunodepressive Activity of Nitrosoureas

	Inhibition humoral antibody			Inhibition cellular reactivity		
Drug	ED$_{50}$[a] (mg/kg)	$\dfrac{LD_{50}}{ED_{50}}$	$\dfrac{OAD^b}{ED_{50}}$	ED$_{50}$[c] (mg/kg)	$\dfrac{LD_{50}}{ED_{50}}$	$\dfrac{OAD^b}{ED_{50}}$
BCNU	7.1	7	4.3	15	3.3	2
Methyl-CCNU	9.2	6	3.2	15	3.1	1.7
CZT	7.6	8	2.6	17.5	3.4	1.1
HECNU	4	10	5	10	4	2

[a] Dose reducing to 50% of controls the number of anti-SRBC AFC in spleen.

[b] Single dose giving optimal activity on ip L1210 Cr.

[c] Dose reducing from 100% to 50% the long-term survivors in preimmunized L1210 Ha-challenged mice.

clinically available NUs tested (Table IX). These data would also seem to indicate an absence of correlation between NUs' capacity to induce bone marrow damage and inhibition of humoral antibody production. Similar results were reported for BCNU and CZT (Panasci *et al.*, 1977).

Schematically, humoral antibodies are believed to be more important in defense against bacterial infections, whereas cell-mediated reactions are involved in resistance to viral infections as well as in neoplastic growth control. As a first model in which to test NUs on the latter type of reactivities, their capacity to reduce in mice the T-lymphocyte-mediated resistance to a challenge with a syngeneic leukemia induced by previous specific immunization with x-irradiated tumor cells was measured. Table X presents data on one typical experiment. It was found that the lowest drug concentration that reduced the number of long-term survivors from 100% in immunized controls to 50% in drug-treated animals was 10 mg/kg for HECNU, 15 mg/kg for BCNU and methyl-CCNU, and 17.5 mg/kg for CZT. When the ratios between the LD_{50}, or the optimal anti-L1210 doses for these compounds, and the $ED_{100-50\%}$ dose in these conditions (Table IX) were calculated, the indexes for HECNU were again not significantly different from those of BCNU and methyl-CCNU and only slightly worse than for CZT. Although a more in-depth analysis of the interaction of these drugs with host defense mechanisms is needed before advancing definitive conclusions on their relative immunodepressive capacity, these findings nevertheless support the contention that for better, or at least comparable antineoplastic effectiveness, no significantly greater price in this toxicity is to be paid in animals with HECNU than with clinically used NUs. In addition, these results, together with those in

TABLE X. Effect of NUs on the Survival of Specially Preimmunized L1210 Ha-Challenged Mice

10^7 x rayed L1210 Ha cells (day −8)	Drug	Dose mg/kg (day −4)	10^7 L1210 Ha cells ip (day 0)	MST (days ± SE)	LTS[a]
−	−	−	+	7.0 ± 0.5	0/20
+	−	−	+	−	20/20
+	BCNU	20	+	15 ± 2.8	4/20
+	BCNU	15	+	24 ± 4.2	9/20
+	BCNU	10	+	−	14/20
+	Methyl-CCNU	20	+	19 ± 33	4/20
+	CCNU	15	+	34 ± 55	11/20
+	CCNU	10	+	−	17/20
+	CZT	20	+	18 ± 4.2	6/18
+	CZT	17.5	+	29 ± 5.5	9/20
+	CZT	15	+	33 ± 4.2	15/20
+	HECNU	15	+	19 ± 4.7	4/20
+	HECNU	10	+	32 ± 5.1	9/20
+	HECNU	7.6	+	−	16/20

[a] Long-term (over 90 days) survivors; in nonpreimmunized controls, drugs had no influence on MST.

Table II, favor the hypothesis that the higher effectiveness of HECNU on various animal cancers is attributable to factors such as intrinsic cytocydal capacity and pharmacokinetics rather than to a relative sparing of host defenses capable of synergizing with the drug's direct cell-killing activity.

In addition or as an alternative to an evaluation of acute toxicity, an obvious basis for selection among analogs is their chronic toxicity. Indeed, considering recent progress in the possibilities of treatment with at least some classes of compounds, chronic toxicity can be a more important factor. The modern clinical tendency to use protracted regimens of aggressive chemotherapy has recently renewed much attention to the well-known carcinogenic potential of many cytotoxic chemicals employed for cancer and immunosuppressive treatment. The significantly higher frequency of malignancies in patients submitted in long-term chemical immunodepression is well established (Harris, 1979). A reduced or absent carcinogenic activity could thus be an important criterion in the selection of novel NU analogs.

Following the above rationale, HECNU and other analogs were compared in rats with the clinically used NUs in terms of influence on life span and induction of neoplasms. The agents were administered iv to male Wistar rats in dosages similar to clinical application schemes (9.5–150 mg/m^2 every 6 weeks), treatment being discontinued when severe or lethal toxicity became apparent or after a maximum of 10 injections had been given. As shown in Table XI, CCNU and methyl-CCNU were the least toxic among the substances tested, the median survival time (MST) of rats given up to 75 mg/m^2 of these drugs being not significantly reduced in respect to vehicle-treated controls. Even at 150 mg/m^2 these agents could be administered for 7 and 8 injections, respectively, before a delayed lethal toxicity became apparent. It may here be mentioned that this toxicity was qualitatively similar for all compounds; the most prominent sign was loss of body weight, death being attributable to terminal infections. Almost invariably, histology reveals signs of liver damage, CTZ showing additionally a high degree of dose-dependent renal toxicity, observable in lesser frequency and severity with CCNU. Morpholion-CCNU was, on the other hand, much more toxic than CCNU and methyl-CCNU, MSTs being significantly reduced with this compound even at the lowest dose of 9.5 mg/m^2. In the case of CZT, doses above 38 mg/m^2 could not be administered since 7/30 rats died within 14 days of a single 75 mg/m^2 injection; MST was significantly reduced at 38 mg/m^2, but no significant influences on life span were observed at lower dosages. BCNU was the most toxic of this series of substances and 75 mg/m^2 could be given only ×4 resulting in a MST of 164 days. Even at 19 mg/m^2, there was a highly significant reduction in life span. The water-soluble HECNU was clearly less toxic than BCNU ($p < .01$). At 75 mg/m^2 a median total dose of 565 mg/m^2 was reached which gave a 50% reduction in life span, whereas lower dose levels did not significantly influence MST. Lastly, HECNU-MS was slightly more toxic than HECNU but was still significantly less toxic than BCNU.

When one considers the capacity of these analogs to induce tumors, which in

TABLE XI. Effects of Repeated Injections with Different NUs on Life Span of Normal Rats

	150 mg/m^2	75 mg/m^2	38 mg/m^2	19 mg/m^2	9.5 mg/m^2
CCNU					
Max. no. injections	7	10	10	10	–
Median total dose (mg/m^2)	1050	750	380	190	–
MST (days)	369	530	601	661	–
(95% range)	(327–617)	(487–617)	(504–619)	(597–702)	
Methyl-CCNU					
Max. no. injections	8	10	10	10	–
Median total dose (mg/m^2)	1200	750	380	190	–
MST (days)	374	599	619	555	–
(95% range)	(304–395)	(558–624)	(563–657)	(509–664)	
Morpholino-CNU					
Max. no. injections	4	4	10	10	10
Median total dose (mg/m^2)	600	300	380	190	95
MST (days)	158	251	467	533	499
(95% range)	(138–168)	(158–462)	(451–476)	(510–632)	(467–523)
CZT					
Max. no. injections	–	–	10	10	10
Median total dose (mg/m^2)	–	–	380	190	95
MSR (days)	–	–	474	590	583
(95% range)			(436–513)	(518–613)	(565–638)
BCNU					
Max. no. injections	–	4	10	10	10
Median total dose (mg/m^2)	–	300	266	152	95
MST (days)	–	164	263	314	568
(95% range)		(157–211)	(230–312)	(295–410)	(514–588)
HECNU					
Max. no. injections	–	10	10	10	10
Median total dose (mg/m^2)	–	565	380	190	95
MST (days)	–	331	541	572	558
(95% range)		(294–456)	(518–627)	(522–617)	(492–606)
HECNUMS					
Max. no. injections	–	4	10	10	10
Median total dose (mg/m^2)	–	300	380	190	95
MST (days)	–	262	502	530	530
(95% range)		(194–485)	(477–525)	(490–554)	(461–547)

[a] Male Wistar rats were used, 30 per experimental group and 120 controls, whose MST was 621 days (range 579–644).

TABLE XII. Carcinogenic Activity of N-Nitroso-(2-chloroethyl)urea Derivatives after Repeated Intravenous Application to Rats

Compound	Median total dose (mg/m²)	Malignant tumors (%)	SUM (%)	Nervous system malignant (%)	Nervous system 1ᵃ (%)	Lung malignant (%)	Lung 1 (%)	Forestomach malignant (%)	Forestomach 1 (%)
CCNU	1050	10	17	–	–	3	7	–	–
	750	7	30	–	–	–	13	–	10
	380	–	10	–	–	–	–	–	7
	190	–	3	–	–	–	–	–	–
Methyl-CCNU	1200	4	19	–	–	4	8	–	8
	750	8	27	–	–	–	–	–	–
	380	18	27	–	–	–	–	–	4
	190	18	23	9	9	–	–	5	5
Morpholino-CNU	600	19	30	4	4	15	19	–	7
	300	7	20	–	–	5	9	–	9
	380	14	21	4	4	7	7	4	8
	190	5	24	–	–	5	9	–	9
	95	–	7	–	–	–	–	–	–
CZT	380	10	15	–	–	5	5	–	5
	190	–	–	–	–	–	–	–	–
	95	8	16	4	4	–	–	4	8
BCNU	300	27	32	5	5	23	23	–	5
	266	–	13	–	–	–	13	–	–
	152	–	3	–	–	–	–	–	–
	95	4	12	4	–	–	8	–	3
HECNU	565	12	32	4	4	–	8	4	16
	380	9	22	9	9	–	4	–	9
	190	8	8	–	–	–	–	–	–
	95	6	12	6	6	–	–	–	6
HECNUMS	300	7	17	7	7	–	–	–	10
	380	7	14	3	3	–	–	–	7
	190	7	10	–	–	–	3	–	–
	95	–	4	–	–	–	–	–	4
Vehicle	–	8	15	1	1	–	1	–	6

ᵃ 1 represents sum of incidence (%) of preneoplastic alterations, benign and malignant tumors, based on numbers of evaluable animals. If an animal had more than one tumor in different organs, these were counted separately.

these experiments were more frequently found in the lung, forestomach, and CNS, CCNU and methyl-CCNU appeared to exert only a comparatively weak effect, no clear dose–response relationship being evident with the latter drug (Table XII). Considering that clear reductions in MST were seen at the higher doses tested, morpholino-CNU appeared to be a stronger carcinogen, inducing malignant, benign, and preneoplastic lesions predominantly in the lung. CZT was also carcinogenic, but the relatively limited number of doses examined due to its toxicity render comparisons more difficult. BCNU appeared to have the strongest effect of this series, inducing 23% malignant tumors at 75 mg/m² ×4 within a MST of only 164 days, whereas HECNU was significantly less carcinogenic ($p = <0.05$) than BCNU at 75 mg/m², and the same was true for HECNU-MS at the same dose level.

IV. CONCLUSION

These findings thus show clear differences in long-term toxic and carcinogenic effects in rats among four clinically employed NUs. In the conditions investigated BCNU emerges as the most toxic in terms both of life span reduction upon repeated administration and of carcinogenic potential. In the absence of undisputed evidence for a clinical superiority of BCNU over CCNU, CZT, and methyl-CCNU (Wasserman *et al.*, 1975), these results can have a direct clinical relevance for a less empirical selection among available agents. With regard to the newer analogs and on a more general level, these results support the contention that chemotherapeutic effectiveness is not necessarily linked with at least certain types of acute toxicity. Of possible greater practical relevance is also the observation that high chemotherapeutic effectiveness can be dissociated from a high chronic toxicity and carcinogenic potential, even relatively minor chemical modifications producing analogs of enhanced cell-killing ability but reduced toxicity, as in the case of HECNU.

ACKNOWLEDGMENT

This work was performed in the frame of the activities of the Screening and Pharmacology Group of the European Organization for Research on the Treatment of Cancer (EORTC) and of the German Drug Development and Testing Working Group.

REFERENCES

Carter, S. K. (1978). *Cancer Chemother. Pharmacol. 1,* 69–72.
Eisenbrand, G., Fiebig, H. H., Zeller, W. J. (1976). *Z. Krebsforsch. 86,* 279–286.
Fiebig, H. H., Eisenbrand, G., Zeller, W. J., Zentgraf, R. (1980). *Oncol. 37,* 177–183.

Harris, C. C. (1979). *J. Natl. Cancer Inst. 63,* 275–277.
Imbach, J. L., Montero, J. L., Moruzzi, A., Serrou, B., Chenn, E., Hayat, M., MaMathé, G. (1975). *Biomed. 23,* 410–413.
Panasci, L. C., Fox, P. A., Schein, P. S. (1977). *Cancer Res. 37,* 3321–3328.
Penn I. (1979). *Transpl. Proc. 11,* 1047–1051.
Schabel, F. M. (1976). *Cancer Treat. Rep. 60,* 665–698.
Spreafico, F. (1980). *In* "Recent Results in Cancer Research" (G. Mathe, ed.), Vol. 75, pp. 209–215. Springer-Verlag, New York.
Wasserman, T. H., Slavik, M., and Carter, S. K. (1975). *Cancer 36,* 1258–1268.
Zeller, W. J., Eisenbrand, G., and Fiebig, H. H. (1978). *J. Nat. Cancer Inst. 60,* 345–348.

Chapter 4

CHEMICAL DECOMPOSITION OF CHLOROETHYLNITROSOUREAS

Michael Colvin
Robert Brundrett

I. INTRODUCTION

The chloroethylnitrosourea antitumor agents were developed on the basis of the observed antitumor activity of methylnitrosoguanidine (Skinner *et al.,* 1960; Schabel *et al.,* 1963), a known methylating agent. The spectrum of antitumor activity was found to be similar to that of the nigrogen mustards, and tumors resistant to the nitrogen mustards were found to be cross-resistant to chloroethylnitrosoureas (Schabel *et al.,* 1963; Wheeler and Bowdon, 1965). These facts suggested that the chloroethylnitrosoureas were acting through an alkylating mechanism. Ludlum and colleagues (Kramer *et al.,* 1974) demonstrated that the reaction of BCNU with polycytidylic acid produced 3-hydroxyethylcytidylic acid and 3,N^4-ethanocytidylic acid. This observation indicated that BCNU was contributing to the target nucleotide a two-carbon fragment capable of a second alkylation, as shown by the formation of 3,N^4-ethanocytidylic acid (Fig. 1).

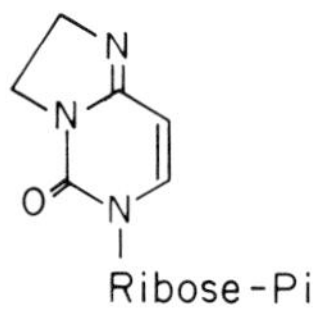

Ribose-Pi

Fig. 1. 3, N^4-ethanocytidylic acid.

II. DECOMPOSITION OF NITROSOUREAS

The aqueous decomposition of the chloroethylnitrosoureas is relatively rapid and is pH dependent (Loo *et al.*, 1966). Representative *in vitro* half-lives of BCNU and CCNU are shown in Table I. As can be seen, the rate of decomposition is markedly accelerated in the presence of plasma.

Levin and colleagues have presented evidence that this increase in the rate of decomposition in plasma is mediated by a heat-labile macromolecule, possibly albumin. These investigators also found that plasma lipids slow the rate of degradation of chloroethylnitrosoureas (Levin *et al.*, 1978).

Montgomery *et al.* (1967) studied the aqueous decomposition of chloroethyl-nitrosoureas and found the major products to be alkyl isocyanates (which undergo further reactions) and acetaldehyde.

On the basis of this finding, these investigators postulated the formation of an oxazolidine intermediate that would decompose to produce an ethylenediazohydroxide alkylating moiety. However, when the decomposition of chloroethylnitrosoureas was studied under buffered conditions at physiological pH, the major product was found to be chloroethanol, as shown in Table II (Colvin *et al.*, 1974). Similar products were shown to be generated by the nitrosative deamination of chloroethylamine (Colvin *et al.*, 1974). On the basis of these findings, a chloroethyl carbonium moiety was postulated to be the critical alkylating entity produced by the spontaneous decomposition of the chloroethylnitrosoureas, as shown in Fig. 2 (Colvin *et al.*, 1976). The production of acetaldehyde was accounted for by the postulation of a hydride shift. Similar products from the buffered aqueous decomposition of BCNU and CCNU have also been found by Reed *et al.*, (1975) and Montgomery *et al.*, (1975).

Evidence for the validity of the proposed decomposition scheme was provided by the synthesis and study of the deuterated BCNU analogs shown in Fig. 3. These compounds were allowed to decompose in buffer at pH 7.4, and the distribution of the deuterium atoms in the products was localized by mass spectrometry. These data were found to be compatible with the postulated decomposi-

TABLE I. Half-Life Periods in Minutes for *In Vitro* Aqueous
Decomposition of Chloroethylnitrosoureas at 37°C

	pH 6.0	pH 7.4
BCNU (0.1 *M* phosphate buffer)	314[a]	52[a]
BCNU (plasma)		17[a]
CCNU (0.1 *M* phosphate buffer)		48[b]

[a] Loo *et al.* (1966).
[b] Reed *et al.* (1975).

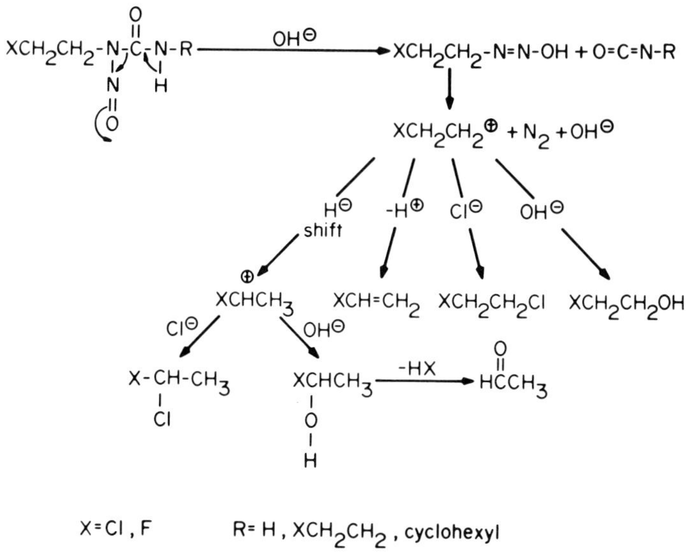
XCH2CH2-N-C-N-R
OH⊖
XCH2CH2-N=N-OH + O=C=N-R
XCH2CH2⊕ + N2 + OH⊖
H⊖ shift
-H⊕
Cl⊖
OH⊖
XCHCH3
XCH=CH2
XCH2CH2Cl
XCH2CH2OH
X-CH-CH3
XCHCH3
-HX
HCCH3
X=Cl, F
R= H, XCH2CH2, cyclohexyl

Fig. 2. Decomposition of chloroethylnitrosoureas.

tion pathway (Fig. 2), but not with alternative pathways, such as vinylcarbonium or diazochlorethane intermediate (Brundrett *et al.*, 1976).

An interesting and potentially important result of these studies was the evidence that the alkylations are occurring through an S_N2 attack on the chloroethyldiazonium hydroxide entity, rather than through a free chloroethyl carbonium ion. If a free chloroethyl carbonium ion were produced, it should undergo substantial transition to the cyclic chloronium ion, which would result in complete scrambling of the carbons bearing the deuterium atoms in the products of decomposition (Fig. 4). Such scrambling appeared to be occurring in less than 10% of the decomposition products, indicating that most of the reactions are occurring through the chloroethyldiazonium moiety. The concerted nature of the substitution is supported by studies on the stereochemistry of the decomposition products from methyl substituted BCNU (Brundrett and Colvin, 1977). This finding may have considerable pharmacologic importance. Chloroethylnitrosoureas bearing a substituent on the alpha carbon, as indicated in Fig. 5, have been found by

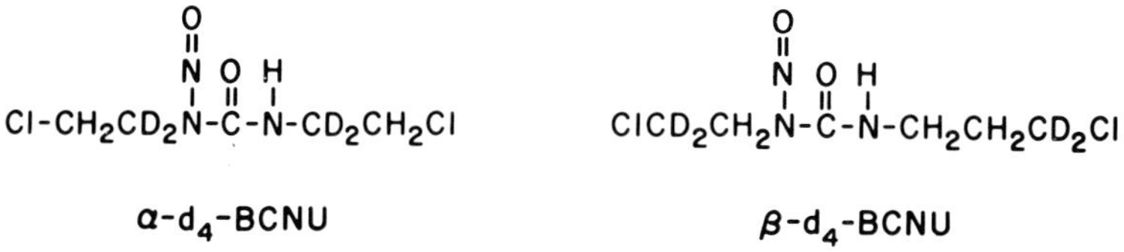
Cl-CH2CD2N-C-N-CD2CH2Cl
α-d4-BCNU
ClCD2CH2N-C-N-CH2CH2CD2Cl
β-d4-BCNU

Fig. 3. Deuterated BCNU analogs.

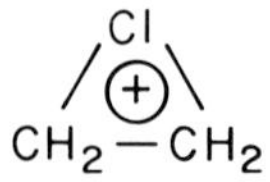

$$\underset{CH_2-CH_2}{\overset{Cl}{\diagup \overset{\oplus}{\diagdown}}}$$

Fig. 4. Cyclic chloronium ion.

Montgomery and colleagues (1976) and by the authors to exhibit no antitumor activity. Since secondary carbonium ions are energetically more stable than primary carbonium ions, the production of the free secondary carbonium ion from the alpha-substituted diazonium intermediate would be favored. Thus the greater alkylation specificity and putative longer half-life of the chloroethyldiazonium molecule may be essential to the antitumor activity of the chloroethylnitrosoureas.

Evidence that the spontaneous decomposition of the chloroethylnitrosoureas is essential to their antitumor activity comes from analog studies. As illustrated in Fig. 2, the initial step in the postulated decomposition scheme is the abstraction of a proton from the 3-nitrogen. If this proton is replaced by an alkyl group (Fig. 6), the molecule becomes much more stable in aqueous solution and the *in vitro* cytotoxicity is markedly reduced (Colvin *et al.*, 1976). This result is consistent with the hypothesis that the base-initiated decomposition of the parent molecule is essential to the characteristic cytotoxic effects of the chloroethylnitrosoureas. Such 3-nitrogen disubstituted derivatives, however, are active *in vivo* since the alkyl groups can be enzymatically removed.

Brundrett has recently compared the decomposition of chloroethylnitrosoureas at pH 5 and 7.4. In contrast to the products at pH 7.4 (see Table II), at pH 5 the major products of decomposition are acetaldehyde (Brundrett, 1980) and ethylene glycol. By studies with deuterium-labeled BCNU it was demonstrated that the acetaldehyde produced at pH 5 arises via a hydride shift. These results are consistent with the formation of an oxadiazole intermediate by cyclization through the nitroso group oxygen resulting from intramolecular cyclization of the molecule and subsequent formation of a hydroxyethyldiazonium moiety, as shown in Fig. 7. Similar products could result from cyclization through the carbonyl group to form an oxazolidine intermediate, and at the present time these two pathways cannot be distinguished.

Thus competing mechanisms for the aqueous decomposition of chloroethyl-

$$\underset{\underset{R_2}{|}}{\overset{\overset{O}{\|}}{\underset{}{N}}} \quad \overset{O}{\underset{}{\|}} \quad H$$

$$ClCH_2\overset{|}{\underset{|}{C}}H-\overset{|}{N}-\overset{|}{C}-\overset{|}{N}-R_1$$

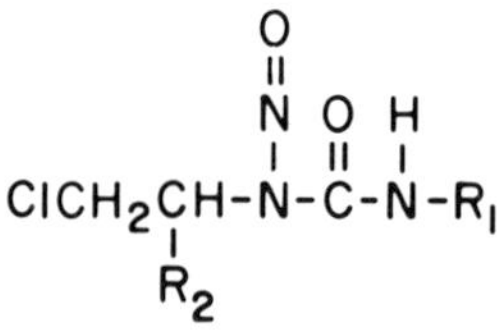

Fig. 5. Alpha-substituted chloronium ion.

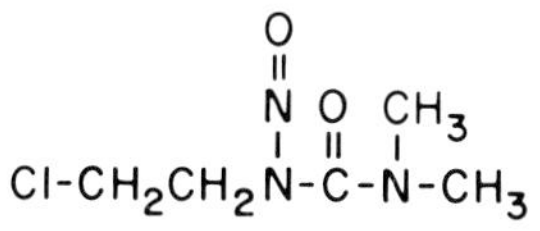

$$
\begin{array}{c}
\text{O} \\
\| \\
\text{N} \quad \text{O} \quad \text{CH}_3 \\
| \quad \| \quad | \\
\text{Cl-CH}_2\text{CH}_2\text{N-C-N-CH}_3
\end{array}
$$

Fig. 6. 1-chloroethyl-3,3-dimethylnitrosourea.

TABLE II. Volatile Products Produced from the Decomposition of
BCNU and Chloroethylamine[a]

	BCNU	Chloroethylamine
Acetaldehyde	31%	18%
Dichloroethane	2%	13%
Chloroethanol	63%	66%
Vinyl chloride	4%	3%
	100%	100%

[a] For experimental details see Colvin et al. (1974). Values expressed
as percentage of total identified volatiles.

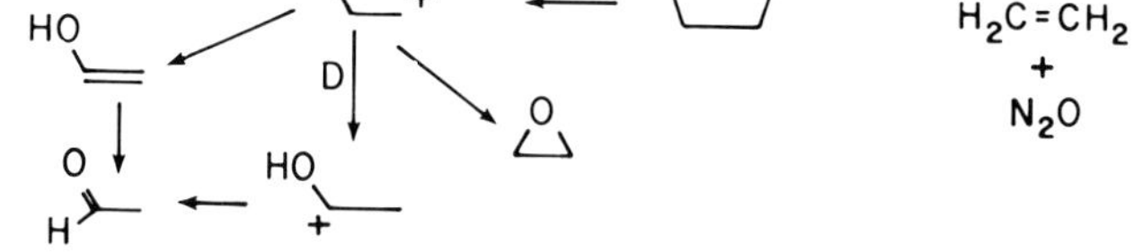

Fig. 7.

nitrosoureas exist. At pHs greater than neutrality, including physiologic pH, proton abstraction to produce chloroethyldiazonium hydroxide clearly predominates. However, as the concentration of hydroxyl ion decreases, the cyclization reaction with production of hydroxyethyldiazonium hydroxide will become predominant and produce hydroxylation, as opposed to chloroethylation, of target nucleophiles.

(a) Chloroethylnitrosocarbamate $CICH_2CH_2N-C-O-CH_3$ (with N=O above the N and C=O)

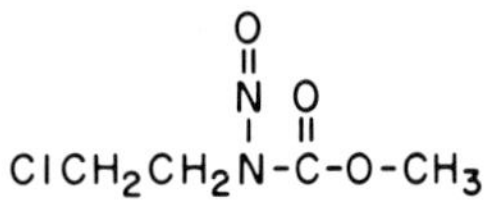

(b) Chloroethylnitrosoacetamide $CICH_2CH_2N-C-CH_3$ (with N=O above the N and C=O)

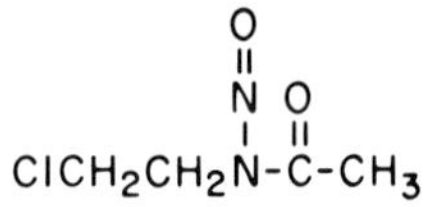

Fig. 8.

III. DECOMPOSITION OF OTHER "NITROSO" COMPOUNDS

There are several nitroso compounds which should decompose to form chloroethyldiazonium hydroxide, in a manner similar to that of the nitrosoureas. Two such types of compounds are chloroethylnitrosocarbamates and chloroethylnitrosoacetamides, shown in Fig. 8. Such compounds are markedly mutagenic, but are inactive against the murine L1210 leukemia *in vivo* (Brundrett *et al.,* 1979). Recently we have found that these compounds are markedly cytotoxic *in vitro* and are approximately 500 times more potent DNA crosslinking agents than are the chloroethylnitrosoureas when exposed to cells *in vitro*. However, the cytotoxic, mutagenic, and crosslinking activities of these compounds are rapidly destroyed by plasma, and this inactivation appears to be mediated by a plasma esterase. This enzymatic destruction probably accounts for the lack of antitumor activity of these compounds in the whole animal.

IV. CONCLUSION

In summary, the alkylating activity of the chloroethylnitrosoureas appears to be mediated by the spontaneous generation of chloroethyldiazonium hydroxide which will chloroethylate nucleophiles, including sites on nucleotides. The formation of a reactive chloroethylamino group *in situ* on nucleotides in DNA is a likely explanation for the crosslinking of DNA that has been observed by Kohn (1977). The nature of the decomposition reaction, and the consequent alkylating entities produced, is very dependent on the pH of the reaction, and relatively subtle chemical alterations in the structure of the nitroso-bearing portion of the molecule can alter the biologic activity of the molecule. Whereas, as would be predicted from chemical considerations, other nitroso compounds have the potential to be cytotoxic by a mechanism similar to that of the nitrosoureas, these compounds are rapidly altered by host enzymatic factors.

REFERENCES

Brundrett, R. B. (1980). *J. Med. Chem. 23,* 1245.
Brundrett, R. B., and Colvin, O. M. (1977). *J. Org. Chem. 42,* 3538.

Brundrett, R. B., Cowens, J. W., Colvin, M., and Jardine, I. (1976). *J. Med. Chem. 19,* 958.

Brundrett, R. B., Colvin, O. M., White, E. H., McKee, J., Hartman, P. E., and Brown, D. L. (1979). *Cancer Res. 39,* 1328.

Colvin, M., Cowens, J. W., Brundrett, R. B., Kramer, B. S., and Ludlum, D. B. (1974). *Biochem. Biophys. Res. Comm. 60,* 515.

Colvin, M., Brundrett, R. B., Cowens, J. W., Jardine, I., and Ludlum, D. B. (1976). *Biochem. Pharmacol. 25,* 695.

Kramer, B. S., Fenselau, C. C., and Ludlum, D. B. (1974). *Biochem. Biophys. Res. Comm. 56,* 783.

Kohn, K. W. (1977). *Cancer Res. 37,* 1450.

Levin, V. A., Hoffman, W., and Weinkam, R. J. (1978). *Cancer Treat. Rep. 62,* 1305.

Loo, T. L., Dion, R. L., Dixon, R. L., and Rall, D. P. (1966). *J. Pharm. Sci. 55:* 492.

Montgomery, J. A. (1976). *Cancer Treat. Rep. 60,* 651.

Montogmery, J. A., James, R., McCaleb, G. S., Kirk, M. C., and Johnston, T. P. (1975). *J. Med. Chem. 18,* 568.

Montgomery, J. A., James, R., McCaleb, G. S., and Johnston, T. P. (1967). *J. Med. Chem. 10,* 668.

Reed, D. J., May, H. E., Boose, R. B., Gregory, K. M., and Beilstein, M. A. (1975). *Cancer Res. 35,* 568.

Schabel, F. M. Jr., Johnston, T. P., McCaleb, G. S., Montgomery, J. A., Laster, W. R., and Skipper, H. E. (1963). *Cancer Res. 23,* 725.

Skinner, W. A., Gram, H. F., Greene, M. O., Greenberg, J., and Baker, B. R. (1960). *J. Med. Pharm. Chem. 2,* 299.

Wheeler, G. P., and Bowdon, B. J. (1965). *Cancer Res. 25,* 1770.

Chapter 5

METABOLISM OF NITROSOUREAS

Donald J. Reed

I. INTRODUCTION

A. Cytochrome P-450-Dependent Monooxygenation of Nitrosoureas

The chemical reactivity of the nitrosoureas would suggest that the initial metabolic fate of this class of drugs in the body would depend primarily upon spontaneous decomposition during body distribution. The main degradation products are reactive intermediates that include carbonium ion alkylating agents and isocyanates capable of carbamoylation reactions. However, as indicated in Fig. 1, certain intact nitrosoureas may undergo several rapid metabolic transformations that include cytochrome P-450-dependent monooxygenation to form monohydroxylated derivatives of the cyclohexyl ring of CCNU, the cyclohexyl ring, the methyl group, and the α-methylene group of MeCCNU and denitrosation to form the parent urea, particularly of BCNU and MeCCNU.

The rates of monooxygenation have been examined using liver microsomes isolated from control and 3-methyl cholanthrene- or phenobarbital (PB)-induced rats (Table I). Hydroxy CCNU metabolite formation from CCNU by microsomes from rats or mice was induced 6 to 8 fold with phenobarbital while being decreased slightly by 3-MC induction. The overall rate of hydroxylation of CCNU was much higher than that of MeCCNU in both control and PB-induced rats and mice.

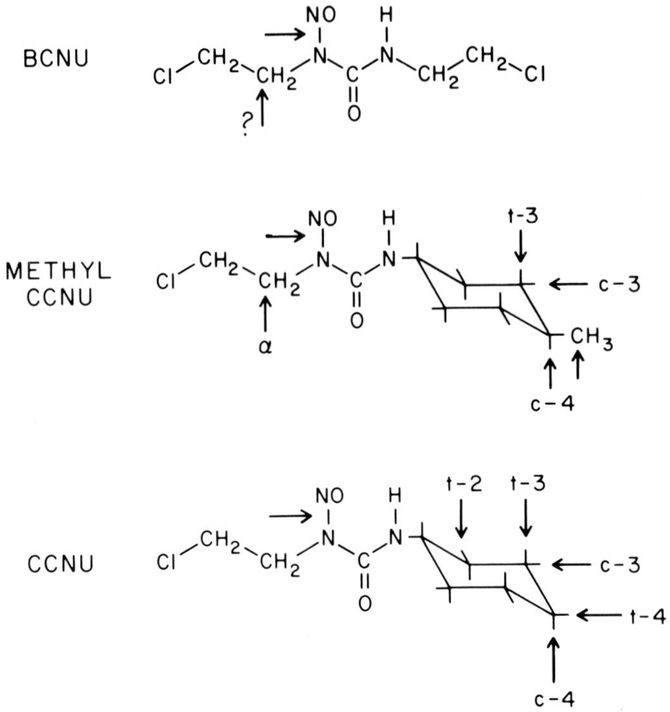

Fig. 1. Sites of cytochrome P-450-dependent hydroxylation and denitrosation.

TABLE I. Rates of Monooxygenation of CCNU and *Trans*-4-Methyl CCNU by Liver Microsomes from Control and Phenobarbital (PB)-Induced Rats and Mice

Nitrosourea	Total OH-CCNU metabolite formation (nmol min^{-1} mg protein^{-1})				
		Rat		Mouse	
	Control	3-MC induced	PB-induced	Control	PB-induced
CCNU	3.3[a]	2.0[a]	19.8[a] 11.7[b] 23.3[c]	2.1[a]	16.3[a]
trans-4-methyl CCNU	1.1[a]		3.3[a]	0.5[a]	4.0[a]

[a] Data of May *et al.* (1975).
[b] Data of Farmer *et al.* (1978).
[c] Data of Hilton and Walker (1975).

The principal hydroxylated metabolite of CCNU is *trans*-3-hydroxy CCNU in control and 3-MC-induced rats and *cis*-4-hydroxy CCNU in PB-induced rats and control mice (Table II).

The data by Hilton and Walker (1975) were obtained by degradation of the hydroxy CCNUs to the corresponding aminocyclohexanols and derivatized to N-(2,4-dinitrophenyl) derivatives for HPLC. The stereochemistry of the *cis*-3-OH-CCNU highly favors intramolecular carbamoylation to give the corresponding cyclic carbamate (urethane) during decomposition of this isomer (May *et al.*, 1975; Montgomery, 1976). Thus, the data obtained by Hilton and Walker (1975) appear to underestimate the quantity of *cis*-3-OH-CCNU formed (Table II).

MeCCNU not only displays a slower rate of microsomal monohydroxylation than CCNU but also a more complex metabolite pattern. Seven metabolites of MeCCNU have been isolated and characterized from control and PB-induced rats and mice (Table III). Not only are the expected ring hydroxylation products observed, but also a metabolite resulting from α-hydroxylation of the ethylene side chain and denitrosation to form the parent urea (Fig. 1).

B. Loss of Stereospecificity during Methyl CCNU Hydroxylation

Most unusual was the formation of a hydroxylated metabolite that could form only from the loss of the methyl group stereospecificity during the cytochrome P-450-dependent monooxygenation of MeCCNU. This metabolite was estimated to be about 30% of the total of metabolites 5A plus 5B (Table III) from control animals and 60% from PB-induced animals. Inversion or flip of the methyl group to the opposite side of the alicyclic ring during hydroxylation at the 4-position yields *trans*-4-hydroxy-*cis*-4-methyl CCNU (Potter and Reed, 1980). These

TABLE II. Microsomal Metabolites of CCNU

	Percentage of total OH-CCNU derivatives						
	Rat						Mouse
Isomer	Control[a]	PB[a]	3-MC[a]	PB[b]	Control[c]	PB[c]	Control[d]
Cis-2-OH				trace			
Trans-2-OH				2	14	3	1
Cis-3-OH	30	16	25	20	trace	3.3	13
Trans-3-OH	39	13	40	6	31	11	12
Cis-4-OH	21	67	30	62	54	77	46
Trans-4-OH	9	5	5	6	3	5	28

[a] Data of May *et al.* (1975).

[b] Data of Farmer *et al.* (1978).

[c] Data of Hilton and Walker (1975).

[d] Data of Wheeler *et al.* (1977).

workers have also shown the rapid hydroxylation of *cis*-4-McCCNU to yield a major metabolite *trans*-4-hydroxy-*cis*-4-MeCCNU and a minor metabolite, *cis*-4-hydroxy-*trans*-4-MeCCNU, which results from a methyl flip in the opposite direction (Potter and Reed, 1980). A possible mechanism for these transformations involves transitory free radical intermediates that permit a limited loss of stereospecificity during the monooxygenation of the carbon atom in the four position (Fig. 2). This mechanism may be applicable to all alicyclic carbon hydroxylation reactions and may account in part for a portion of the ring *cis*, *trans* hydroxylated products of CCNU as well as the *cis* and *trans*-4-MeCCNUs.

C. Denitrosation of Nitrosoureas

Certain nitrosoureas undergo a denitrosation reaction that is catalyzed by liver microsomes in the presence of NADPH. The rates of parent urea formation are

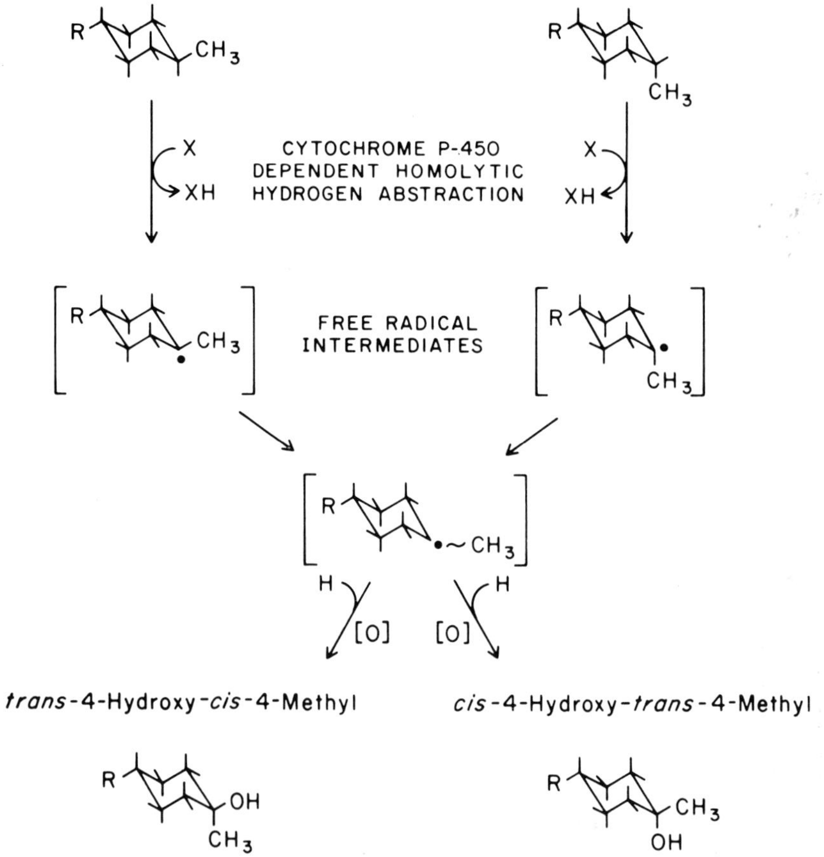

Fig. 2. Postulated mechanism for the loss of methylcyclohexyl stereospecificity during cytochrome P-450-dependent monooxygenation.

TABLE III. Comparison of the Rate of Formation of Methyl CCNU Metabolites by Liver Microsomes from Control and PB-Induced Rats and Mice[a]

	Metabolite	nmol/min/mg protein			
		Rat		Mouse	
		Control	PB	Control	PB
1	trans-3-OH-trans-4-methyl-CCNU	0.11	0.15	0.04	0.29
2	α-OH-trans-4-methyl-CCNU	0.07	0.48	0.04	0.40
3	cis-4-OH-trans-4-methyl-CCNU	0.16	1.11	0.12	1.85
4	cis-3-OH-trans-4-methyl-CCNU	0.13	0.37	0.08	0.17
5A & 5B	Methyl-CCU + trans-4-OH-cis-4-methyl-CCNU	0.51	1.45	0.06	0.91
6	trans-4-hydroxymethyl-CCNU	0.50	0.35	0.21	0.77
	Total	1.48	3.91	0.55	4.39

[a] Data of May et al. (1979).

increased by PB induction of both rats and mice (Tables III and IV). Limited evidence to date suggests a possible role for the cytochrome P-450 enzyme system in this reaction which leads to a possible nitroso complex with an absorption maximum at 445 nm (Reed and May, 1977).

Certain nitrosoureas can undergo denitrosation *in vivo* which results in decreased antitumor activity, especially after induction of the endoplasmic reticulum. Phenobarbital pretreatment of rats bearing intracerebral 9L tumors eliminated the antitumor activity of BCNU and reduced the activity of PCNU and CCNU (Levin et al., 1979). Reduction of systemic toxicity was noted which agreed with a microsomal catalyzed rate of BCNU disappearance that was doubled by phenobarbital pretreatment.

D. Plasma Half-Lives of Nitrosoureas

Every nitrosourea examined, with the possible exception of RFCNU (Godenèche et al., 1980), has been found to have a shorter *in vivo* plasma

TABLE IV. Rates of Chloroethylnitrosourea Denitrosation by Liver Microsomes

Nitrosourea	Rate of denitrosation (nmol min^{-1} mg protein^{-1})			
	Rat		Mouse	
	Control	PB-induced	Control	PB-induced
BCNU	1.7[a]	4.2[a]	0.7[b]	3.4[b]
Trans-4-methyl-CCNU	0.4[c]	0.6[c]	0.1[c]	0.4[c]

[a] Data from Levin et al. (1979).

[b] Data from Hill et al. (1975).

[c] Data from May et al. (1979).

Table V. *In vivo* Plasma Half-Lives of 2-Chloroethylnitrosoureas

| | Plasma half-life (min) | | | | | |
| | Nitrosourea | | | | | |
Species	BCNU	CCNU	MeCCNU	ACNU	CZ	RFCNU
Rat				12^f		$<5^c$
Mouse	$<5^a$	6^a	5^a		5^d	
	ca 15^b	ca 5^b				
Monkey	$<5^a$	$<15^a$				
Dog	$<5^a$	$<15^a$				
	$<15^b$	ca 5^b				
Man	11.6^e		$<10^b$			$<5^c$
	15.6^e					

[a] Data from Oliverio (1976).
[b] Data from Wheeler (1974).
[c] Data from Godenèche *et al.* (1980).
[d] Data from Mhatre *et al.* (1978).
[e] Data from Levin *et al.* (1978).
[f] Data from Shigehara and Tanaka (1978).

half-life than that predicted from the chemical half-life in buffer at physiological pH (Table V; and Colvin and Brundrett, this volume). Thus, rapid *in vivo* transformations of nitrosoureas appear dependent not only upon their chemical stability but also upon protein-mediated decomposition (Weinkam *et al.*, 1980), cytochrome P-450-dependent hydroxylation, and denitrosation reactions.

Whole-body disposition of nitrosoureas is characterized by a rapid and widespread distribution, including the central nervous system. Except where intramolecular carbamoylation limits the extent of cellular carbamoylation, the ethyl moiety is retained to about the same degree as the isocyanate moiety. An example of the effects of intramolecular carbamoylation is shown with chlorozotocin (Mhatre *et al.*, 1978; Plowman *et al.*, 1978). Protein binding, however, has been shown to be greater for the cyclohexyl moiety than the ethyl moiety of CCNU, whereas the reverse was true for biliary excretion (Oliverio *et al.*, 1970). Rapid disappearance of both the intact nitrosoureas and their degradation products from the body results in high yields of urinary metabolites within 24–48 hr. However, some covalent labeling may persist for much longer periods (Wheeler *et al.*, 1964; DeVita *et al.*, 1967; Oliverio *et al.*, 1970; Sponzo *et al.*, 1973; Plowman *et al.*, 1978; Mhatre *et al.*, 1978; Shigehara and Tanaka, 1978 and Godenèche *et al.*, 1980).

E. Protein-Mediated Decomposition of Nitrosoureas

Weinkam *et al.* (1980) have shown that the rates of chemical degradation of certain nitrosoureas in serum are significantly higher than in aqueous buffer at the same pH and temperature. This effect appears to be due to nonspecific interactions between lipophilic nitrosoureas and serum proteins. Half-life change for BCNU in serum was 49–14 min, 53–30 min for CCNU, and 49–29 min for MeCCNU, with no change for PCNU. Lack of effect upon PCNU may be due to its having a much greater water solubility than BCNU or CCNU (Levin and Kabra, 1974).

F. *In Vivo* **Metabolism of Nitrosoureas**

Reed and May (1977) have described rapid *in vivo* formation of hydroxy CCNU metabolites within 2 min after iv administration of CCNU to PB-induced rats. Transformation products included unidentified products that were more polar than the hydroxy CCNU metabolites (Reed and May, 1977).

Comparison of the distribution of ^{14}C labeling during metabolism of either ethylene or cyclohexyl-labeled CCNU indicates that blood and liver tissue underwent more extensive interaction with the cyclohexyl moiety than the ethylene moiety (Table VI). With [chx-1-^{14}C] CCNU, 30% of the ^{14}C was unextractable from blood by hexane, ether, or ether and methanol (Table VI). Liver tissue, while being labeled with 17–20% of the administered ^{14}C, contained only small amounts of unchanged CCNU as indicated by only 6–12% of the ^{14}C being extractable with hexane. More than one-half of the ^{14}C was extracted with ether and methanol after the hydroxy CCNU metabolites had been extracted with ether. ^{14}C labeling of the small intestine resulted in a low level of unextractable ^{14}C whereas brain ^{14}C was about 70% unchanged [chx-1-^{14}C] CCNU. [Ethylene-U-^{14}C] CCNU was not transformed into unextractable product(s) in the blood within 2 min after administration (Table VI). A total of 99% of the ethylene ^{14}C was extractable from blood, and 93% was extractable from liver tissue. The small intestine and brain were ^{14}C labeled largely with intact CCNU and hydroxy CCNU metabolites (Table VI) as indicated by analysis of the hexane and ether extracts respectively.

The low levels of unextractable ^{14}C during metabolism of ethylene-labeled CCNU can be related to the extensive role of sulfur in the formation of urinary metabolites. Reed and May (1975) isolated and identified thiodiacetic acid and S-carboxymethyl L-cysteine as major metabolites. These metabolites may form via 2-chloroethanol or by direct alkylation of glutathione or cysteine (Fig. 3).

G. **Specific Protein Carbamoylation Reactions**

Much effort has been expended to correlate the biological effects of individual nitrosoureas to their alkylation and/or carbamoylation capabilities as measured by *in vitro* assay methods (Wheeler *et al.,* 1974). Carbamoylation of lysine has been used as a quantitative measure of nitrosourea decomposition to yield a reactive isocyanate. Extrapolation of the results from this *in vitro* test to *in vivo*

TABLE VI. [Ethylene-U-^{14}C]CCNU or [Chx-1-^{14}C]CCNU Distribution Two Minutes after IV Administration to Phenobarbital-Induced Rats[a]

Organ	[^{14}C]CCNU		Percentage of total ^{14}C[b]	Percentage extracted[b]			
	Ethylene	Cyclohexyl		Hexane	Ether	Ether & methanol	Residue
Blood	+		14.7 ± 3.5	34.3 ± 3.5	36.0 ± 2.0	28.7 ± 3.2	<1
		+	9.5 ± 2.4	26.2 ± 6.2	20.3 ± 3.0	23.7 ± 6.5	29.8 ± 5.0
Liver	+		20.3 ± 5.0	12.0 ± 8.5	28.7 ± 4.7	52.0 ± 1.4	7.0 ± 0.0
		+	16.9 ± 6.4	5.5 ± 6.2	16.2 ± 5.1	56.4 ± 9.3	21.9 ± 2.0
Small intestine	+		7.0 ± 1.0	55.3 ± 7.2	32.0 ± 3.6	11.7 ± 3.8	1.0 ± 0.0
		+	5.6 ± 0.4	47.4 ± 24.8	16.7 ± 4.0	33.4 ± 19.2	2.4 ± 0.6
Brain	+		3.1 ± 0.5	68.3 ± 7.5	22.3 ± 6.7	8.3 ± 5.5	1.0 ± 0.0
		+	2.1 ± 0.1	71.3 ± 2.5	19.3 ± 0.4	8.5 ± 2.2	1.1 ± 0.1

[a] Dose of CCNU was 30 mg/kg body weight. Rats were induced with phenobarbital at a dose of 80 mg/kg body weight per day for 4 days followed by a 24 hr period prior to CCNU administration.
[b] Data presented as the mean of three experiments with ± standard error.

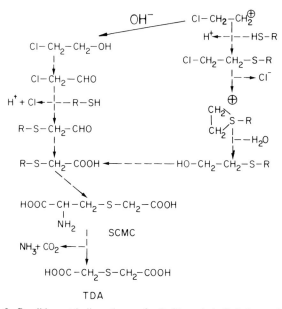

Fig. 3. Possible metabolic pathways for 2-chloroethyl alkylating moiety.

fate and effects of nitrosoureas is highly questionable. Patients are known to undergo extensive loss of red cell glutathione reductase activity immediately after BCNU therapy (Frischer and Ahmad, 1977). Glutathione reductase inactivation was shown to be due to a highly specific active site-directed inactivation by 2-chloroethylisocyanate which is released upon BCNU decomposition (Babson and Reed, 1978). *In vitro* studies showed that lysine at concentrations up to 10 mM did not affect this stoichiometric inactivation. Specific and stoichiometric carbamoylation of proteins has been shown also for chymotrypsin (Babson *et al.*, 1977) and tubulin (Brodie *et al.*, 1980). We now show that those urinary metabolites of CCNU and MeCCNU that are derived from the respective isocyanates are carbamoylated peptides. These peptides were found to contain major amounts of cysteine and serine residues and only a small amount of lysine residues.

II. MATERIALS AND METHODS

1. Chemicals

CCNU was synthesized by Parke Davis; [chx-1-^{14}C]CCNU, and [chx-1-^{14}C]methyl CCNU were supplied by the National Cancer Institute, Bethesda, Maryland. Emulphor EL-620 was obtained from GAF Corp., New York, New York. Column chromatography packings were LiChrosorb SI60, 5μ particle size (EM Laboratories, Elmsford, New York); Cellex-CM, 0.68 meq/g, and BioGel P-2, 200–400 mesh (BioRad Laboratories, Richmond, California). All solvents used in HPLC were spectral grade purchased from J. T. Baker (Phillipsburg, New York) or Burdick and Jackson (Muskegon, Michigan). All other chemicals were made of reagent grade material.

2. Urinary Metabolite Synthesis

Synthesis of hydroxycyclohexylamines and monohydroxylated methylcyclohexylamines was accomplished by previously published procedures (May *et al.*, 1979).

3. Urinary Metabolites of [Chx-1-^{14}C]CCNU and [Methylchx-1-^{14}C]Methyl CCNU

Sprague-Dawley male rats were injected ip with [chx-1-^{14}C] CCNU or [methylchx-1-^{14}C]methyl CCNU in an Emulphor-absolute ethanol vehicle, and the urines collected over dry ice for at least two consecutive 24-hr periods. The urines were centrifuged through Aminco Centriflo cones with 25,000 daltons molecular weight cutoff. The filtrates were stored at $-80°$ until required for analysis. Urine from animals that did not receive CCNU were pooled and used as a control.

4. Chromatography of Urine

Urine was first fractionated on Cellex CM (3.2 × 250 mm column) using 0.005 *M* ammonium carbonate, pH 8.5–9.0, followed by a 20 min linear gradient to 0.5 *M* ammonium carbonate. A nonretained fraction and the urinary ammonia eluted with the solvent front; a retained fraction eluted as a single peak between 0.13 *M* and 0.23 *M* ammonium carbonate. The retained fraction which contained basic components was derivatized directly using DNBS, separated by HPLC on Lichrosorb, and a mass spectrum determined for each peak to determine the presence of cyclohexyl-containing DNP-amines.

Fractionation of filtered urine samples was also performed with a BioGel P-2 on a 1.0 × 35 cm column eluted with 0.5 *M* ammonium carbonate, pH 7.0. The urine ^{14}C-labeled metabolites eluted as four fractions for which molecular weights were determined.

5. Hydrolysis Procedures for Urinary Metabolites

Twenty-four-hour hydrolysis of urine was carried out with equal volumes of sample and 6 *N* HCl or 6 *N* NaOH in sealed ampuls at 110°. The ampuls were chilled and opened, and the pH of the acid hydrolyzed sample was adjusted to 11 with 6 *N* NaOH to permit formation of DNP derivatives of the amines as described below. Following derivatization with DNBS, the amines in the urines were isolated using sequential extractions, and their quantities were determined from ^{14}C and independently by absorbance of HPLC peaks. Authentic amines were used as standards to monitor derivatization yields, extraction recovery, and identification by HPLC, and mass spectrometry of amine metabolites.

Urine samples were incubated with Sigma β-glucuronidase (from Lipets, Lot 113c-62201) containing glucuronidase and aryl sulfatase at 37°C to hydrolyze the conjugated amines. The glucuronidase and sulfatase activities were determined using α-napthyl-D-glucuronic acid and α-napthylsulfate according to the method of van Asperen (1962).

Sodium dihydrogen phosphate (0.1 M) was added to inhibit aryl sulfatase without affecting the activity of the glucuronidase, thereby allowing analysis of the amines derived from glucuronides or sulfates. After enzymatic hydrolysis, the urines were derivatized with DNBS and quantitated as described above.

6. DNP Derivatives of Amine Standards and Urinary Metabolites

The amine, amine oxalate, or urinary sample was incubated for 60 min at 100°C with an equal volume of 0.6 M DNBS in saturated sodium tetraborate according to a modified method of Smith and Jepson (1967). The DNP amines were extracted with benzene until the benzene layer was colorless.

7. HPLC of DNP Amines

All HPLC was performed on a Spectra Physics model 3500 liquid chromatograph equipped with a model 770 spectrophotometric detector at 254 nm. A mixture of DNP derivatives of cyclohexyl amine, cis-2, trans-3, cis-3, cis-4, and trans-4-hydroxycyclohexyl amines was readily separable on LiChrosorb (3.2 × 250 mm 5μ particle size) isocratically using isooctane:dichloromethane:2-propanol, with volume ratio of 930:63:50, and a flow rate of 1.5–1.7 ml/min. The 2-propanol content was 0.8 parts instead of 50 when DNP cyclohexylamine was chromatographed separately. The relative retention times (minutes) were rather constant and were cis-2, 0.32; trans-3, 0.58; cis-3, 0.86; cis-4, 1.00; trans-4, 1.08; 2,4-dinitroaniline, 0.71. Similar procedures were utilized to characterize the [chx-1-^{14}C]methyl CCNU metabolites.

All quantitation of urinary cyclohexyl metabolites was based on the specific radioactivity of the injected [chx-1-^{14}C]CCNU. Optical traces of chromatograms at 254 nm were used to confirm these quantities.

8. Mass Spectral Analysis of DNP Amines

Mass spectra were obtained with a Varian Model M7 mass spectrometer using a direct probe, temperature programming and an ionizing potential of 70 V.

9. Amino Acid Analysis

A modified Beckman Model 120B Amino Acid Analyzer equipped with a single 6 × 510 mm column system was used for amino acid analyses of acid hydrolyzed metabolites.

10. Liquid Scintillation Counting

Radioactivity was quantitated in either toluene or emulsion fluors assayed with a Packard Model 2425 liquid scintillation counter.

III. RESULTS

A. Urinary Metabolites

Administration of [chx-1-^{14}C] CCNU to control and PB-induced rats resulted in 44 and 51% respectively of the ^{14}C dose being excreted in the urine within 24 hr. PB induction decreased the portion of ^{14}C-labeled free cyclohexylamine of total urinary ^{14}C from 12 to 6% compared to control rats. The remaining ^{14}C metabolites lacked free amino groups and were considered conjugated amine metabolites. Molecular weight distribution of the ^{14}C metabolites indicated that more than one-half of the total urinary ^{14}C was present with a molecular weight of 628, 20% each with molecular weights of 413 and 329, and 5% with a molecular weight of 243. Sequential extraction of urine aliquots with hexane, methylene chloride, and ether demonstrated that less than 1% of the urinary ^{14}C was either unchanged CCNU or the parent urea, CCU.

B. Urinary Free Amine Metabolites

The fraction of ^{14}C free cyclohexylamines was analyzed by 1-fluoro-2,4-dinitrobenzene derivatization, extraction with benzene and HPLC (Fig. 4). Less than 20% of the ^{14}C was present as cyclohexylamine; the remaining ^{14}C was the *cis*- and *trans*-3 and 4-hydroxycyclohexylamines (Fig. 4). As expected, PB induction of rats before CCNU administration, enhanced the relative quantities of the *cis* and *trans*-4-hydroxy isomers while decreasing *cis*-3-hydroxy cyclohexylamine excretion.

C. Urinary Amine Conjugate Metabolites

Hydrolysis of urinary metabolites with glucuronidase failed to release free amines. Hydrolysis with glucuronidase plus sulfatase released only about 10% of the conjugate ^{14}C as free amines (Table VII). Rigorous acid or base hydrolysis released up to 40% of the conjugate ^{14}C as free amines from [Chx-1-^{14}C] CCNU and up to 55% from [Chx-1-^{14}C] *trans*-4-methyl CCNU urinary metabolites. Some destruction of free hydroxy-cyclohexylamines was shown to occur during these hydrolysis reactions, causing reduced recovery yields.

No significant change was seen in the distribution of hydroxy isomer metabolites between free and conjugated amines after acid or base hydrolysis for both CCNU and MeCCNU. An example is shown in Fig. 5.

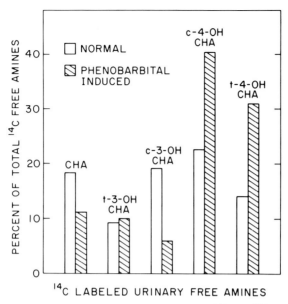

Fig. 4. ^{14}C urinary free amines after metabolism of [chx-1-^{14}C]CCNU by rats for 24 hr.

D. Peptide Conjugates of Urinary Amines

Fractions of urinary metabolites from [Chx-1-^{14}C] CCNU metabolism represented by molecular weights of 628, 413, and 329 were each collected, dried, and individually passed through a P-2 column (1.0 × 100 cm) prior to acid hydrolysis and amino acid analysis. Major amino acids present after acid hydrolysis in decreasing amounts were cysteine, serine, glutamic acid, alanine, glycine, aspartic acid, leucine, threonine, lysine, and isoleucine. The presence of these amino acids along with ^{14}C cyclohexylamine and ^{14}C *cis* and *trans*-3 and 4-hydroxy-cyclohexylamines suggest that these conjugates are a complex mixture of peptide conjugates containing ^{14}C-labeled cyclohexyl moieties. It should

TABLE VII. Effect of Hydrolytic Procedures upon Urinary ^{14}C from Rats after [Chx-1-^{14}C]CCNU or [Chx-1-^{14}C]*Trans*-4-Methyl CCNU Metabolism for 24 Hr[a]

Urine treatment	% of total urinary ^{14}C as extractable free amines	
	[^{14}C]CCNU	[^{14}C]*trans*-methyl-CCNU
None	9.9	11.9
Glucuronidase	7.6	6.7
Glucuronidase plus sulfatase	19.3	21.9
Acid hydrolysis (6 *N* HCl, 110°, 24 hr)	42.1	62.6
Alkaline hydrolysis (6 *N* NaOH, 110°, 24 hr)	49.0	66.4

[a] CCNU or MeCCNU dose was 30 mg/kg body weight.

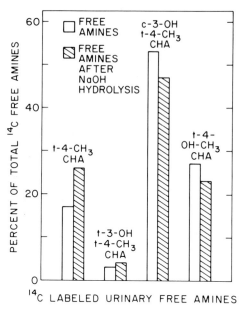

Fig. 5. ^{14}C Urinary amines after metabolism of [chx-1-^{14}C]methyl-CCNU by rats for 24 hr.

be noted that fractions with molecular weights of 628 and 413 underwent partial degradation under mild alkaline conditions (pH 8.5) at ambient temperatures, releasing small quantities of free cyclohexylamines. Similar chemical instability was observed by Twu and Wold (1973) in their study of the properties of S-(butylcarbamoyl)-L-cysteine and by Gross *et al.* (1975) upon methylisocyanate inactivation of transglutaminase.

IV. DISCUSSION

The nitrosoureas have received considerable attention in terms of their antitumor efficacy compared to their toxicity, particularly bone marrow toxicity (Hansch *et al.*, 1972; Levin and Kabra, 1974; Wheeler *et al.*, 1977; and Heal *et al.*, 1979). Hansch *et al.* (1972) speculated that the more hydrophilic nitrosoureas may be more potent and less toxic if the nitrosoureas tend to bind to serum proteins or other macromolecules, possibly causing alkylation at such sites rather than at sites for antitumor activity. Heal *et al.* (1979) noted, however, that upon comparing seven water-soluble nitrosoureas with CCNU as a reference lipid-soluble analog, there was no correlation between relative myelotoxicity, octanol-water partition coefficients (log P) and *in vitro* assays of alkylation activity or carbamoylation activity.

Hydroxylation of CCNU occurs within essentially a single pass through the liver (Walker and Hilton, 1976; Reed and May, 1977) yielding hydroxy CCNU isomers which are the major immediate precursors of the therapeutic moieties (Reed and May, 1975). Methyl CCNU also appears to express many of its pharmacological properties via hydroxy derivatives.

An apparent accelerated rate of nitrosourea degradation *in vivo* when compared to the rates of degradation in buffer at physiological pH has been demonstrated. This effect, which is very pronounced with CCNU, results in extensive protein carbamoylation in the blood and liver. Alkylation by the ethyl moiety was much less than was carbamoylation by the isocyanate moiety, possibly indicating a high degree of protection by intracellular thiols such as glutathione. Main urinary products of the ethyl moiety include S-carboxymethyl L-cysteine and thiodiacetic acid. Glutathione conjugates formed in the liver are in some instances excreted almost exclusively via the bile (Wahlländer and Sies, 1979). In agreement with this observation, biliary excretion of ethyl ^{14}C after [2-chloroethyl-^{14}C]CCNU administration to rats was extensive (>40% in 24 hr) and significantly greater than the cyclohexyl moiety (Oliverio *et al.*, 1970). Also, BCNU has been shown to decrease the glutathione content in liver of mice (McConnell *et al.*, 1979) and in isolated hepatocytes (Reed and Babson, 1980).

Extensive carbamoylation of proteins by BCNU and CCNU *in vivo* has resulted in glutathione reductase inactivation (Frischer and Ahmad, 1977) and formation of carbamoylated urinary peptides. These *in vivo* effects combined with carbamoylation effects *in vitro* on DNA polymerase (Baril *et al.*, 1975), DNA repair (Kann *et al.*, 1980), chymotrypsin (Babson *et al.*, 1977), and tubulin (Brodie *et al.*, 1980) give strong support for rather specific biological carbamoylation effects. Thus, the mounting evidence indicates that the *in vitro* criteria for carbamoylation capability is probably misleading and may result in assumptions that are incorrect for *in vivo* situations. A specific example is the observation that ACNU has virtually no carbamoylating activity *in vitro* with lysine but is a potent carbamoylating agent against glutathione reductase (Babson and Reed, 1978). In addition, Tanaka *et al.* (1977) have identified an intramolecular carbamoylation product that forms nonenzymically. Therefore, one can speculate that other nitrosoureas, including chlorozotocin, may be capable of some *in vivo* carbamoylation, even though intramolecular carbamoylation is suggested as the reason for the lack of lysine carbamoylation *in vitro* (Montgomery, 1976).

V. CONCLUSIONS

In addition to more rapid decomposition *in vivo* than in buffer solution at physiological pH, this decomposition is complex and involves proteins both enzymatically and nonenzymatically.

We may conclude that cytochrome P-450-catalyzed reactions are capable of rapid hydroxylation of CCNU and methyl CCNU with some loss of stereospecificity, indicating transitory free radical intermediates.

Denitrosation of nitrosoureas is inducible *in vitro* by phenobarbital, and it appears to be related to the cytochrome P-450 enzyme system.

Urinary metabolite patterns of CCNU and methyl CCNU strongly indicate extensive carbamoylation of proteins at sites other than lysine. In addition, evidence has been presented that questions the carbamoylation capability of nitrosoureas *in vivo*.

ACKNOWLEDGMENT

The author wishes to acknowledge the major contributions of Drs. Sue Kohlhepp and Gene May to this work. The assistance of Donald A. Griffin in mass spectral analyses is gratefully acknowledged.

REFERENCES

Babson, J. R., and Reed, D. J. (1978). *Biochem. Biophys. Res. Comm. 83*, 754–762.
Babson, J. R., Reed, D. J., and Sinkey, M. A. (1977). *Biochem. 16*, 1584–1589.
Baril, B. B., Baril, E. F., Laszlo, J., and Wheeler, G. P. (1975). *Cancer Res. 35*, 1–5.
Brodie, A. E., Babson, J. R., and Reed, D. J. (1980). *Biochem. Pharmacol. 29*, 652–654.
DeVita, V. T., Denham, C., Davidson, J. D., and Oliverio, V. T. (1967). *Clin. Pharmacol. Therapeut. 8*, 566–577.
Farmer, P. B., Foster, A. B., Jarman, M., and Oddy, M. R. (1978). *J. Med. Chem. 21*, 514–520.
Frischer, H., Ahmad, T. (1977). *J. Lab. Clin. Med. 89*, 1080–1091.
Godenèche, D., Madelmont, J.-C., Moreau, M.-F., Montoloy, D., and Plagne, R. (1980). *Cancer Res. 40*, 3351–3356.
Gross, M., Whetzel, N. K., and Folk, J. E. (1975). *J. Biol. Chem. 250*, 7693–7699.
Hansch, C., Smith, N., Engle, R., Wood, H. (1972). *Cancer Chemother. Rep. 56*, 443–456.
Heal, J. M., Fox, P., and Schein, P. S. (1979). *Biochem. Pharmacol. 28*, 1301–1306.
Hill, D. L., Kirk, M. C., and Struck, R. F. (1975). *Cancer Res. 35*, 296–301.
Hilton, J., and Walker, M. D. (1975). *Biochem. Pharmacol. 24*, 2153–2158.
Kann, H. E., Jr., Blumenstein, B. A., Petkas, A., and Schott, M. A. (1980). *Cancer Res. 40*, 771–775.
Levin, V. A., and Kabra, P. (1974). *Cancer Chemother. Rep. Part 1 58*, 787–792.
Levin, V. A., Hoffman, W., and Weinkam, R. J. (1978). *Cancer Treat. Rep. 62*, 1305–1312.
Levin, V. A., Stearns, J., Byrd, A., Finn, A., and Weinkam, R. J. (1979). *J. Pharmacol. Exptl. Therapeut. 208*, 1–6.
May, H. E., Boose, R., and Reed, D. J. (1974). *Biochem. Biophys. Res. Comm. 57*, 426–433.
May, H. E., Boose, R., and Reed, D. J. (1975). *Biochem. 14*, 4723–4730.
May, H. E., Kohlhepp, S. J., Boose, R. B., and Reed, D. J. (1979). *Cancer Res. 39*, 762–772.
McConnell, W. R., Kari, P., and Hill, D. L. (1979). *Cancer Chemother. Pharmacol. 2*, 221–223.
Mhatre, R. M., Green, D., Panasci, L. C., Fox, P., Woolley, P. V., and Schein, P. S. (1978). *Cancer Treat Rep. 62*, 1145–1151.
Montgomery, J. A. (1976). *Cancer Treat. Rep. 60*, 651–664.
Oliverio, V. T., (1976). *Cancer Treat. Rep. 60*, 703–707.
Oliverio, V. T., Vietzke, W. M., Williams, M. K., and Adamson, R. H. (1970). *Cancer Res. 30*, 1330–1337.
Plowman, J., Dingell, J. V., Adamson, R. H. (1978). *Cancer Treat. Rep. 62*, 31–44.

Potter, D. W., and Reed, D. J. (1980). *In* "Microsomes, Drug Oxidations, and Chemical Carcinogenesis," pp. 371–374. Academic Press, New York.

Reed, D. J., and Babson, J. R. (1980). *Proc. Am. Assoc. Cancer Res. 21*, 307.

Reed, D. J., and May, H. E. (1975). *Life Sci. 16*, 1263–1270.

Reed, D. J., and May, H. E. (1977). *In* "Microsomes and Drug Oxidations" (V. Ullrich, I. Roots, A. Hildebrandt, R. W. Estabrook, and A. H. Conney, eds.), pp. 680–687. Pergamon Press, New York.

Shigehara, E., and Tanaka, M. (1978). *Gann. 69*, 709–714.

Smith, A. D., and Jepson, J. B. (1967). *Anal. Biochem. 18*, 36–45.

Sponzo, R. W., DeVita, V. T., and Oliverio, V. T. (1973). *Cancer 31*, 1154–1159.

Tanaka, M., Nakajima, E., Nishigaki, T., Shigehara, E., and Nakao, H. (1977). *In* "Proceedings of the Tenth International Congress of Chemotherapy," Abstr. 570. Zurich, Switzerland.

Twu, J., and Wold, F. (1973). *Biochem. 12*, 381–386.

Van Asperen, K. J. (1962). *Insects Physiol. 8*, 401–416.

Wahlländer, A., and Sies, H. (1979). *Eur. J. Biochem. 96*, 441–446.

Walker, M. D., and Hilton, J. (1976). *Cancer Treat. Rep. 60*, 725–728.

Weinkam, R. J., Liu, T.-Y. J., and Lin, H.-S. (1980). *Chem.-Biol. Interact. 31*, 167–177.

Wheeler, G. P., Bowdon, B. J., and Herren, T. C. (1964). *Cancer Chemother. Rep. 42*, 9–12.

Wheeler, G. P., Bowdon, B. J., Grimsley, J. A., and Lloyd, H. H. (1974). *Cancer Res. 34*, 194–200.

Wheeler, G. P., Johnston, T. P., Bowdon, B. J., McCaleb, G. S., Hill, D. L., and Montgomery, J. A. (1977). *Biochem. Pharmacol. 26*, 2331–2336.

Chapter 6

DNA CROSSLINKING AND THE ORIGIN OF SENSITIVITY TO CHLOROETHYLNITROSOUREAS

Kurt W. Kohn
Leonard C. Erickson
Guy Laurent
Jonathan Ducore
Nancy Sharkey
Regina A. Ewig

I. INTRODUCTION

A. DNA Interstrand Crosslinking

From the chemistry and structure–activity relations of the nitrosoureas, it seems clear that the cell-killing and antitumor activities arise from alkylation reactions (Montgomery, 1976). Other classes of alkylating agents, such as the nitrogen mustards, generally have potent activities only if the molecule has two or more alkylating groups. Bifunctional alkylations can produce covalent crosslinks between nucleophilic sites on biomolecules, and it has long been suspected that the critical reaction is some form of crosslinking of DNA (Stacey *et al.*, 1958). The nitrosoureas, however, have only a single alkylating group. Indeed, compounds such as methylnitrosourea and streptozotocin can transfer methyl groups to DNA but cannot form crosslinks; these compounds have relatively weak cell-killing and antitumor activities, but are strongly mutagenic (Bradley *et al.*, 1980). The chloroethylnitrosoureas, on the other hand, do produce DNA crosslinks (Kohn 1977; Lown *et al.*, 1978) and are highly effective against a variety of rodent tumors.

The sequence of reactions leading to crosslink formation by chloroethylni-trosoureas is thought to begin with the generation of highly reactive intermediates that transfer chloroethyl groups to nucleophilic sites (Fig. 1). The resulting chloroethyl monoadducts can react with a second nucleophilic site through the displacement of chloride to form a crosslink consisting of an ethylene bridge between the two nucleophilic sites. This is illustrated in Fig. 1 for the formation of interstrand crosslinks in DNA.

Interstrand crosslinking is thought to involve essentially 2 steps: (*a*) chloroethylation of a nucleophilic site on one DNA strand; and (*b*) reaction of the chloroethyl monoadducts with a nucleophilic site on the opposite DNA strand to form an ethylene bridge between the strands. The second step, which converts monoadducts to interstrand crosslinks, is slow, requiring several hours for completion, both in purified DNA (Kohn 1977; Lown *et al.*, 1978) and in living cells (Ewig and Kohn, 1978). Hence there would be ample time for a DNA repair process to remove chloroethyl monoadducts before they convert to potentially lethal interstrand crosslinks.

It is not known which DNA site is capable of becoming chloroethylated and of then converting to interstrand crosslinks. It was proposed, however, that this site could be the guanine-O^6 position (Kohn, 1977) on the basis of the demonstrated ability of nitrosoureas to alkylate this site (Singer, 1975; Lawley *et al.*, 1975) and the possibility of the chloroethyl group reaching to the adjacent cytosine amino group with which it could react to complete the crosslinks (Fig. 2).

Evidence from model reactions (e.g., the reaction of chloroethylnitrosoureas with poly C) shows that an ethylene bridge reaction with the cytosine amino group is possible (Kramer *et al.*, 1974; Lown and McLaughlin, 1979). The cytosine amino group is, however, a minor alkylation site (Singer, 1975); this could account for the slowness of the conversion of chloroethyl monoadducts to interstrand crosslinks.

Another factor to be considered is that alkylation of the guanine-O^6 position would disturb the hydrogen bonding of the guanine ring and interfere with base pairing to cytosine. The DNA helix could thereby be locally distorted so that

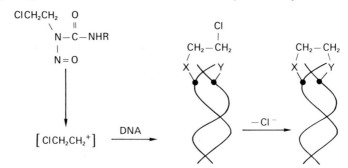

Fig. 1. Proposed mechanism of DNA interstrand crosslinking by chloroethylnitrosoureas (X and Y are nucleophilic sites located on opposite DNA strands).

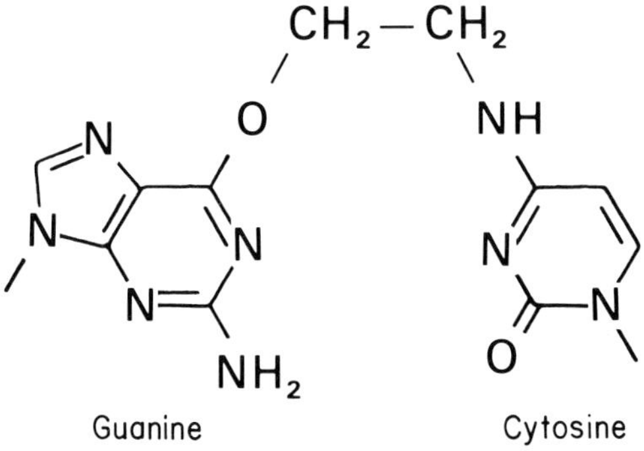

Fig. 2. Proposed structure of an ethylene-bridge crosslink in a GC base pair.

other nucleophilic sites, which normally would be inaccessible, could at times come within range of the guanine-O^6-chloroethyl group. Crosslinks might then connect guanine-O^6 with a variety of possible DNA sites. The chemical identity of the interstrand crosslinks thus remains to be determined.

B. DNA–Protein Crosslinking

As is the case with most agents that form interstrand crosslinks, chloroethyl-nitrosoureas also produce DNA–protein crosslinks (Kohn, 1979; Ewig and Kohn, 1978). DNA–protein crosslinks, however, form relatively quickly following treatment of cells with chloroethylnitrosoureas, as opposed to the delayed formation of interstrand crosslinks. Although the chemical details of DNA–protein crosslinking are as yet unknown, it seems likely that the initial chloroethylation would usually be at a protein amino or sulfhydryl group. The resulting 2-chloroethylamine or 2-chloroethyl-sulfide would constitute a nitrogen mustard or sulfur mustard substituent on the protein. These mustard groups would be likely to react rapidly with the most nucleophilic DNA sites, such as the guanine-N_7 position.

C. Monoadduct Repair

The central problem in cytotoxic chemotherapy, as we now understand it, is how a selective toxic action can be delivered to tumor cells while sparing critical normal cells. A clue to the origin of selective cytotoxicity by nitrosoureas was derived from recent work on DNA methylating agents. Day and his co-workers (Day and Ziolkowski, 1979; Day et al., 1980) found that human tumor cell strains differ in their ability to support the growth of adenovirus that had been treated with DNA methylating agents such as 1-methyl-1-nitrosourea (MNU) or 1-methyl-1-nitroso-3-nitroguanidine (MNNG). The tumor cell strains were assumed to differ in their ability to repair the methylated adenovirus DNA. The cell strains that were deficient in the ability to support the growth of methylated adenovirus also were deficient in the ability to remove O^6-methylguanine from their nuclear DNA (Day *et al.*, 1981). The cell types generally fell into two groups, differing in their proficiency to support the growth of methylated adenovirus and to remove O^6-methylguanine from DNA. The discrete differences between the cell strains suggested the existence of a DNA methylation repair phenotype (denoted "Mer") in which some of the cell strains were deficient. The deficient cells are designated as having the Mer⁻ phenotype. Normal fibroblasts, even from patients with Mer⁻ tumors, always exhibited the Mer⁺ phenotype.

The removal of O^6-methyl as well as O^6-ethylguanine from DNA probably occurs by destruction rather than by excision of these residues from DNA (Robins and Cairns, 1979; Karran and Lindahl, 1979; Renard and Verly, 1980). The tautomeric structure of the guanine ring is altered by O^6 alkylation; this could be a recognition feature for an enzyme that acts on the O^6-alkylguanine ring, regardless of the identity of the alkylating group. Therefore the enzyme system that repairs O^6-methyl and O^6-ethylguanines might also repair O^6-chloroethylguanine monoadducts.

If the monoadducts lead to interstrand crosslinks, and if either of these are the critical lethal lesions, then the above argument leads to the prediction that the sensitivity of human tumor cells to chloroethylnitrosoureas should be critically dependent on the Mer phenotype.

D. Hypothesis

In the work that follows (Erickson *et al.*, 1980; 1981), we hypothesized that human tumor cells having the Mer⁺ phenotype would be capable of repairing O^6-chloroethylguanine monoadducts and thereby would prevent the formation of chloroethylnitrosourea-induced interstrand crosslinks. DNA–protein crosslinks, on the other hand, would not be affected. Human tumor cells having the Mer⁻ phenotype would exhibit more interstrand crosslinking than Mer⁺ cells and would have greater sensitivity to chloroethylnitrosoureas.

II. MATERIALS AND METHODS

A. Cells

IMR-90, normal human embryo cells, were obtained from the Institute of Medical Research, Camden, New Jersey. VA-13 is an SV-40 transformed derivative of normal human embryo strain WI-38. Human colon tumor cell lines, BE and HT29, were obtained from B. Giovanella, St. Joseph's Hospital, Houston, Texas, and E. Jensen, Mason Research Institute, Rockville, Maryland, respectively. All other cell strains were obtained from R. Day, National Cancer Institute, Bethesda, Maryland.

Cells were cultured in Eagle's basal medium with 10% calf serum and 7.5% CO_2 or in Dulbecco's modified Eagle medium with 10% heat inactivated fetal calf serum and 0.02 M HEPES buffer.

B. Cytotoxicity Assays

Approximately 5×10^5 cells were seeded in 25-cm^2 culture flasks and incubated for 24 hr. Cells were then exposed to drug for 1 or 2 hr in fresh medium containing 0.02 M HEPES buffer. After further incubation for the appropriate period of time, the cells were suspended by trypsinization, and cell number was determined using an electronic cell counter.

C. DNA Crosslinking Assays

DNA interstrand and DNA–protein crosslinking were assayed by means of the alkaline elution technique (Kohn *et al.*, 1976; 1981). The cells to be studied were labeled with 0.01 μCi/ml [2-^{14}C]-thymidine for 24 hr. After drug treatment and postincubation, as described in individual experiments, the cells were scraped from the surface of the flask and mixed with standard cells that served as internal reference in the assay (for the latter purpose, ^{3}H-thymidine-labeled L1210 cells were used). The mixed cells were exposed to 300 rads of γ-irradiation at 0°C in order to introduce a suitable frequency of DNA single-strand breaks. The cells were deposited on 2 μm pore-size polyvinylchloride filters (Millipore Corp.) and lysed with 2% sodium dodecyl sulfate, 0.02 M EDTA, 0.1 M glycine, pH 10, with or without 0.5 mg/ml proteinase K. A solution of 0.02 M H_4 EDTA, tetrapropylammonium hydroxide, pH 12.1, was then pumped through the filter at a rate of 2 ml/hr, and fractions were collected for radioactivity determination.

Crosslinking was calculated in terms of the "crosslink index," defined as $\sqrt{(1-R_0)/(1-R_1)} - 1$, where R_0 and R_1 are the "relative retention" for untreated and drug-treated cells, respectively (Ewig and Kohn, 1978). Relative retention is the fraction of the ^{14}C-DNA remaining on the filter when 30% of the internal standard ^{3}H-DNA remained on the filter. The crosslink index was found to be proportional to drug concentration under standard conditions of cell treatment and therefore is taken to be a proportional measure of DNA crosslinking; this was true both for interstrand and for DNA–protein crosslinks.

III. RESULTS

A. Comparison between Normal and Transformed Human Embryo Cells

The initial objective was to compare a normal with a presumably neoplastic human cell line. For this purpose we utilized the normal human embryo cells strain, IMR 90, and the SV40-transformed human embryo cell line, VA-13

(Erickson *et al.*, 1980). The two cell lines were proliferating at approximately equal rates at the time of drug treatment.

1. Cytotoxicity

Studies of the effects of several chloroethylnitrosoureas on cell proliferation consistently showed that the transformed VA-13 line was more sensitive than the normal IMR 90 cells (Figs. 1 and 2). The magnitude of the differential toxicity between the 2 cell types, however, differed, depending on the drug used (Fig. 2). The differences between the two cell types were also shown in experiments in which colony formation was used as the assay for cytotoxicity (Erickson *et al.*, 1980).

2. DNA Alkaline Elution Assays

DNA interstrand and DNA–protein crosslinking were measured by means of the alkaline elution technique (see section II). This technique measures effective DNA single-strand size by determining the rate at which DNA from lysed cells is able to pass through membrane filters at pH 12. The longer the DNA single strands, the longer is the average time required for the strands to elute from the filters. For the measurement of crosslinks, a known frequency of DNA single-strand breaks is introduced into the DNA by exposing the cells to a suitable x ray dose at 0°C. This causes the DNA to elute at a clearly measurable rate (filled

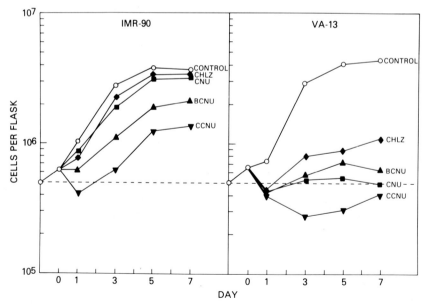

Fig. 3. Effect of various chloroethylnitrosoureas on the proliferation of IMR-90 (left) and VA-13 (right) cells. Cells were treated with 50 μM of the indicated drug for 2 hr on day 0 (from Erickson *et al.*, 1980).

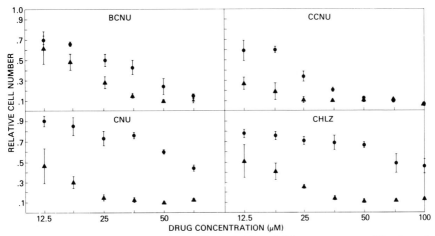

Fig. 4. Concentration–dependence of the effect of chloroethylnitrosoureas on proliferation of IMR-90 (●) or VA-13 (▲) cells. Cells were exposed to drug for 2 hr, and cell number relative to control was determined on day 3 (from Erickson *et al.*, 1980).

circles in Figs. 3 and 4). The presence of crosslinks is then gauged by a reduction of the DNA elution rate (filled symbols other than circles in Figs. 3 and 4). Interstrand crosslinks reduce the elution rate by increasing the effective size of the DNA strands. DNA–protein crosslinks can also reduce elution rate because proteins adsorb to the filters under the conditions used for these assays. The DNA strands linked to adsorbed protein are thus also retained by the filters. The effect of DNA–protein crosslinks can be eliminated by digesting the lysed cells with proteinase-K and by including a detergent in the alkaline eluting solution so as to further reduce adsorptive effects; the retardation of elution that remains is assumed to be a measure of interstrand crosslinks (Fig. 4). The assay without proteinase measures the combined effect of interstrand and DNA–protein crosslinks; however, in these experiments the major contribution is from DNA–protein crosslinks (Fig. 3).

Assays were also conducted without the use of x ray in order to determine whether the drugs themselves produced DNA strand breaks (open symbols in Figs. 3 and 4). A low frequency of DNA strand breaks appears transiently following chloroethylnitrosourea treatment (Erickson *et al.*, 1977), but the magnitude of this effect did not significantly alter the crosslink estimates.

3. Interstrand and DNA–Protein Crosslinks

The major finding was that all four chloroethylnitrosoureas used in this work produced interstrand crosslinks in VA-13 cells but not in IMR-90 cells (Figs. 4 and 5). DNA–protein crosslinks, on the other hand, appeared equally in both cell types (Fig. 3), indicating that reactive drug gained equal access to the DNA in both cell types.

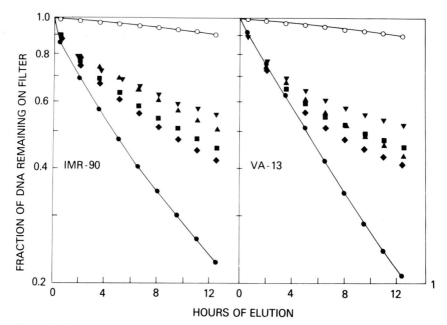

Fig. 5. Alkaline elution assays without proteinase K (primarily showing DNA–protein crosslinking). IMR-90 (left) and VA-13 (right) cells were treated with 50 μM drug for 2 hr and then incubated in the absence of drug for 12 hr. ○ represents controls assayed without x ray (in all other assays, cells were exposed to 300 rads at 0°C); ●, no drug; ▲, BCNU; ▼, CCNU; ■, CNU; ◆, CHLZ (from Erickson *et al.*, 1980).

The results with VA-13 cells cannot, however, be generalized for all SV-40 transformed human embryo cells because Day *et al.* (1980) have found that although some cell lines of this type have the Mer⁻ phenotype, others have the Mer⁺ phenotype.

B. Dependence on the Mer Methylation Repair Function

The absence of interstrand crosslinking in IMR-90 cells suggested that these cells were capable of removing chloroethyl monoadducts before they could react with the opposite DNA strand. The IMR-90 and VA-13 cells were then tested by Dr. Rufus Day in order to determine whether the interstrand crosslinking difference between these cells could be related to the Mer phenotype. The expected result was obtained: IMR-90 cells were found to be Mer⁺ and VA-13 cells were Mer⁻.

We then tested for DNA crosslinking and cytotoxic sensitivity 9 additional cell strains which had been determined to be Mer⁺ or Mer⁻ (Erickson *et al.*, 1981).

TABLE I. Alkaline Elution Assays of DNA Interstrand and DNA–Protein Crosslinking by 1-(2-Chloroethyl)-1-Nitrosourea (CNU)
(Crosslink Index $\times 10^3$)[a]

Symbol	Cell strain	Origin	Mer phenotype	Interstrand crosslinking (proteinase assay)		Total crosslinking (no proteinase)	
				50 µM CNU	100 µM CNU	50 µM CNU	100 µM CNU
■	HT29	colon ca.	+	0 (−13–13)[3]	7 (0–17)[3]	126, 135	268, 410
●	IMR90	embryonic lung	+	3 (−3–7)[3]	23 (23–24)[3]	210, 216	485
◆	A2182	lung ca.	+	14 (8–20)[3]	31	154, 218	788
▨	A673	rhabdomyosarcoma	+	16 (12–20)[3]	16	186, 261	523
◆	A549	lung ca.	+	33 (20–54)[4]	52 (19–91)[4]	203, 216	504, 545
◀	HT1080	fibrosarcoma	+	62 (20–107)[4]	127 (51–230)[3]	221, 251	437, 462
●	A431	epidermoid ca. (vulva)	−	69 (60–76)[3]	143 (142–144)[2]	190, 190	385, 446
⬡	A1336	ovarian ca.	−	69 (38–177)[3]	221	135, 152	692
▷	A427	lung ca.	−	72 (61–84)[4]	149 (101–247)[4]	122, 208	295, 471
□	BE	colon ca.	−	81 (71–93)[3]	201 (182–224)[4]	166, 207	438, 530
◇	A172	astrocytoma (grade IV)	−	84 (62–113)[3]	217 (209–225)[2]	193, 204	432, 457
△	VA13	SV40-transformed WI-88	−	88 (50–146)[3]	232 (145–320)[2]	148, 152	384
○	A875	melanoma	−	112 (82–142)[3]	240 (201–281)[3]	250, 281	540

[a] Cells previously labeled with [2-^{14}C]-thymidine were treated with 50 or 100 µM CNU for 1 hr and then incubated for 6 hr in drug-free medium. Alkaline elution assays were performed with or without the use of proteinase-K (see Section II). The parentheses contain the range of values obtained in independent experiments; the superscript indicates the number of experiments. Symbols in the first column on the left are used to represent the cell strains in Figures 6 and 7 (from Erickson et al., 1981).

1. DNA Crosslinking

Cells were treated with 50 or 100 μM chloroethylnitrosourea (CNU) for 1 hr and then incubated for 6 hr in drug-free medium to allow delayed interstrand crosslink formation. The results showed a clear correlation between interstrand crosslinking (proteinase assay) and methylation repair capacity (Table I). The 6 strains deficient in this capacity (Mer⁻ phenotype) produced consistently higher interstrand crosslinking levels than did 5 of the 7 Mer⁺ strains.

The remaining 2 Mer⁺ strains (A431 and HT1080) gave intermediate or variable results. It may be that cells can lose the Mer⁺ function during culture or that some cell strains have a limited capacity for this function.

In contrast to the interstrand crosslinking results, assays without the use of proteinase (Fig. 6) showed that DNA–protein crosslinking did not differ significantly between Mer⁺ and Mer⁻ cells (Table I). Hence the differences in interstrand crosslinking cannot be attributed simply to differences in drug uptake or inactivation.

The absence of a correlation between DNA–protein crosslinking and Mer status indicates that the O⁶-methylguanine repair mechanism does not act on

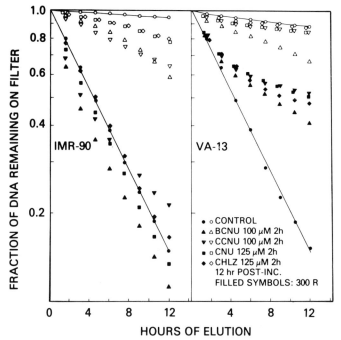

Fig. 6. Alkaline elution assays with proteinase K (showing DNA interstrand crosslinking). IMR-90 (left) and VA-13 (right) cells were treated with drug for 2 hr and then incubated in the absence of drug for 12 hr. Open symbols, no x ray; filled symbols, 300 rads at 0°C. ○,●, No drug; △,▲, 100 μM BCNU; ▽,▼, 100 μM CCNU; □,■, 125 μM CCNU; ◇,◆, 125 μM CHLZ (from Erickson *et al.*, 1980).

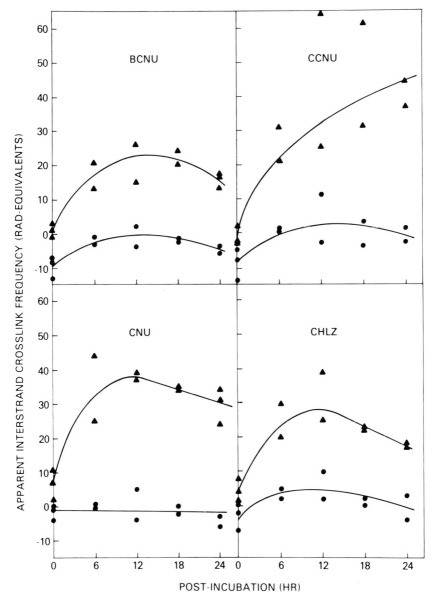

Fig. 7. DNA interstrand crosslinking as a function of time after treatment of IMR-90 (●) or VA-13 (▲) cells with various chloroethylnitrosoureas at 50 μM for 2 hr (from Erickson *et al.*, 1980).

these lesions. Since DNA–protein crosslinks form rapidly, it is unlikely that a monoadduct repair process could be significant for these lesions. There also appeared to be little or no repair of the crosslinks themselves during the time of the experiments (Fig. 7). Hence it is reasonable to suppose the DNA–protein

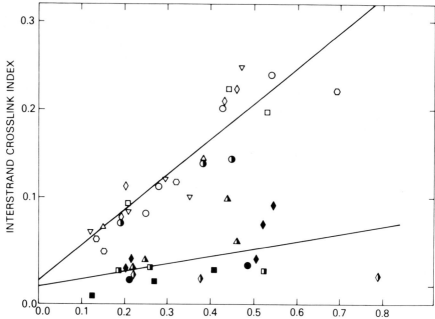

Fig. 8. Interstrand crosslinking (assay with proteinase) plotted against DNA–protein crosslinking (assay without proteinase) in cells treated with CNU. Open symbols, Mer⁻ strains; closed and half closed symbols, Mer⁺ strains. The symbols representing individual cell strains are shown in Table I. The experiments were as in Table I (from Erickson *et al.*, 1981).

crosslinking can be taken as a measure of the amount of chemically reactive drug that gains access to the DNA.

We applied this concept by plotting the values for interstrand crosslinking (proteinase assay) against the values for DNA–protein crosslinking (no proteinase) obtained within the same experiment (Fig. 8). The cell strains are seen to fall into 2 distinct groups, the Mer⁻ strains (open symbols) being clearly separated from most of the Mer⁺ strains (closed and half-closed symbols). The A431 (Mer⁺) strain again falls within the Mer⁻ group (half closed circles). The segregation of the cell strains into 2 distinct groups is analogous to the distinct grouping with respect to methylation repair noted by Day *et al.* (1980). This gives additional support to the idea that the cells differ with respect to a discrete phenotype.

2. Cytotoxicity

The cytotoxic sensitivity of the cell strains to CNU was determined by inhibition of cell proliferation (Fig. 9). The six Mer⁻ strains (open symbols) were clearly more sensitive to CNU than were five of the seven Mer⁺ strains. The other two Mer⁺ strains gave intermediate (half-closed circles) or deviant (half-closed triangles) results; these were the same two strains that gave intermediate

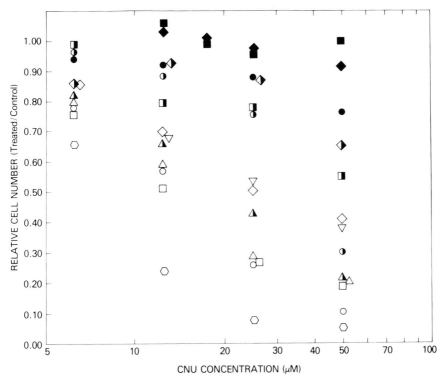

Fig. 9. Effect of CNU treatment on proliferation of various cell strains. Cells were treated with CNU for 1 hr. After drug removal, cells were allowed to proliferate in fresh medium for 3 days (3 to 5 doubling times for control cells). Symbols are as in Table I; open symbols, Mer⁻ strains; closed and half-closed symbols, Mer⁺ strains.

or variable interstrand crosslinking results. The correlation between interstrand crosslinking and cytotoxicity thus was excellent.

IV. DISCUSSION

Recent results with nitrosoureas, including the results that have been reviewed here, suggest a unifying hypothesis relating DNA repair deficiency, carcinogenesis, and chemotherapeutic sensitivity of human tumors. The hypothesis emerges from the following considerations. It is known that DNA repair defects of various kinds predispose individuals to cancer (Cetlow, 1978). A DNA lesion that is particularly likely to be mutagenic and carcinogenic is alkylation of the guanine-O^6 position (Singer, 1975). Normal human cells have a repair mechanism that eliminates these lesions. The finding that perhaps 20 to 30% of human tumor cell strains are deficient in this repair process (Day et al., 1980; 1981) suggests that such tumors may arise from repair-deficient tissue cells. It may be that repair-deficient cells accumulate with time in various tissues, perhaps by somatic mutation. Such cells, deficient in the ability to repair O^6-alkylguanine lesions in DNA, would have an increased probability of undergoing neoplastic transformation due to the action of chemical carcinogens that alkylate at this position. It is possible that a significant fraction of human tumors arise by this mechanism. Our findings suggest that this class of tumors would be vulnerable to chemotherapy with chloroethylnitrosoureas. We hypothesize that these drugs react at the DNA guanine-O^6 position to form an alkylation product that slowly is converted to potentially lethal interstrand crosslinks. The O^6-alkylguanine repair mechanism present in normal human cells would be capable of removing the initial drug-induced DNA lesions before they react further to produce the crosslinks. Those tumor cells that are deficient in this repair mechanism would not be able to remove the initial chloroethylnitrosourea-induced lesions and would therefore be killed by the formation of interstrand crosslinks. (Alternatively, it may be the O^6-chloroethylguanine adducts themselves that are the major cytotoxic lesions.) The results that we have presented suggest that the potentially sensitive tumors could be identified by DNA interstrand crosslink measurement following exposure of the cells to chloroethylnitrosourea *in vitro*.

These considerations lead us to the viewpoint that cancer chemotherapy with nitrosoureas, and perhaps also with a variety of other DNA-damaging agents, may be effective only against those tumors that have some specific vulnerability, such as a particular DNA-repair defect. According to this viewpoint, the further development of nitrosourea therapy would require (*a*) development of *in vitro* tests to identify the potentially sensitive tumors; and (*b*) structure–activity studies aimed at finding new nitrosoureas having the greatest possible selective action against the sensitive tumor cells.

REFERENCES

Bradley, M. O., Sharkey, N. A., and Kohn, K. W. (1980). *Cancer Res. 40*, 2719–2725.

Day, R. S., III, and Ziolkowski, C. H. J. (1979). *Nature 279*, 797–799.

Day, R. S., III, Ziolkowski, C. H. J., Scudiero, D. A., Meyer, S. A., and Mattern, M. R. (1980). *Carcinogenesis 1*, 21–32.

Day, R. S., III, Ziolkowski, C. H. J., Scudiero, D. A., Meyer, S. A., Lubiniecki, A., Girardi, A., Galloway, S. M., and Bynum, G. D. (1981). *Nature*. In press.

Erickson, L. C., Bradley, M. O., and Kohn, K. W. (1977). *Cancer Res. 37*, 3744–3750.

Erickson, L. C., Bradley, M. O., Ducore, J. M., Ewig, R. A. G., and Kohn, K. W. (1980). *Proc. Nat. Acad. Sci. 77*, 467–471.

Erickson, L. C., Laurent, G., Sharkey, N. A., and Kohn, K. W. (1981). *Nature*. In press.

Ewig, R. A. G., and Kohn, K. W. (1978). *Cancer Res. 38*, 3197–3203.

Karran, P., and Lindahl, T. (1979). *Nature 280*, 76–77.

Kohn, K. W. (1977). *Cancer Res. 37*, 1450–1454.

Kohn, K. W. (1979). *In* "Methods in Cancer Research" (H. Busch and V. DeVita, eds.), Vol. 16, pp. 291–345. Academic Press, New York.

Kohn, K. W., Erickson, L. C., Ewig, R. A. G., and Friedman, C. (1976). *Biochem. 15*, 4629–4637.

Kohn, K. W., Ewig, R. A. G., Erickson, L. C., and Zwelling, L. A. (1981). *In* "DNA Repair: A Laboratory Manual of Research Techniques" (E. C. Friedberg and P. C. Hanawalt, eds.). Marcel Dekker, New York. In press.

Kramer, B. S., Fenselau, C. C., and Ludlum, D. B. (1974). *BBRC 56,* 783–788.

Lawley, P. D., Orr, D. J., and Jarman, M. (1975). *Biochem. J. 145,* 73–84.

Lown, J. W., and McLaughlin, L. W. (1979). *Biochem. Pharmacol. 28,* 2123–2128.

Lown, J. W., McLaughlin, L. W., and Chang, Y-M. (1978). *Bioorganic Chem. 7,* 97–100.

Montgomery, J. A. (1976). *Cancer Treat. Rep. 60,* 651–664.

Renard, A., and Verly, W. G. (1980). *FEBS Lett. 114,* 98–102.

Robins, P., and Cairns, J. (1979). *Nature 280,* 74–76.

Setlow, R. B. (1978). *Nature 271,* 713–717.

Singer, B. (1975). *Prog. Nucleic Acid Res. Mol. Biol. 15,* 219–284.

Stacey, K. A., Cobb, M., Cousens, S. F., and Alexander, P. (1958). *Ann NY Acad. Sci. 68(3),* 682–701.

Chapter 7

MODIFICATION OF DNA AND RNA BASES[1]

David B. Ludlum
William P. Tong

I. INTRODUCTION

Early studies on the mechanism of action of the therapeutic nitrosoureas demonstrated that these agents have both alkylating and carbamoylating activities (Bowdon and Wheeler, 1971; Schmall et al., 1973; Wheeler and Chumley, 1967). In 1972, Cheng et al. found that a radioactive label contained in the chloroethyl group of chloroethyl cyclohexyl nitrosourea (CCNU) was transferred to nucleic acids, while a label in the cyclohexyl group was transferred to proteins. The former reaction was attributed to nucleic acid alkylation and the latter to protein carbamoylation.

The first indication as to the nature and significance of nucleic acid modifications was contained in a paper by Kramer et al. (1974) which described two derivatives of cytidylic acid: 3-hydroxyethylcytidylic acid and 3,N[4]-ethanocytidylic acid. These workers suggested that such derivatives were the result of a more general two-step process in which a haloethyl group was first transferred to a DNA or RNA base. The haloethyl nucleoside thus formed would then be free to react a second time with displacement of the chlorine atom to yield hydroxyethyl nucleosides, ethanonucleosides, or, possibly, interstrand crosslinks.

[1]This research was supported by grant CA 20129 from the National Cancer Institute, DHEW.

Subsequent studies have supported many of these conclusions, but additional data discussed below indicate that the situation is probably somewhat more complex than this.

In any case, physical studies described by Kohn (this volume) have shown that interstrand crosslinks are definitely formed in DNA that has been exposed to haloethyl nitrosoureas. Commonly held hypotheses associate such interstrand crosslinking reactions with cytotoxicity and relate the enzymatic repair of these crosslinks to cellular resistance. Accordingly, as the nature of the interstrand crosslinks are elucidated, enzymes that repair them will probably also be discovered.

Since the crosslinks formed by the nitrosoureas are necessarily different from those formed by other DNA-modifying agents, it is tempting to suggest that details of their structure may explain the unique cytotoxic actions of these agents. In what follows, we will describe those base modifications that have been identified in our laboratory and speculate on their mode of formation and probable significance.

II. MATERIALS AND METHODS

The first step in our investigations was to determine which DNA and RNA bases were modified by the therapeutic nitrosoureas. This was done by reacting homopolyribonucleotides, obtained from Miles Laboratory or Schwarz Bio-Research, with BCNU labeled in the chloroethyl groups with ^{14}C. Labeled drug was supplied by the Division of Cancer Treatment, Drug Research and Development, National Cancer Institute.

Polynucleotides were incubated with labeled BCNU at 37° in pH 7 cacodylate buffer, and aliquots were withdrawn at various time periods. These aliquots were applied to filter paper discs, washed free of unbound radioactivity, and counted to determine the amount of bound radioactivity. The results indicated that cytidine was the most reactive nucleoside under these conditions while guanosine and adenosine reacted to a lesser extent (Ludlum et al., 1975). There was practically no labeling of uridine.

The next step was to digest the polyribonucleotides—and, later, DNA—to the monomeric level so that modified nucleosides and bases could be separated and identified. Polyribonucleotides were hydrolyzed with 1 N HCl at 100° for 30 min or with venom phosphodiesterase for 14 hr at 37°. DNA was digested to its component nucleosides with a combination of deoxyribonuclease 1, venom phosphodiesterase, spleen phosphodiesterase, and bacterial alkaline phosphatase.

Modified purines were separated from DNA very effectively by a controlled depurination procedure without complete digestion. DNA was dissolved in 25 mM sodium cacodylic buffer, pH 7, and heated at 100° in a water bath for 15 min. At the end of this time, the solution was cooled in an ice bath. Partially depurinated DNA and oligonucleotides were precipitated by the addition of 1/10 volume of 1 N HCl and removed by centrifugation. Purines in the supernatant were concentrated by lyophilization and redissolved in water.

High-pressure liquid chromatography, with its superior resolving power for nucleosides and bases, was used to isolate modified nucleosides from both DNA and RNA. Hydrolysates were separated by reverse phase chromatography on a C_{18} analytical column eluted with 5 mM KH_2PO_4, pH 4.5, containing a few percent acetonitrile. Eluents were monitored with a Perkin Elmer LC-55B ultraviolet detector interfaced with a Sigma 10 Data System.

In most cases, structures of the modified nucleosides were determined from ultraviolet and mass spectral data. Additional quantities of several modified nucleosides were prepared by known synthetic methods to verify structural assignments (Gombar et al., 1980; Ludlum and Tong, 1978; Tong and Ludlum, 1978 and 1979).

Ultraviolet spectra were obtained in 0.1 N HCl, 0.1 N sodium cacodylate buffer, pH 7, and in 0.1 N NaOH on a Beckman Model 35 spectrophotometer. In most cases, mass spectrometry was performed on underivatized bases and trimethylsilylated nucleoside derivatives using an electron impact technique. Recently, through the courtesy of Mr. Marion Kirk at the Southern Research Institute, we have obtained direct mass spectral data on nonsilylated nucleosides and bases by field desorption mass spectrometry.

III. RESULTS

The modified bases that have been isolated from polyribonucleotides or DNA treated with therapeutic nitrosoureas are shown in Fig. 1. As noted above, their structures have been deduced from a combination of ultraviolet and mass spectrometry and shown to be correct by independent organic synthesis. Based on the reaction mechanisms proposed below, it is predictable that other modified nucleosides will be found in the future.

As illustrated in Fig. 1, derivatives of all the bases except uracil have been identified. Uracil, and the corresponding DNA base, thymine, are evidently much less reactive. The modifications themselves fall into two general classes: those consisting of a single substitution at a nucleophilic site and those that necessarily involve a two-step modification. Mono-substituted bases include the haloethyl, hydroxyethyl, and aminoethyl derivatives, whereas the ethano derivatives and diguanylethane are examples of more extensive modifications.

Although the haloethyl derivatives are probably somewhat unstable in DNA, trace quantities of the chloroethyl derivatives have been found in hydrolysates of DNA that has been reacted with BCNU. Similarly, trace amounts of the fluoroethyl derivatives have been found in hydrolysates of DNA reacted with BFNU.

Although it was expected that the haloethyl derivatives would be precursors of

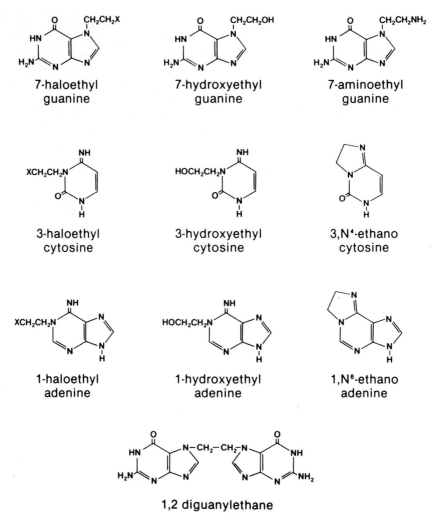

Fig. 1. Bases modified by BCNU.

the hydroxyethyl derivatives, hydrolysis studies have failed to demonstrate this. Thus, when 3-fluoroethylcytidine was incubated in neutral aqueous solution at 37°, this compound was converted quantitatively to ethanocytidine; no detectable hydroxyethylcytidine was formed. Similarly, 3-chloroethylcytidine and 7-chloroethylguanosine failed to yield the hydroxyethyl derivatives when they were incubated in aqueous solution. Other possible origins of the hydroxyethyl derivatives will be discussed below.

7-Aminoethylguanine has been identified in hydrolysates of DNA treated with BCNU (Gombar *et al.*, 1980). This product is not found when CCNU is the reactant, indicating that it is the chloroethyl group in the 3 position ("right-hand end'') of BCNU (1,3-bis[2-chloroethyl]-1-nitrosourea) that is involved in this reaction. Aminoethylguanine is also a product of the reaction between guanosine and chloroethylisocyanate or chloroethylamine, both of which are formed from the chloroethyl group in the 3 position during the decomposition of BCNU (Montgomery et al., 1967). This, of course, explains why CCNU, which contains a cyclohexyl group in the 3 position, does not form aminoethylguanine. Similarly, methyl CCNU would not be expected to yield this derivative.

Since aminoethylguanine formation is unique to BCNU, it is probably not involved in the cytotoxic action of the nitrosoureas. However, this DNA modification might be a source of unwanted mutagenic or carcinogenic side effects. On the other hand, CCNU does form chloroethyl and hydroxyethyl derivatives. Thus, it is evident that the chloroethyl group in the 1 position (''left-hand end'') of the nitrosourea is involved in these reactions.

In many ways, the most interesting nucleoside derivative shown in Fig. 1 is diguanylethane. This structure was first isolated as a product of the reaction between guanosine and BCNU (Gombar et al., 1979) and has since been found in DNA treated with BCNU (Tong and Ludlum, 1981). The presence of this derivative in DNA clearly indicates that BCNU can crosslink nucleosides through a two-carbon bridge. Diguanylethane can be isolated with relative ease since it is released intact from BCNU-treated DNA by mild depurination. We are

TABLE I. High-Pressure Liquid Chromatographic Retention Times (Minutes) of Modified Bases and Nucleosides

Compound	System[a] A	B	Compound	System A	B
Cytidine	8.5	4.8	Adenosine		23.2
3-(β-hydroxyethyl)cytidine	10.3	5.2	1-(β-fluoroethyl)adenosine		11.8
3-(β-fluoroethyl)cytidine	16.5	7.0	1-(β-hydroxyethyl)adenosine		6.4
4-(β-hydroxyethyl)cytidine	13.6	6.1	N^6-(β-hydroxyethyl)adenosine		46.0
3,N^4-ethanocytidine	12.3	5.7	1,N^6-ethanoadenosine		8.9
Guanosine		10.0	Adenine		14.7
7-(β-chloroethyl)guanosine		24.5	1-(β-fluoroethyl)adenine		15.7
7-(β-hydroxyethyl)guanosine		8.0	1-(β-hydroxyethyl)adenine		5.8
O^6-(β-hydroxyethyl)guanosine		41.2	1,N^6-ethanoadenine		9.6
7-(β-aminoethyl)guanosine	7.5	4.5			
1,2-(diguanosin-7-yl)ethane		13.4			
Guanine		6.8			
7-(β-chloroethyl)guanine		35.7			
7-(β-hydroxyethyl)guanine		9.1			
O^6-(β-hydroxyethyl)guanine		22.3			
7-(β-aminoethyl)guanine		5.9			
1,2-(diguan-7-yl)ethane		29.5			

[a] System A: Spherisorb ODS 5μ 4.6 × 250 mm eluted isocratically with 0.05 M KH$_2$PO$_4$, pH 4.5, at 1 ml/min. System B: Same column eluted isocratically with 3% acetonitrile 0.05 M KH$_2$PO$_4$, pH 4.5, at 1 ml/min.

TABLE II. Ultraviolet Spectral Characteristics of Modified Bases and Nucleosides

Compound	Acid		pH 7		Base	
	λmax	λmin	λmax	λmin	λmax	λmin
3-(β-hydroxyethyl)cytidine	280	246	280	244	268	245
3-(β-fluoroethyl)cytidine	280	246	280	244	268	246
4-(β-hydroxyethyl)cytidine	283	245	271	248	271	250
			238	228	238	232
3,N^4-ethanocytidine	286	244	283	251	279	256
7-(β-hydroxyethyl)guanosine	258	232	284	270	266	246[a]
			257	240		
7-(β-chloroethyl)guanosine	258	232	284	270	266	246[a]
			257	240		
7-(β-aminoethyl)guanosine	258	232	284	270	266	246[a]
			257	240		
O^6-(β-hydroxyethyl)guanosine	288	261	280	262	280	262
	244	232	248	227	248	232
1,2-(diguanosin-7-yl)ethane	258	232	284	270	266	246[a]
			257	240		
7-(β-hydroxyethyl)guanine	250	228	284	260	280	257
			245	235		
7-(β-chloroethyl)guanine	250	230	284	260	280	257
			245	235		
7-(β-aminoethyl)guanine	250	230	284	260	280	257
			243	234		
O^6-(β-hydroxyethyl)guanine	288	257	282	258	284	260
			240	228		
1,2-(diguan-7-yl)ethane	250	230	284	260	280	257
			245	235		
1-(β-hydroxyethyl)adenosine	258	234	259	234	259	235
1-(β-fluoroethyl)adenosine	260	235	260	234	260	236
1,N^6-ethanoadenosine	262	235	262	235	269	239
1-(β-hydroxyethyl)adenine	262	235	262	235	269	239
1-(β-fluoroethyl)adenine	262	235	262	235	269	239
1,N^6-ethanoadenine	262	234	265	242	274	245

[a] All 7-alkyl purine nucleosides are unstable in alkali, and the imidazole ring opens.

investigating the presence of crosslinked bases and nucleosides in DNA that has been digested enzymatically to release pyrimidines as well as purines.

Since HPLC separations have proved to be vital in isolating these derivatives, we have included information on typical separations of nucleosides and bases in Table I. This table gives the retention times for various derivatives on a Spherisorb ODS 5μ column (4.6 × 250 mm) eluted with phosphate buffers.

Ultraviolet spectral data on these derivatives in acid, base, and neutral solution are given in Table II. If these data are combined with chromatographic properties that have been obtained in comparison with known derivatives, they are usually sufficient for identification purposes.

IV. DISCUSSION

A scheme indicating how BCNU could modify bases is shown in Fig. 2. There is good evidence that chloroethyl carbonium (or chloronium) ions are generated from the left-hand side of BCNU and that they can add to nucleosides to form chloroethyl derivatives. Chloroethylisocyanate and chloroethylamine are known to be generated from the right-hand side of BCNU and to form aminoethyl nucleosides. The oxadiazoline intermediate shown in the center of the figure has been postulated to explain the formulation of hydroxyethyl derivatives (Lown and McLaughlin, 1979; Tong and Ludlum, 1979).

In our quantitative studies, hydroxyethyl derivatives have always constituted a major fraction of the modified nucleosides. Yet, as described above, the corresponding haloethyl derivatives generate these products very slowly. Thus, it seems likely that there is an alternate route to the hydroxyethyl nucleosides that would be explained by the postulated oxadiazoline intermediate.

If the oxadiazoline intermediate is formed, it could also lead to crosslink formation. As shown in Fig. 2, attack by a nucleoside at C-1 of this cyclic intermediate would again generate a nitrosourea—this time bearing a nucleoside rather than a halogen substituent on the ethyl group. If this first nucleoside were contained in a DNA strand and if the "nucleoside nitrosourea" reacted with a second nucleoside, a crosslink would be produced. Thus, either the secondary reaction of a haloethyl nucleoside or this mechanism would explain the delayed crosslinking observed by others (Kohn, 1977; Lown et al., 1978).

As mentioned above, other modified nucleosides will almost certainly be uncovered. Since all nucleophilic sites in the DNA bases apparently react with ethyl nitrosourea (Singer, 1975), it seems likely that the same reactions will occur with the haloethyl nitrosoureas. Furthermore, Lown and McLaughlin (1979) have shown that the phosphate groups in DNA are alkylated by BCNU. We are inclined to focus attention on dinucleotide formation, however, because it seems probable that crosslinking reactions are related to cytotoxicity.

If one examines a geometric model of double helical DNA, an interstrand link could be formed most readily between guanine and cytosine (Kohn, 1977; Lown et al., 1978). Although such a derivative has not yet been described, we have preliminary evidence from this laboratory for its existence. Of course, other crosslinks are possible if the helical structure is denatured, and this may occur as a result of the initial alkylating reaction.

The possible consequences of a crosslinking reaction are shown in Fig. 3. An interstrand crosslink can be formed between opposing bases or neighboring ones. Alternately, an intrastrand crosslink can be formed between neighboring bases in one strand of DNA or, possibly, between more widely separated bases if there is looping out of the DNA strand. In any case, because of geometric constraints, it seems very likely that there would be considerable distortion of the DNA. This situation is, perhaps, not unlike that postulated for the cis-platinum compounds.

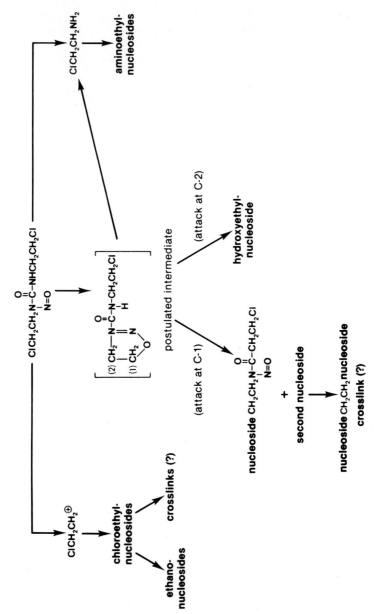

Fig. 2. Scheme for modification of bases by BCNU.

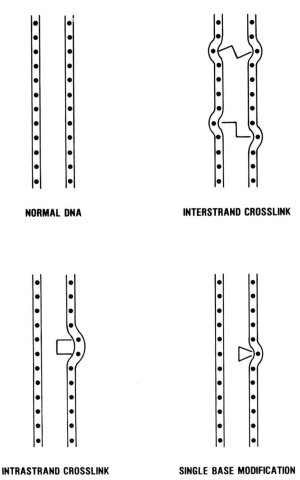

NORMAL DNA **INTERSTRAND CROSSLINK**

INTRASTRAND CROSSLINK **SINGLE BASE MODIFICATION**

Fig. 3. Changes in structure of DNA induced by base modification.

The increase in bulk in a single base produced by either an ethano bridge or a halo- or hydroxyethyl substituent would also cause local denaturation.

The ultimate mechanism by which these base modifications produce cytotoxicity has yet to be established. Interstrand crosslink formation would probably be cytotoxic because it would almost certainly interfere wtih DNA replication. However, certain of the other modifications could also be cytotoxic by interfering with replicative and transcriptive phenomena, or by causing DNA fragmentation.

ACKNOWLEDGMENTS

The authors would like to acknowledge the scientific contributions of their associates, particularly Barnett S. Kramer and Charles T. Gombar; also, the editorial assistance of Suzanne Wissel.

REFERENCES

Bowdon, B. J., and Wheeler, G. P. (1971). *Proc. Am. Assoc. Cancer Res. 12,* 67.

Cheng, C. J., Fujimura, S., Grunberger, D., and Weinstein, I. B. (1972). *Cancer Res. 32,* 22–27.

Gombar, C. T., Tong, W. P., and Ludlum, D. B. (1979). *Biochem. Biophys. Res. Comm. 90,* 878–882.

Gombar, C. T., Tong, W. P., and Ludlum, D. B. (1980). *Biochem. Pharmacol.* 29: 2639–2643.

Kohn, K. W. (1977). *Cancer Res. 37:* 1450–1454.

Kramer, B. S., Fenselau, C. C., and Ludlum, D. B. (1974). *Biochem. Biophys. Res. Comm. 56,* 783–788.

Lown, J. W., and McLaughlin, L. W. (1979). *Biochem. Pharmacol. 28,* 1631–1638.

Lown, J. W., McLaughlin, L. W., and Chang, Y. M. (1978). *Bioorg. Chem. 7,* 97–110.

Ludlum, D. B., and Tong, W. P. (1978). *Biochem. Pharmacol. 27,* 2391–2394.

Ludlum, D. B., Kramer, B. S., Wang, J., and Fenselau, C. (1975). *Biochem. 14,* 5480–5485.

Montgomery, J. A., James, R., McCaleb, G. S., and Johnston, T. P. (1967). *J. Med. Chem. 10,* 668–674.

Schmall, B., Cheng, C. J., Fujimura, S., Gersten, N., Grunberger, D., and Weinstein, I. B. (1973). *Cancer Res. 33,* 1921–1924.

Singer, B. (1975). *In* "Progress Nucleic Acid Research and Molecular Biology" (W. E. Cohn, ed.), Vol. 15, pp. 219–280. Academic Press, New York.

Tong, W. P., and Ludlum, D. B. (1978). *Biochem. Pharmacol. 27,* 77–81.

Tong, W. P., and Ludlum, D. B. (1979). *Biochem. Pharmacol. 28,* 1175–1179.

Tong, W. P., and Ludlum, D. B. *Cancer Res.* In press.

Wheeler, D. G., and Chumley, S. (1967). *J. Med. Chem. 10,* 259–261.

Chapter 8

CARBAMOYLATING ACTIVITY OF NITROSOUREAS

Herbert E. Kann, Jr.

I. INTRODUCTION

The carbamoylation reaction occurring between an isocyanate and a reactant capable of losing a proton involves formation of a covalent bond between the isocyanate and its reactant (Fig. 1). Isocyanates are highly reactive compounds; the carbamoylation reaction is therefore spontaneous and rapid. The number and range of possible sites for carbamoylation in biologic systems is vast.

Isocyanates capable of reacting with cellular constituents via the carbamoylation reaction are formed on decomposition of most nitrosoureas (Montgomery *et al.*, 1967). An abbreviated scheme for decomposition of nitrosoureas is shown in Fig. 2. It is quite clear that carbamoylation is the basis for many of the biochemical actions that have been observed with nitrosoureas, including inhibition of macromolecular syntheses (Kann *et al.*, 1974a), inhibition of RNA processing (Kann *et al.*, 1974b), and inhibition of DNA repair (Kann *et al.*, 1980a). Carbamoylation reactions between the decomposition products of nitrosoureas and cellular constituents are, however, not essential to the antitumor activity of nitrosoureas. Chlorozotocin, for example, is a nitrosourea with very weak carbamoylating activity for cellular constituents (Wheeler, 1976) but with potent antitumor activity. The isocyanate formed on decomposition of chlorozotocin carbamoylates internally (Fig. 3), and this is the basis for its limited ability to carbamoylate other compounds (Montgomery, 1976).

At about the time carbamoylation was being recognized as nonessential to

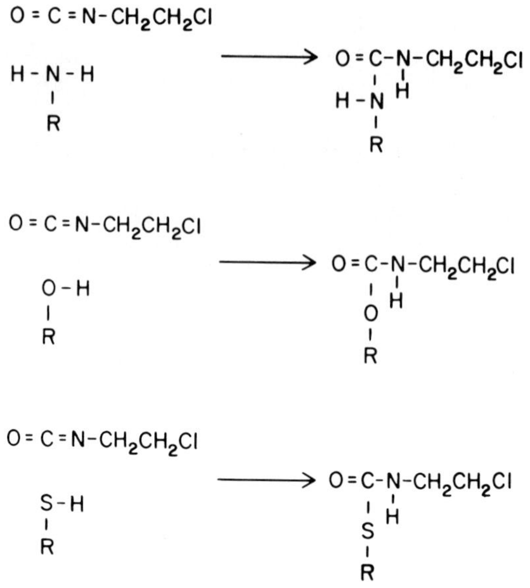

$$O = C = N - CH_2CH_2Cl$$

$$H - N - H \quad\longrightarrow\quad O = C - N - CH_2CH_2Cl$$

with R below, yielding $H - N - H$ and R.

$$O = C = N - CH_2CH_2Cl$$

$$O - H \quad\longrightarrow\quad O = C - N - CH_2CH_2Cl$$

$$O = C = N - CH_2CH_2Cl$$

$$S - H \quad\longrightarrow\quad O = C - N - CH_2CH_2Cl$$

Fig. 1. Carbamoylation reactions.

antitumor activity, Kohn and his coworkers were demonstrating that N-1-chloroethylnitrosoureas are capable of bifunctional alkylation and crosslinking of DNA (Kohn, 1977; Ewig and Kohn, 1978). The evidence now strongly favors the view that alkylation is the principal basis for antitumor activity of nitrosoureas, and a reasonable concern exists regarding the possibility of unwanted, unneeded, toxic side effects on normal host tissues from the strong carbamoylating activity associated with most nitrosoureas.

In spite of the above, one of the actions of strongly carbamoylating nitrosoureas—inhibition of DNA repair—has been a point of continuing interest and research (Kann *et al.*, 1980a; Kann *et al.*, 1980b). Repair inhibition is, at

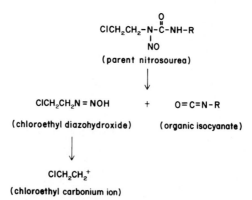

Fig. 2. Abbreviated scheme for decomposition of a nitrosourea (Brundrett *et al.*, 1976).

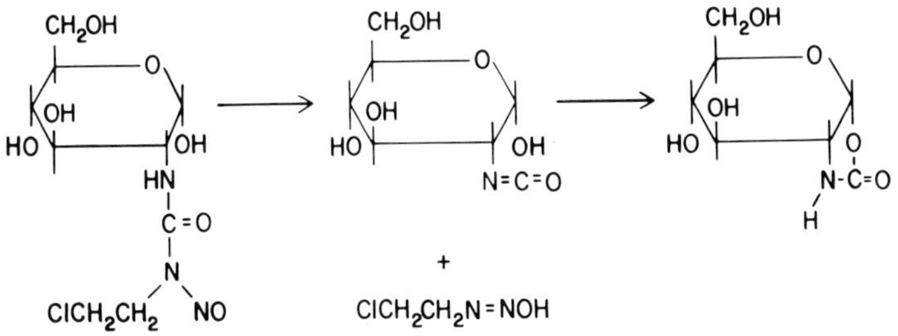

Fig. 3. Decomposition of chlorozotocin followed by intramolecular carbamoylation.

least in theory, a significant action because damage to DNA is such a common mode of action for oncolytic agents. In this chapter are discussed the results of studies dealing with the basis for inhibition of DNA repair by nitrosoureas and the biologic significance of repair inhibition. The drugs studied are listed in Fig. 4. The principal findings from this work are that (a) strongly carbamoylating nitrosoureas (but *not* the more weakly carbamoylating chlorozotocin) inhibit the repair of DNA damaged by ionizing radiation; and (b) repair-inhibiting nitrosoureas (and *only* the repair-inhibiting nitrosoureas) function as radiation synergists, enhancing the cytotoxic effects of ionizing radiation by reducing the mean lethal dose of radiation for cells in culture.

Drug	$$R-N-C-NHR'$$ (R=)	(R'=)	Relative to CCNU[*] Carbamoylating activity	Alkylating activity
Chlorozotocin	$CICH_2CH_2-$	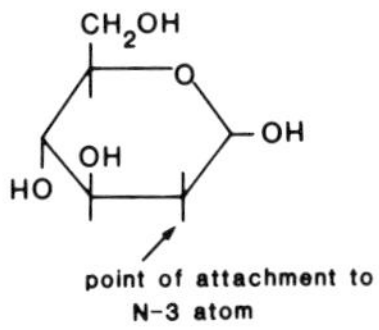 [structure] point of attachment to N-3 atom	0.02	4.52
CCNU	$CICH_2CH_2-$	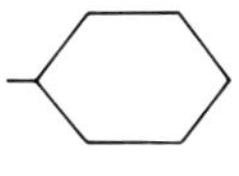 [structure]	1.0	1.0
BCyNU	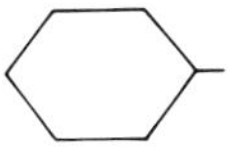 [structure]	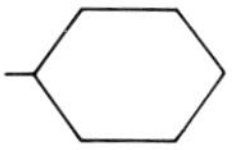 [structure]	0.87	0.01

[*] Data are from the work of Wheeler

Fig. 4. Comparison of alkylating and carbamoylating activity of nitrosoureas (Wheeler *et al.*, 1977).

II. MATERIALS AND METHODS

Mouse leukemia L1210 cells (Moore *et al.*, 1966) in suspension culture were used in this study. Cells were maintained in log-phase growth, under conditions that have been described previously (Kann *et al.*, 1974b). Cells were checked periodically and were found to be free of mycoplasma contamination.

Drugs were obtained from the Drug Synthesis and Chemistry Branch, Division of Cancer Treatment, National Cancer Institute, and from the Southern Research Institute, Birmingham, Alabama. The exact conditions of drug storage, preparation, and addition have been given previously (Kann *et al.*, 1980a).

Radiation was delivered from a cobalt source at a rate of 950 rads/min. Cells were irradiated in plastic culture flasks; following irradiation, they were analyzed for frequency of unrepaired strand breaks and for viable cell number.

A. Evaluation for Inhibition of DNA Repair

Repair inhibition was estimated in cells exposed to radiation by determining the frequency of strand breaks immediately after and 1 hr after irradiation. Strand breaks were detected using the alkaline elution technique developed by Kohn *et al.* (1976), with modifications that have been described previously (Fornace *et al.*, 1978). Proteinase K digestion of cellular lysates was performed by the method described by Ewig and Kohn (1977).

B. Cell Survival

Survival of cells exposed to drug alone, to radiation alone, or to drugs plus radiation was assessed from soft-agar cloning efficiency (Chu and Fischer, 1968).

III. RESULTS

For evaluation of drug effects on DNA repair, alkaline elution profiles of DNA from cells irradiated with or without drug were compared (Fig. 5). Results of a control experiment, in which no drug was present, are shown in the left-hand panel. DNA from cells not exposed to irradiation elutes slowly (circles), with greater than 90% of the DNA remaining on the filter at the end of a 15 hr elution period. DNA from cells irradiated and given no time for postirradiation repair elutes quite rapidly (squares), with only about 6% of the DNA being retained on the filter at the end of 15 hr. The more rapid elution is a reflection of the large number of unrepaired, radiation-induced strand breaks. DNA from cells irradiated and given 1 hr for postirradiation repair (triangles) elutes at an intermediate rate, with approximately 80% of the DNA being retained on the mem-

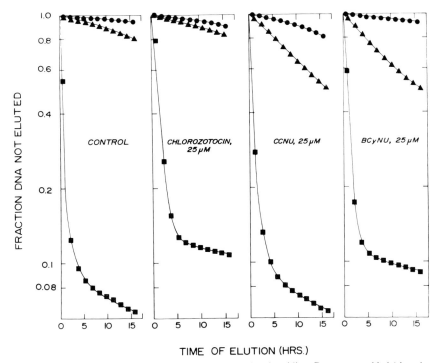

TIME OF ELUTION (HRS.)

Fig. 5. Alkaline elution of DNA prelabeled with radioactive thymidine. Drugs were added 1 hr prior to irradiation. ●, no irradiation; ■, 1000 rads, no repair; ▲, 1000 rads, 60 min repair.

brane filter at the end of a 15 hr elution experiment. The slower rate of elution in this sample (compared to the radiation-with-no-repair sample) is a consequence of repair of the radiation-induced strand breaks; significant strand rejoining occurs during the 1-hr postirradiation repair period. The more rapid elution of the radiation-with-repair sample compared to the unirradiated control is an indication that not all of the strand breaks have been rejoined at the end of a 60-min period. Drugs that inhibit rejoining of radiation-induced strand breaks will cause a more rapid rate of elution in the radiation-with-repair samples, the more rapid elution being a consequence of the presence of a larger-than-normal number of unrejoined strand breaks. In cells exposed to chlorozotocin (Fig. 5), the rate of elution in the radiation-with-repair sample is no greater than that of the radiation-with-repair sample from the culture not exposed to drug. Chlorozotocin, therefore, does not inhibit repair. For both CCNU and BCyNU the rate of elution in the radiation-with-repair sample is increased, due to the larger number of unrejoined strand breaks in DNA from these cells compared to the cells not preincubated with drug. Thus CCNU and BCyNU, the two drugs with strong carbamoylating activity, are both potent inhibitors of repair.

Differences exist in the rates of elution of DNA from unirradiated cells (circles) from some of the drug-treated cultures compared to the control culture.

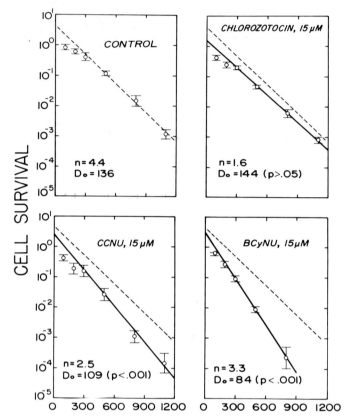

Fig. 6. Radiation survival curves for L1210 cells with and without drug incubation. Either drug or solvent was added 60 min prior to irradiation. Thirty minutes after radiation, cells were washed to remove extracellular drug. Survival was then assessed from soft-agar cloning efficiency. Cell survival in unirradiated cultures was taken to be 1.0, and survival of irradiated cells was defined in relation to survival of unirradiated cells. Values shown are mean ± S.D. for 6 or more determinations from 3 or more independent experiments. Straight-line regressions were fitted by the method of least squares to survival values at 300 rads and above. The broken line in each panel provides a reference point and represents the results obtained in the no-drug control cultures. Calculation of D_0 and n were performed as described in Kann *et al.* (1980b). The *p* values refer to differences in D_0 for drug-treated versus control cultures.

These differences are a consequence of drug effects on the DNA, with production of strand breaks and/or alkali-labile sites (the alkaline elution technique does not distinguish between strand breaks and alkali-labile sites). Note also the differences in the rates of elution of DNA from the radiation-with-no-repair samples of the drug-treated cultures compared to the control culture; the relative slowing of elution in the drug-treated samples may be a reflection of drug-induced crosslinking of DNA. The production of strand breaks, alkali-labile sites, and crosslinks by the various nitrosoureas is interesting and significant and has been analyzed in detail elsewhere (Kann *et al.*, 1980a). However, the information of central importance from these experiments is obtained by comparing the differences between the unirradiated (circles) and the radiation-with-repair samples (triangles) from each experiment; from this comparison the drug effects on the rate of rejoining of radiation-induced strand breaks can be discerned.

To assess the biologic significance of repair inhibition, each of the nitrosoureas was evaluated for cytotoxic synergism with ionizing radiation. A conclusion in favor of a biologically significant role for repair inhibition would be based on the finding of significantly greater cytotoxic synergism between radiation and the repair-inhibiting nitrosoureas (CCNU and BCyNU) than between radiation and the non-repair-inhibiting nitrosourea (chlorozotocin). Results of the experiment pertaining to this question are shown in Fig. 6. A radiation survival curve for L1210 cells not exposed to drug is shown in the upper left panel. The general configuration of the radiation survival curve is typical of that generally observed with mammalian cells. There is a "shoulder region" to the curve at low radiation dosages, in which increments in dosage have a relatively slight effect on cell survival. However, once a threshold of radiation dosage is exceeded, survival decreases by a fixed amount with each increment in radiation dosage. In this "linear portion" of the survival curve, a straight line fits the results when the data are displayed on a semilogarithmic plot. Typically, radiation and survival curves are analyzed by fitting a straight line to the linear portion of the survival curve, using the method of least squares. The straight line then can be defined by its slope and intercept. Drugs that affect only the intercept have principally an effect on the shoulder region of the survival curve. Drugs that affect the slope of the survival curve, however, reducing the mean lethal dose of radiation, are by definition radiation synergists. The mean lethal dose of radiation is determined from the slope of the survival curve. The mean lethal dose (D_0) for L1210 cells not preincubated with drugs is 136 rads. The intercept, or extrapolation number (n), for these cells is 4.4.

Cells preincubated with chlorozotocin show no increase in the slope of the survival curve. The survival curve is affected by a decrease in the extrapolation number, but the mean lethal dose (D_0) of radiation is not decreased by the chlorozotocin preincubation (D_0 = 144 rads for cells preincubated with chlorozotocin versus 136 rads for cells not exposed to drug). Both of the repair-inhibiting nitrosoureas, however, do change the slope of the survival curve, producing a significant decrease in D_0 (109 rads for cells preincubated with CCNU; 84 rads for cells preincubated with BCyNU).

IV. DISCUSSION

The first hypothesis tested in this work stated that carbamoylation is the sole basis for inhibition of DNA repair by nitrosoureas. If the hypothesis were correct, nitrosoureas without carbamoylating activity should not inhibit repair, and carbamoylating nitrosoureas, with or without alkylating activity, should inhibit repair. The demonstration that repair inhibition does not occur with chlorozotocin but does occur with CCNU and BCyNU provides unambiguous confirmation of the hypothesis. Clearly, carbamoylating activity is the sine qua non for repair inhibition by nitrosoureas; alkylating activity is irrelevant to this action.

The second hypothesis tested in this work stated that repair inhibition is a biologically significant action. If the hypothesis were correct, then a nitrosourea that does not inhibit DNA repair should not enhance the cytotoxic effect of the DNA-damaging agent (ionizing radiation) whereas nitrosoureas that do inhibit repair should enhance radiation cytotoxicity. The finding that chlorozotocin does not increase the slope of the radiation survival curve whereas both CCNU and BCyNU do, the latter two compounds significantly reducing the mean lethal dose (D_0) for the cells in culture, provides clear evidence in support of this second hypothesis.

The extent of the significance of these findings is currently uncertain. At a minimum, it can be said that the repair-inhibiting nitrosoureas represent a new class of radiation synergists, whose mechanism of action apparently differs from that of other previously studied radiation synergists. Practical significance may be attached to the finding that radiation synergism is retained by a nitrosourea with no alkylating activity (BCyNU); the absence of alkylating activity makes this a less cytotoxic compound (Kann et al., 1980b), and the diminished cytotoxicity should allow for greater flexibility in the design of synergism experiments. The nonalkylating drug can be used at higher concentrations and could, in all probability, be administered more frequently than the more toxic alkylating nitrosoureas.

The design of the radiation synergism experiments was patterned after the work of Wheeler et al. (1977), who had previously studied cell killing by BCNU plus radiation. The findings reported here, however, differ from their results. They found an enhancing action of BCNU on the cell-killing effect of radiation, but that effect involved primarily the shoulder region of the survival curve. Here the principal finding is an increase in the slope of the straight-line portion of the survival curve. This effect has been observed with all repair-inhibiting nitrosoureas tested so far, including BCNU (Kann et al., 1980b).

Several questions remain after this work. One question concerns the link between repair inhibition and radiation synergism. So far a cause-and-effect relationship has not been proven; conceivably some other consequence of strong carbamoylating activity could be the basis for radiation synergism. Another question is whether cytotoxic synergism by repair-inhibiting nitrosoureas will occur only with ionizing radiation or be demonstrable with other DNA-damaging agents as well, such as ultraviolet radiation and alkylating drugs. A third question would pertain to studies of radiation synergism to be performed in vivo: Hydroxylation of the N-3 cyclohexyl ring of BCyNU can be anticipated in vivo (Reed and May, 1975); there will be a resulting decrease in carbamoylating activity (Montgomery, 1976) and, perhaps, an associated decline in repair inhibition and radiation synergism. The extent of the decline is the point of uncertainty. Finally, there is no way of predicting for the *in vivo* studies whether radiation synergism will be, to any appreciable degree, a selective action; effects on tumor cells and normal cells may prove to be identical.

A number of additional studies might reasonably follow from this work. There might be some value in repeating the radiation synergism experiments, adding the nitrosourea immediately after (rather than before) radiation. This would eliminate the question of any possible radiation–drug–DNA interaction and could help clarify the status of repair-inhibiting nitrosoureas as compounds whose mechanism of action is distinctly different from other radiation synergists. The objective would be to establish more firmly that nitrosoureas are not acting by increasing the number of strand breaks produced by radiation.

Further evidence either for or against a connection between repair inhibition and cytotoxic synergism with DNA-damaging agents might be obtained from studies comparing normal and repair-deficient cells. If repair inhibition (and not some other effect of carbamoylation) were the basis for the radiation synergism, then cytotoxic synergism should be lost when cells genetically incapable of repairing DNA damage are used in place of normal cells. The best characterized system for this kind of study would involve ultraviolet light as the DNA-damaging agent and Xeroderma Pigmentosum cells as the repair-deficient line. A prerequisite for this study would be a demonstration that repair-inhibiting nitrosoureas do produce cytotoxic synergism in cells whose DNA has been damaged by ultraviolet light.

One of the problems that can be anticipated with studies *in vivo* has to do with the limited water solubility of BCyNU (it was dissolved in acetone in the

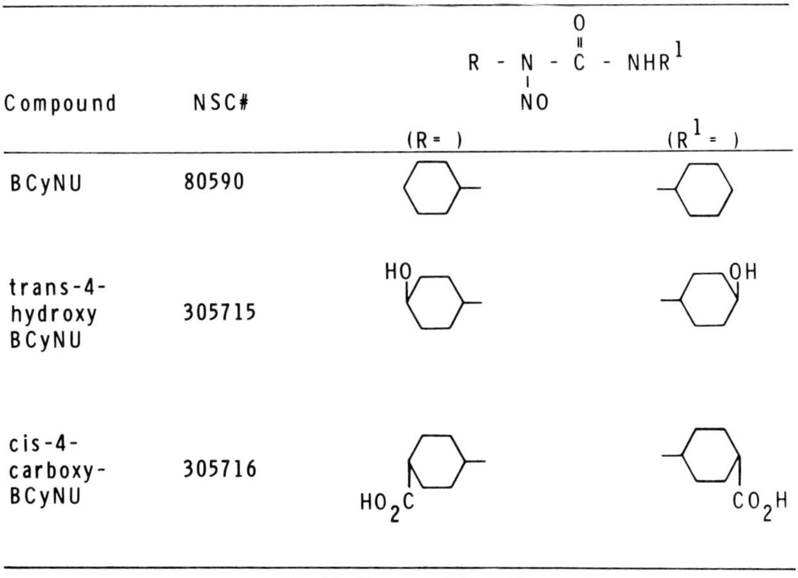

Compound	NSC#	$R - N(NO) - C(=O) - NHR^1$ (R =)	$(R^1 =)$
BCyNU	80590	⬡—	—⬡
trans-4-hydroxy BCyNU	305715	HO—⬡—	—⬡—OH
cis-4-carboxy-BCyNU	305716	HO_2C—⬡—	—⬡—CO_2H

Fig. 7. Structure of BcyNU.

studies performed *in vitro*). To obtain more water-soluble derivatives capable of producing the same biochemical and biologic effects, Tom Johnston at Southern Research Institute in Birmingham, Alabama, has synthesized analogs of BCyNU. Two that have been tested for repair-inhibiting ability are shown in Fig. 7. Both of these compounds inhibit DNA repair. Neither has been tested so far for radiation synergism.

A great deal of work by others has shown that isocyanates are capable of selective inhibition (or selective sparing) of certain enzymes on the basis of stereochemical factors (Brown and Wold, 1973, and Babson *et al.*, 1977). From that work is suggested the possibility, following the identification of a specific enzyme whose inhibition is the basis for radiation synergism, of designing a nitrosourea which would form on decomposition an isocyanate capable of inhibiting that enzyme in a relatively specific fashion. Carbamoylation at other, unwanted sites might thereby be decreased.

Since isocyanates are the active agents involved in carbamoylation, the question of using these compounds directly in studies *in vivo* should be addressed. This approach is unworkable because of the volatility and toxicity of the isocyanates themselves. The parent nitrosourea can be regarded as a useful precursor to the isocyanate. When the nitrosourea is nonalkylating, no obvious added toxicity is incurred by using the nitrosourea as a precursor.

ACKNOWLEDGMENT

The author thanks Terri Burnham for her assistance in preparation of this manuscript.

REFERENCES

Babson, J. R., Reed, D. J., and Sinkey, M. A. (1977). *Biochem. 16*, 1584.
Brown, W. E., and Wold, F. (1973). *Biochem. 12*, 835–840.
Brundrett, R. B., Cowens, J. W., and Colvin, M. (1976). *Chem. 19*, 958–961.
Chu, M. Y., and Fischer, G. A. (1968). *Biochem. Pharmacol. 17*, 753–767.
Ewig, R. A. G., and Kohn, K. W. (1977). *Cancer Res. 37*, 2114–2122.
Ewig, R. A. G., and Kohn, K. W. (1978). *Cancer Res. 38*, 3197–3202.
Fornace, A. J., Jr., Kohn, K. W., and Kann, H. E., Jr. (1978). *Cancer Res. 38*, 1064–1069.
Kann, H. E., Jr., Kohn, K. W., and Lyles, J. M. (1974a). *Cancer Res. 34*, 398–402.
Kann, H. E., Jr., Kohn, K. W., Widerlite, L., and Gullion, D. (1974b). *Cancer Res. 34*, 1982–1988.
Kann, H. E., Jr., Schott, M. A., and Petkas, A. (1980a). *Cancer Res. 40*, 50–55.
Kann, H. E., Jr., Blumenstein, B. A., Petkas, A., and Schott, M. A. (1980b). *Cancer Res. 40*, 771–775.
Kohn, K. W., Erikson, L. C., Ewig, R. A. G., and Friedman, C. A. (1976). *Biochem. 15*, 4629–4637.
Kohn, K. W. (1977). *Cancer Res. 37*, 1450–1454.
Montgomery, J. A. (1976). *Cancer Treat. Rep. 60*, 651–664.

Montgomery, J. A., James, R., McCaleb, G. S., and Johnston, T. P. (1967). *J. Med. Chem. 10,* 668–674.

Moore, G. E., Sandberg, A. A., and Ulrich, K. (1966). *J. Nat. Cancer Inst. 36,* 405–421.

Reed, D. J., and May, H. E. (1975). *Life Sci. 16,* 1263–1270.

Wheeler, G. P. (1976). *In* "Cancer Chemotherapy" (A. Sartorelli, ed.), pp. 87–119. American Chemical Society, Washington, D.C.

Wheeler, K. T., Deen, D. F., Wilson, C. B., Williams, M. E., and Sheppard, S. (1977). *Int. J. Radiat. Oncol. Biol. Phys. 2,* 79–88.

Chapter 9

CHROMATIN AND ASSOCIATED NUCLEAR COMPONENTS AS POTENTIAL DRUG TARGETS

Kenneth D. Tew

I. INTRODUCTION

A. The Concept of Nuclear Target Specificity

Although there is much evidence to suggest that ultimate cytotoxic events are mediated through alterations in cellular DNA, less consideration has been given to a comparison of the relative importance of nuclear macromolecules as potential targets for drug interaction. The chloroethylnitrosoureas are a highly reactive group of compounds capable of producing a variety of nuclear modifications (Kohn, 1979). These include (*a*) monofunctional adducts (alkylations) within nucleic acids and proteins; (*b*) single-strand breaks; (*c*) bifunctional crosslinkage of nucleic acids (inter- or intrastrand) or (*d*) of nucleic acids with proteins; and (*e*) carbamoylation of proteins. A relationship between DNA crosslinking and cytotoxicity has been demonstrated (Kohn, 1979). However, each of these drug interactions has the potential to interfere with nuclear functional integrity and thereby to perpetrate cytotoxicity. Moreover, it is possible that a combination of these lesions may be responsible for ultimate cell death or an increased cell-killing potential.

The complexity of chromatin organization and the presence of regions of functional redundancy provide the basis for the theory that the site of drug interaction within the nucleus will be as important, or more important, than the

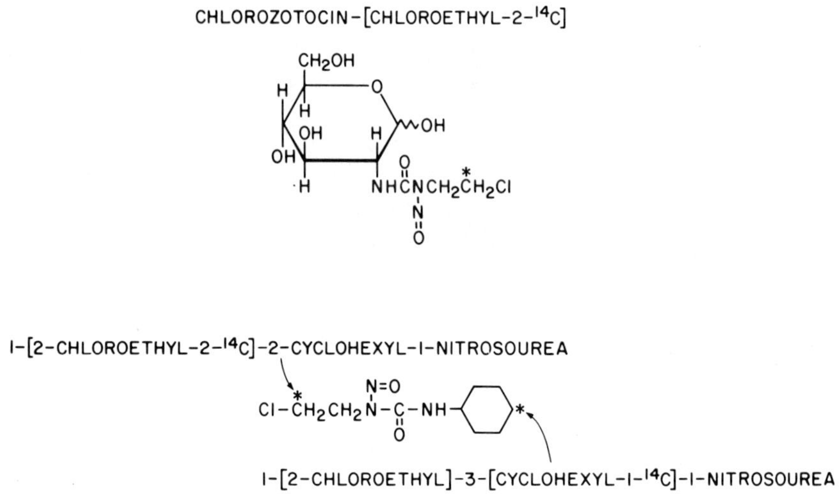

CHLOROZOTOCIN−[CHLOROETHYL−2−^{14}C]

I−[2−CHLOROETHYL−2−^{14}C]−2−CYCLOHEXYL−I−NITROSOUREA

I−[2−CHLOROETHYL]−3−[CYCLOHEXYL−I−^{14}C]−I−NITROSOUREA

Fig. 1. Structural formulae of ^{14}C radiolabeled nitrosoureas. The asterisks represent possible positions of ^{14}C label.

type of lesion. Indeed, the potential for a cell to survive may be determined by the type, number, or diversity of drug lesions; their location within the chromatin; and the speed and efficiency of their repair. These factors may be closely interrelated, and it is important to understand which regions of the nucleus are susceptible to drug attack in order that specific macromolecules may be considered as potential sites for drug-mediated cytotoxicity. Because chloroethylnitrosoureas have clinical applicability, this chapter will emphasize these compounds, thereby excluding the methylnitrosoureas which possess similar pharmacologic characteristics (Tew *et al.*, 1980e; Jump *et al.*, 1980) but lack the potential to crosslink macromolecules. Chlorozotocin (CLZ) and 1-(2-chloroethyl)-3-cyclohexyl-1-nitrosourea (CCNU) were used in all the studies described. The presence of ^{14}C labels (Fig. 1) in the chloroethyl groups of both drugs and in the cyclohexyl moiety of CCNU were used to monitor alkylation or carbamoylation of nuclear macromolecules.

II. RESULTS

A. Subnucleosomal DNA

The basic repeat subunit of chromatin, the nucleosome, consists of a double tetrameric core of histones (H2a, H2b, H3 and H4) with which approximately 170 base pairs of DNA are associated. Micrococcal nuclease initially cleaves linker DNA between nucleosomes. This linker DNA is 40–60 base pairs in length and has associations with H1 and other nonhistone proteins. Figure 2 illustrates a

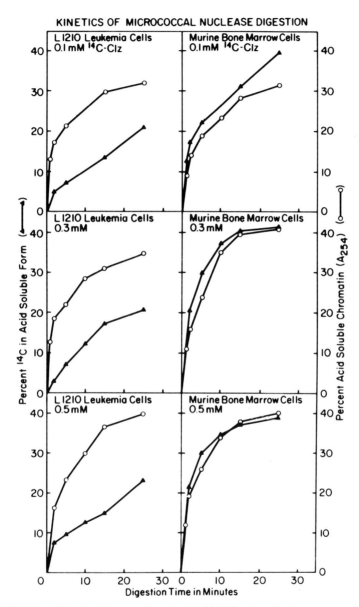

Fig. 2. Limit digest kinetics (micrococcal nuclease) of L1210 and murine bone marrow cells pre-treated for 2 hr with various concentrations of (^{14}C-chloroethyl)CLZ. The enzymic release of acid soluble ^{14}C (▲) was compared with the digestion of bulk chromatin, measured at A_{254} (○) (from Tew et al., 1980a).

micrococcal nuclease limit digest of chromatin from L1210 and murine bone marrow cells treated with three concentrations of radiolabeled CLZ. In L1210 cells, the digestion of chromatin, measured at A_{254}, was greater than the concomitant release of [14C]-labeled material (i.e., alkylated macromolecules). This suggested a preferential alkylation of the DNA which was inaccessible to micrococcal nuclease, namely, the DNA associated with the nucleosome core particle. Conversely, the digestion of alkylated chromatin was greater than that of bulk chromatin in murine bone marrow, suggesting a preferential alkylation of

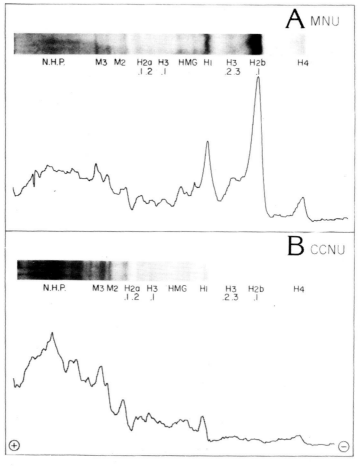

Fig. 3. Modification of nuclear proteins by [carbonyl-14C]MNU and [cyclohexyl-14C]CCNU. HeLa cells (5×10^7) in 1 ml were incubated with 30 μCi of [carbonyl-14C]MNU (7.9 Ci/mol) and [cyclohexyl-14C] CCNU (8.3 Ci/mol) for 1 and 2 hr, respectively. Nuclei were isolated, and proteins were extracted with $0.4N$ H_2SO_4 and analyzed on triton-acetic acid-urea slab gel as described by Alfagame *et al.* (1976). A, autoradiograph of MNU-modified histones and densitometer scan; B, same as in A, of CCNU-modified histones.

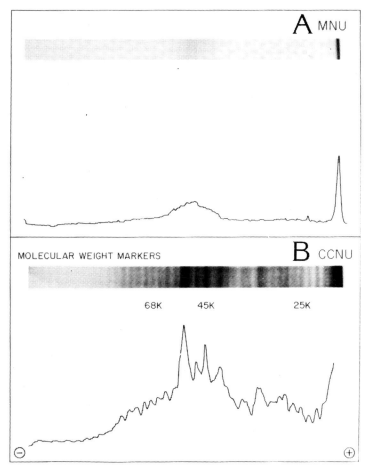

Fig. 4. SDS gel electrophoresis of nonhistone proteins from [carbonyl-^{14}C] MNU and [cyclohexyl-^{14}C]CCNU-treated cells. Proteins insoluble in 0.4N H$_2$SO$_4$ from the experiment described in Fig. 3 were analyzed on a standard SDS slab gel as described by Laemmli (1970). The pellet of proteins was solubilized in the gel buffer and electrophoresed on a 10% acrylamide slab gel containing 1% SDS at 200 V until the tracking dye, bromophenol blue, reached the bottom of the gel. A, autoradiograph of MNU-modified proteins and densitometer scan of the negative of the photograph shown. B, same as in A, of CCNU-modified proteins (Figs. 3 and 4 from Sudhakar *et al.*, 1979).

the linker DNA. A similar methodological approach has demonstrated that CCNU, at the same concentrations as CLZ, alkylated the core particle DNA of both L1210 and murine bone marrow (Tew *et al.*, 1978b). Because there is potential for differential repair of lesions within these chromatin regions (Ramanathan *et al.*, 1976; Bodel and Banerjee, 1979), the comparative non-myelotoxic properties of CLZ may relate to the observed qualitative differences in subnucleosomal alkylation sites.

B. Histone and Nonhistone Proteins

Although cytoplasmic proteins were heavily alkylated by chloroethylnitrosoureas, at the nuclear level histone and nonhistone proteins were not substantially alkylated at physiological drug concentrations (Tew *et al.*, 1978a). In
L1210 cells, at high drug concentrations MNU was found to alkylate histones
H2a and H1, in addition to a number of nonhistone proteins (Pinsky *et al.*,
1979). Carbamoylation of nuclear proteins from HeLa cells occurred at lower
drug concentrations, as shown in Figs. 3 and 4. Significant carbamoylation of
core histones, especially H2b, was detected with MNU (Fig. 3A). This con-
trasted with CCNU, where negligible core histone carbamoylation was found.
The internucleosomally located histone H1 was modified to a minor extent (Fig.
3B), but the prevalent interaction of CCNU was with the nonhistone protein
components of the nucleus. A wide spectrum of proteins within the molecular
weight range of 40,000 to 60,000 was extensively carbamoylated by CCNU (Fig.
4B). At these drug concentrations, the covalent binding of the carbonyl group of
MNU to nonhistone proteins was significantly lower (Fig. 4A). The fact that
CLZ has negligible carbamoylating activity, while maintaining cytotoxic properties, provides evidence that carbamoylation of proteins contributes little to the
overall cell-killing potential of the chloroethylnitrosoureas.

C. Nuclear RNA

Because of the constant flux of intranuclear and cytoplasmic pools of RNA, it
is difficult to gain an absolute quantitation of chloroethylnitrosourea alkylation of
nuclear RNA. By using isolated nuclei from drug-treated HeLa cells, it has been
possible, employing T2 RNase as a probe, to estimate that 20% of the total
nuclear alkylation was of RNA (Tew *et al.*, 1980b). Identification of the extent
of modification and the functional integrity of specific types has not been determined. The extent of duplication and the rate of turnover of these RNA species
predisposes that such modifications should contribute minimally to the cytotoxic
properties of the nitrosoureas. It is interesting, however, that residual ribonuclear
protein complexes are effective targets for drug modifications (see Section E).

D. Transcriptional Chromatin: Manipulation of Drug Targets with
Corticosteroids And Butyrate

It has been possible, using the techniques of ECTHAM–cellulose chromatography (Simpson, 1977) or DNase II digestion, followed by solublization in
magnesium buffer (Gottesfeld, 1977), to demonstrate that transcriptional
chromatin acts as a preferential target for nitrosourea interaction. The precise
degree of selective binding is dependent upon the methodology employed. Figure
5 shows the separation of transcriptional chromatin in the latter-eluting fractions

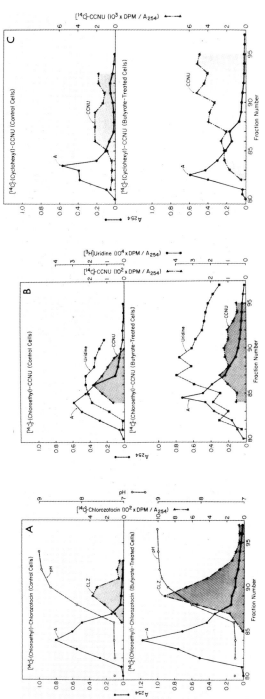

Fig. 5. A, ECTHAM-cellulose column separation of chromatin from normal or butyrate-treated HeLa cells after 2 hr incubation with 30 uM [^{14}C]CLZ. Butyrate treatment was at 5 mM for 24 hr prior to drug incubation. The change in pH after buffer change (Simpson, 1977) was monitored for representative 0.5 ml fractions (○) The elution of chromatin was measured on an ISCO flow cell and UA-5 recorder at A_{254} (●). Drug incorporation in each 0.5 ml fraction was measured by scintillation counting, and these data are expressed as $10^2 \times$ dpm/A_{254} (▲). B, ECTHAM-cellulose column separation of chromatin from normal or butyrate-treated HeLa cells after 2 hr incubation with 30 uM [chloroethyl-^{14}C] CCNU. Data are the same as in A except that a 30 min pulse of [5-^{3}H] uridine was given, and its uptake is represented as dpm/A_{254} (■). C, ECTHAM-cellulose column separation of chromatin from normal or butyrate-treated HeLa cells after 2 hr incubation with 30 uM [cyclohexyl-^{14}C] CCNU. Data as for A (from Tew et al., 1978a).

TABLE I. Acid-Insoluble Cellular, Nuclear, and Chromatin Drug Incorporation Following Hydrocortisone[a]

Drug	Hydrocortisone (μM)	[b]Cellular incorporation (%)	[b]Nuclear incorporation (%)	Acid insoluble incorporation (pmol/A_{260})		
				Whole chromatin	Mg^{2+} soluble chromatin	[c]Mg^{2+} soluble ^{14}C as percent of control (0μM HC)
CLZ (chloroethyl) alkylation	0	0.92	0.024	132	41	100
	0.1	0.88	0.023	126	66	159
	1.0	1.0	0.014	114	85	206
	2.0	1.1	0.016	107	89	216
CCNU (chloroethyl) alkylation	0	0.75	0.036	106	46	100
	0.1	0.88	0.057	82	64	139
	1.0	0.66	0.044	63	59	129
	2.0	0.60	0.038	109	65	140
CCNU (cyclohexyl) carbamoylation	0	2.21	0.29	2,265	816	100
	0.1	2.50	0.37	2,593	980	120
	1.0	2.25	0.31	2,419	935	115
	2.0	2.05	0.23	2,345	1,242	152

[a] Log-phase HeLa cells (4×10^5/ml) were incubated for 22 hr with HC at the concentrations shown. Cell cultures were concentrated and treated with 0.6 mM concentrations of all labeled drugs.

[b] Values are mean of three experiments. There is no significant difference between HC-treated and untreated cells ($p > 0.1$).

[c] The amount of drug bound to chromatin from cells without HC pretreatment was standardized at 100%, and each of the HC treated values was expressed as a function of this control. All values for HC-treated cells differ significantly from control ($p < 0.05$), except CCNU carbamoylation with HC concentrations of 0.1 and 1.0 μM (from Tew et al., 1980b).

(final 10–20% of the bulk A_{254}) of an ECTHAM–cellulose column. Transcriptional activity was measured by the uptake of a pulse of ^{3}H-uridine which localized within these fractions (Fig. 5B). The stippled areas of each panel represent CLZ alkylation (Fig. 5A), CCNU alkylation (Fig. 5B), and CCNU carbamoylation (Fig. 5C). In each case the covalent (acid-precipitable) drug binding was preferentially localized within these latter fractions. Stimulation of transcriptional activity with sodium butyrate (bottom panels) increased the drug binding approximately twofold, while maintaining the increase within the transcriptional fraction of the chromatin. The extended structural configuration of transcriptional chromatin predicates that it be a convenient target for drug attack. Further confirmation was gained by using DNase II digestion, followed by differential solubilization in magnesium buffer (Table I). Approximately 10% of the total nuclear DNA was magnesium soluble. Approximately 30–50% of the drug binding was localized in this fraction. Treatment with hydrocortisone stimulated alkylation of the magnesium soluble fraction, and had no significant effect on cellular or nuclear uptake, or whole chromatin binding (Table I).

E. The Nuclear Matrix

The nonhistone protein nuclear matrix is responsible for many facets of nuclear structure and function (Fig. 6). It has, to this time, received only cursory

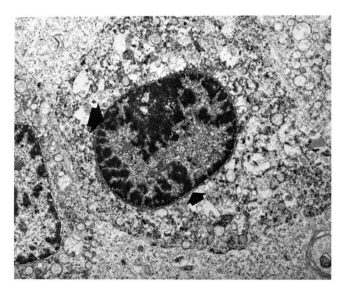

Fig. 6. Untreated HeLa cells ($\times$ 15,000 plate magnification). Electron-lucent euchromatin is surrounded by more densely stained membrane-associated heterochromatin (large arrow). The small arrow indicates the nuclear membrane, showing a nuclear pore and both inner and outer lamellae. The nucleolus is the dark staining area juxtaposed to the membrane on the right center of the nucleus (from Tew *et al.*, 1980b).

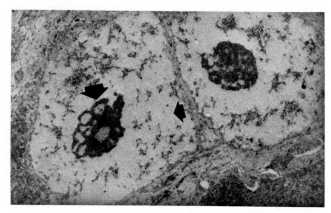

Fig. 7. Electron micrograph of isolated nuclear matrices (× 15,000 plate magnification). The spongelike protein matrix resulted from treating isolated HeLa cell nuclei as described in the legend of Fig. 8. Remnants of the nucleolus (large arrow) are surrounded by the residual nonhistone protein fibrillar matrix (small arrow) (from Tew *et al.*, 1980e).

attention as a possible site of drug interaction (Hemminki and Vainio, 1979). The matrix is a composite structure of fibrillar nonhistone proteins, ribonucleoproteins (Berezney, 1980), residual nucleolar RNA (<5%), and small amounts of DNA (approximately 1% of total).

The basic spongelike structure seen by electron microscopy (Fig. 7) was unaffected by prior treatment with nitrosoureas. However, the contiguous association of the matrix with the nuclear membrane (and also the peripheral chromatin) has been shown to break down during early stages of cell death (Tew *et al.*, 1980b). Since the matrix constitutes just 5% of the total nuclear protein and less than 5% and 1% of the nuclear RNA and DNA (Berezney and Coffey, 1977), Table II demonstrates that a preferential alkylation and carbamoylation of the matrix occurred following exposure to CLZ or CCNU. These drug interactions

TABLE II. Nitrosourea Interaction with the Nuclear Matrix[a]

	Protein (μg%)	Total drug bound	Acid-precipitable drug bound
CLZ alkylation	4.7	26.7	27.0
CCNU alkylation	5.0	31.3	31.7
CCNU carbamoylation	5.2	33.1	38.8

[a] 4×10^8 HeLa cells were treated with 50 μCi of each of the ^{14}C-labeled drugs. Total nuclear drug associations were compared with isolated nuclear matrix fractions. Protein levels (protein μg%) were calculated similarly using colorimetric assays. Data are expressed as (nuclear matrix × 100)/total nucleus.

occurred in both the fibrillar and ribonuclear protein components of the matrix (Tew *et al.*, 1980e) and may have potential significance in disrupting the structural and organizational functions of the matrix.

F. Nitrosourea Interference with Chromatin Organization: Implications for DNA Replication

The presence of replicating chromatin attached to the nuclear matrix (Pardoll *et al.*, 1980) implies that specific DNA–protein attachment sites may be required to facilitate normal DNA synthesis. The potential for nitrosoureas to interfere with these reassociations was demonstrated by a nitrocellulose filter assay (Fig. 8). Normal reassociation kinetics provided a plateau of 80–90% at an approximate protein:DNA ratio of 100:1. Pretreatment of the DNA with either CLZ or CCNU did not alter the retention of bulk ^{3}H-labeled DNA (compare buffer to CLZ-^{3}H and CCNU ^{3}H). However, when alkylated DNA was labeled with (^{14}C-chloroethyl)-CLZ or (^{14}C-chloroethyl)-CCNU (Fig. 8: CCNU-^{14}C or CLZ-^{14}C), the alkylated DNA did not reassociate with the matrix proteins and was not retained on the nitrocellulose filter. If one predicts that specific DNA base sequences (Comings and Wallack, 1978) recognize specific matrix protein-binding sites, alkylation of these base sequences will be responsible for the observed inhibition of reassociation. The stoichiometric straight-line relationship for the ^{14}C-labeled DNA can be explained by the presence of alkylations or crosslinks at sites other than the recognition base sequences. Since the average size of the DNA was 10^4 base pairs, it is likely that much of this DNA will not be involved in reassociation recognition. In addition, the conversion of DNA–nitrosourea monoadducts to DNA–matrix protein crosslinks during the 1 hr *in vitro* incubation could contribute to the observed nonspecific filter retention. Figure 9 shows a diagrammatic representation of the proposed mechanism by which drug interactions with specific base regions may prevent reassociation with nuclear matrix proteins.

III. CONCLUSION AND PERSPECTIVES

A broad spectrum of chloroethylnitrosourea interactions with DNA, nuclear RNA, and nuclear proteins have been demonstrated. Transcriptionally active, consequently extended regions of chromatin were attacked preferentially. By manipulation with transcriptional inducers, it was possible to increase drug interaction within these functional regions. Such drug-induced perturbations have the potential to alter transcriptional activity and corrupt its fidelity. In cases of critical gene products, these effects may be critically deleterious to the cell. Since steroids and alkylating agents in combination are known to increase tumor cell kill (Wilkinson *et al.*, 1979) drug interactions with transcriptional chromatin may have important implications for cytotoxicity.

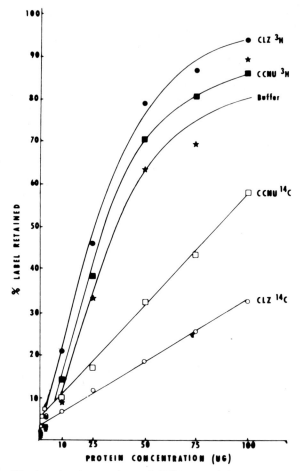

Fig. 8. The modification of nuclear matrix protein/DNA reassociation properties by treatment with CLZ or CCNU. The following methodology was employed:

Isolation of ^3H-DNA. Log-phase HeLa cells were pretreated with 50 μCi of ^{3}H-dThd for 24 hr, and DNA was isolated by a modified phenol extraction. Single stranded regions were removed by treatment with S_1 nuclease and dialysis (>3500 MW) against 10 mM Tris pH 7.4, 1 mM EDTA for 24 hr. Final specific activity of ^{3}H-DNA = 18,000 dpm/μg.

Isolation of nuclear matrix. Essentially by the method of Berezney and Coffey (1977). Nuclei were isolated from log-phase HeLa cells (5 × 10^8 total) by the method of Sporn *et al.* (1969) and resuspended in 5 ml isotonic solution (0.32 M sucrose, 2 mM MgCl$_2$, 1 mM KH$_2$PO$_4$ pH 6.8). Nuclei were treated with 5 μg/ml DNase 1 for 15 mins and resuspended in a 0.2 mM MgCl$_2$ buffer; the pellet was resuspended in a 2 M NaCl buffer (× 3). The resultant pellet was treated with a 1% triton solution and finally treated with 200 μg/ml RNase A and DNase 1 (60 min at 22°). The pellet was resuspended in 2 M NaCl/5 M urea and dialysed against 0.1 M NaCl, 0.1 mM EDTA, 10 mM Tris pH 7.4 (× 4 changes).

Drug Treatments. In some cases isolated DNA was treated with (^{14}C-chloroethyl)CLZ or (^{14}C-chloroethyl)CCNU to yield a final specific activity of ≈5000 dpm/μg DNA.

Reassociation. Varying concentrations of matrix proteins were added to 1 μg of DNA in 1.5 ml of TN buffer (0.1 M NaCl, 0.1 mM EDTA, 10 mM Tris pH 7.4, 3 mM MgCl$_2$, 0.1 mM DTT and 1

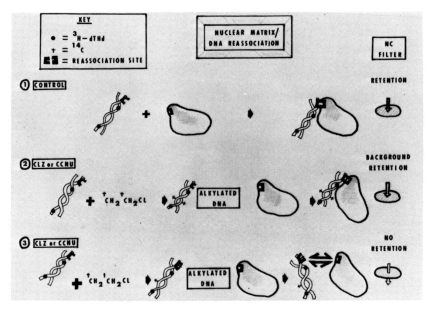

Fig. 9. Schematic representation of the proposed mechanism of nitrosourea-mediated inhibition of DNA–nuclear matrix associations.

Although studies with bone marrow cells are fraught with the problem of cellular heterogeneity, the qualitative differences in subnucleosomal alkylation between the myelotoxic CCNU and nonmyelotoxic CLZ were significant. These data have been further strengthened by recent findings (Green *et al.*, 1980) that ACNU (myelosuppressive) preferentially binds to the DNA associated with the core histones in both L1210 and murine bone marrow, whereas GANU (non-myelosuppressive), like CLZ, binds preferentially to linker DNA in bone marrow and core DNA in L1210. Consistent with the possible reduced cytotoxic potential of drug associations with linker DNA is evidence that methylated products were more efficiently repaired when they occurred in linker DNA following carcinogen treatment (Ramanathan *et al.*, 1976; Bodell and Banerjee, 1979).

Although nuclear RNA and nuclear proteins were modified heavily by the chloroethylnitrosoureas, their importance as crucial targets is questionable because of the turnover rate and predominant duplication of many of these molecules. There is evidence, however, that carbamoylation of the DNA ligase

---

mM PMSF) and incubated for 1 hr at 22°. The mixture was filtered through nitrocellulose HA 0.45 μm filters and washed with 5 ml TN buffer. In a glass scintillation vial, the DNA bases were hydrolysed by boiling in 0.5 ml of 0.5 N HCl. Finally, the filter was dissolved in 1 ml ethyl acetate and counted.

Background filter retention of DNA <2%. ★ no drug treatment; ● pretreatment with unlabeled CLZ; ■ pretreatment with unlabeled CCNU; ○ pretreatment with (^{14}C-chloroethyl)CLZ; □ pretreatment with (^{14}C-chloroethyl)CCNU (from Tew *et al.*, 1980e).

involved in DNA repair will inhibit the repair process (Kann *et al.*, 1974). Other enzyme systems have been identified as sensitive to carbamoylation (Babson *et al.*, 1977). Such disturbances must presumably result from drug interaction with amino acids involved at the active site of the enzyme. Since this process must possess characteristics of antimetabolite kinetics, it is possible that the expression of cytotoxicity through protein carbamoylation (or alkylation) is not achieved because the high concentration of nitrosourea required to cause total enzyme inhibition would primarily cause extensive DNA alkylation. However, it remains difficult to draw definite conclusions when the functional properties, rates of syntheses, and pool sizes of the more obscure nonhistone proteins (many of which are heavily carbamoylated by CCNU) are unclear. At present, it is probable that an interference with the nuclear protein apparatus could produce cytotoxic potential indirectly through an interference with protein–nucleic acid interactions.

Perhaps the most speculative assessment of nitrosourea-induced cytotoxicity relates to the interference of DNA associations with the nuclear matrix. The evidence that DNA replication occurs at a large number of different sites upon the matrix (Pardoll *et al.*, 1980) predisposes that chromatin organization will not be totally random. Since the appearance of chromatin under the electron microscope can be radically altered by treatment with steroids and/or nuclear-reactant drugs (Wilkinson *et al.*, 1979; Tew *et al.*, 1980b), the nuclear matrix may exert a chromatin-organizing function that permits fluidity while maintaining some restrictive influence upon chromatin arrangement. Although the nuclear matrix was found to be a prominent target for nitrosourea interaction, protein modifications per se were shown not to influence nuclear matrix interactions with DNA (Tew *et al.*, 1980e). However, alkylation of DNA did interfere with the matrix protein reassociation process. Since AT-rich regions have been shown to associate preferentially with matrix proteins (Comings and Wallack, 1978), it is possible that the nitrosoureas interact with these base sequences, thereby preventing recognition of the requisite binding proteins. If these associations are indeed critical to replication (Berezney and Coffey, 1975), this drug-induced interference (especially during, or just prior to, S phase) may provide great cytotoxic potential. The data were generated under *in vitro* conditions, and it remains to be shown that this effect can occur under physiological circumstances. However, it is interesting to speculate that the failure to replicate an important region of the genome could result from one alkylation (or crosslink) of a base sequence that would normally be the recognition site for matrix association. As a result, an inhibition of DNA synthesis would occur, since the DNA (or more correctly, chromatin fiber) would be prevented from associating with the replicational machinery associated with the nuclear matrix.

REFERENCES

Alfagame, R. C., Zweidler, A., Mahowald, A., and Cohen, L. H. (1976). *J. Biol. Chem. 219*, 3729–3736.

Babson, J. R., Reed, D. J., and Sinkey, M. A. (1977). *Biochem. 16*, 1584-1589.

Berezney, R. (1980). *J. Cell Biol. 85*, 641-650.

Berezney, R., and Coffey, D. S. (1975). *Science 189*, 291-293.

Berezney, R., and Coffey, D. S. (1977). *J. Cell Biol. 73*, 616-637.

Bodell, W. J., and Banerjee, M. R. (1979). *Nucleic Acids Res. 6*, 359-370.

Comings, D. E., and Wallack, A. S. (1978). *J. Cell Sci. 34*, 233-246.

Gottesfeld, J. M. (1977). *Methods Cell Biol. 16*, 421-436.

Green, D., Tew, K. D., Hisamatsu, T., and Schein, P. S. (1980). *Biomed. Pharmacol.* Submitted.

Hemminki, K., and Vainio, H. (1979). *Cancer Lett. 6*, 167-173.

Jump, D. B., Sudhakar, S., Tew, K. D., and Smulson, M. E. (1980). *Chem. Biol. Int. 30*, 35-52.

Kann, H. E., Kohn, K. W., and Lyles, J. M. (1974). *Cancer Res. 34*, 398-402.

Kohn, K. (1979). *In* "Effects of Drugs on the Cell Nucleus" (H. Busch, S. T. Crooke, and Y. Daskal, eds.), pp. 207-239. Academic Press, New York.

Laemmli, U. K. (1970). *Nature (London) 227*, 680-685.

Pardoll, D. M., Vogelstein, B., and Coffey, D. S. (1980). *Cell 19*, 527-536.

Pinsky, S., Tew, K. D., Smulson, M. E., and Woolley, P. V. (1979). *Cancer Res. 38*, 3371-3378.

Ramanathan, R., Rajalakshmi, S., Sarma, D. S. R., and Farber, E. (1976). *Cancer Res. 36*, 2073-2079.

Simpson, R. T. (1977). *Methods Cell Biol. 16*, 437-446.

Sporn, M. B., Berkowitz, D. M., Glinki, R. P., Ash, A. B., and Stevens, C. L. (1969). *Science 164*, 1408-1410.

Sudhakar, S., Tew, K. D., Schein, P. S., Woolley, P. V., and Smulson, M. E. (1979). *Cancer Res. 39*, 1411-1417.

Tew, K. D., Sudhakar, S., Schein, P. S., and Smulson, M. E. (1978a). *Cancer Res. 38*, 3371-3378.

Tew, K. D., Green, D., and Schein, P. S. (1978b). *Proc. Amer. Assoc. Cancer Res. 19*, 113.

Tew, K. D., Smulson, M. E., and Schein, P. S. (1980a). *In* "Human Cancer. Its Characterization and Treatment" (W. Davis, K. R. Harrap, and G. Stathopolous, eds.), Vol. 5, pp. 332-341. Excerpta Medica, Amsterdam.

Tew, K. D., Schein, P. S., Lindner, D., Wang, A. L., and Smulson, M. E. (1980b). *Cancer Res 40*, 3697-3703.

Tew, K. D., Wang, A. L., and Schein, P. S. (1980c). *Proc. Amer. Assoc. Cancer Res. 21*, 297.

Tew, K. D., Smulson, M. E., and Schein, P. S. (1980d). *Recent Results Cancer Res. 70*.

Tew, K. D., Wang, A. L., and Schein, P. S. (1980e). *J. Cell Biol.* Submitted.

Wilkinson, R., Birbeck, M., and Harrap, K. R. (1979). *Cancer Res. 39*, 4256-4261.

Chapter 10

ADP-RIBOSYLATION AND DNA REPAIR AS INDUCED BY NITROSOUREAS[1]

Mark Smulson
Tauseef Butt
Swaroop Sudhakar
Don Jump
Nancy Nolan

I. INTRODUCTION

Elucidation of the molecular mechanisms of chemical carcinogenesis has awaited a greater understanding of both the chemical interactions of these compounds with nuclear components and the organization of chromatin. Our laboratory has undertaken a detailed study of one specific chromatin associated enzyme, poly (ADP-rib) polymerase. These studies have recently taken on great significance for those workers concerned with nitrosourea interaction with cells, since it appears that this enzymatic system is directly responsive to nitrosourea and other alkylating agent damage to DNA, and participates in repair reaction mechanisms, as first shown by our group (Smulson et al., 1977).

The reaction catalyzed by poly (ADP-rib) polymerase is shown in Fig. 1. The substrate for the enzyme is nuclear NAD. The enzyme transfers successive units of ADP-rib covalently to histones (Fig. 1). Chains of poly(ADP-rib) thus are generated on nucleosomal histones, with the concomitant depletion of NAD in cells. It is significant that NAD is synthesized within the eukaryotic nucleus and

[1]This work was supported by Public Health Service Grants CA 13195 and CA 25344.

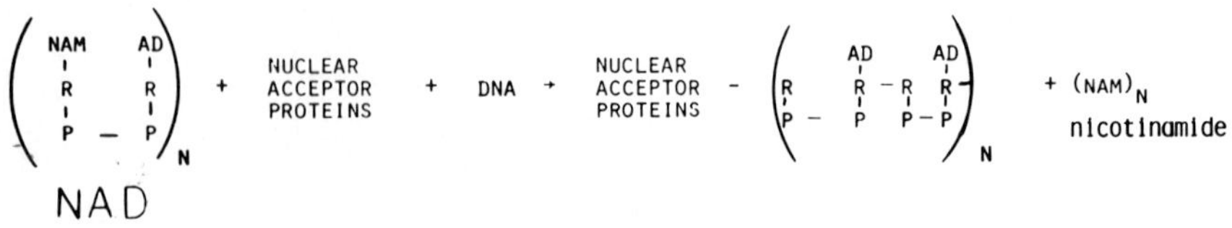

Fig. 1. The reaction catalyzed by poly(ADP-rib) polymerase.

that the large majority (90%) of this compound remains in the cellular nucleus, presumably for biosynthesis of poly ADP-ribose. The important point for this present volume is that a variety of cellular DNA strand breakers, including a number of nitrosoureas, cause dramatic lowering of nuclear NAD pools. Thus, a simple NAD determination could well be developed into a sensitive assay for nitrosourea potency.

The enzyme poly(ADP-rib) has a strict dependency on DNA for activity (Fig. 1), and it is probable that the enzyme might represent a nuclear "sensor" for DNA strand breaks. Data described below will show that the above-mentioned NAD lowering in eukaryotic cells due to exposure to nitrosoureas is a direct consequence of the activation of poly (ADP-rib) polymerase.

Two recently published works provide compelling evidence that the poly ADP-ribosylation modification of chromatin histones is an important event in DNA repair. Juarez-Salinas *et al.*, (1979) have shown, using a highly sensitive new assay for poly ADP-ribose polymer, a direct correlation between MNNG-induced NAD lowering in 3T3 cells and accumulation of poly (ADP-rib). In a second series of experiments, Berger *et al.* (1980) studied MNNG alkylation effects in control fibroblasts and those derived from the repair disease xeroderma pigmentosum. Both cell lines responded to MNNG by lowering of NAD and stimulation of poly (ADP-rib) synthesis. However, when DNA strand breaks induced by UV radiation were examined in the two cells, only normal cells, and not xeroderma cells, responded with NAD lowering and poly (ADP-rib) activation. It would appear that the lack of strand breaks, due to inadequate thymine dimer repair, in the latter cells abolished the poly (ADP-rib) effect.

Because of this apparent close relationship between poly (ADP-rib) and DNA breakage as induced by a variety of agents, including nitrosoureas, our primary approach toward elucidating the mechanism of ADP-ribosylation involvement in DNA repair and/or replication has been directed at determining the effect that this reaction elicits on the structure of chromatin. We have mainly directed our attention at the unit nucleosomal or simple polynucleosomal level of chromatin, since only limited information is available on the effects of cytotoxic agents such as nitrosoureas on these structures. Data presented below shows that chains of newly generated poly(ADP-rib) appear to promote condensation, via nucleosome crosslinking, of localized and distal regions of chromatin. A hypothesis explaining a putative mechanism by which nitrosoureas indirectly alter the eukaryotic genomic environment is thus possible.

II. RESULTS AND DISCUSSION

A. Effects of Nitrosoureas on Poly ADP-Ribosylation

To demonstrate the basic influence of nitrosoureas on nuclear polymerase activity, HeLa cells were incubated in the presence or absence of methylnitrosourea (MNU), and consequently nuclei were prepared and assayed *in vitro* for enzymatic activity (Table I). The data show that an apparent twofold stimulation of specific activity resulted from cellular incubation with the drug. As indicated in the previous section, considerable evidence suggests that this stimulation is due to MNU-induced DNA fragmentation. The data in Table I also show that nuclear control activity is effectively inhibited by thymidine, nicotinamide, and caffeine. It is of interest that while thymidine and nicotinamide also severely reduce the stimulated activity in MNU-derived nuclei, much less inhibitory activity was noted for caffeine. We have no explanation for these results at this time; however, it is of interest that caffeine has been reported to inhibit DNA repair reactions.

Another agent known to cause breaks in DNA was analyzed to determine its effect on enzyme activity. Bleomycin, a polypeptide antibiotic, causes extensive

TABLE I. Effect of Inhibitors on the Stimulation of Poly (ADP-Ribose) Polymerase Activity in Nuclei Derived from Cells Incubated in the Presence or Absence of MNU[a]

Inhibition (mM)	Nuclei from control cells		Nuclei from MNU-treated cells	
	Spec. act. of poly (ADP-ribose) polymerase (CPM/μg prot.) $\times 10^{-3}$	Inhibition (%)	Spec. act. of poly (ADP-ribose) polymerase (CPM/μg prot.) $\times 10^{-3}$	Inhibition (%)
None	12.0	–	22.0	–
Thymidine				
1	4.0	67	10.2	46
2	2.5	80	7.5	75
Nicotinamide				
5	2.0	83	4.0	82
10	0.9	92	2.6	88
Caffeine				
5	7.5	47	19.0	14
10	5.4	55	16.0	27
20	3.6	70	14.0	37

[a] Log-phase HeLa cells (2×10^6/ml) were treated with MNU (7.5 mM) with an appropriate control for one hr at 37°C. Poly (ADP-rib) polymerase activity was assayed in nuclei in the presence and absence of its inhibitors, thymidine, nicotinamide, and caffeine. Percentage of inhibition of enzyme activity was calculated by the following relationship:

$$\frac{\text{sp. activity of enzyme in the presence of inhibitor}}{\text{sp. activity of enzyme in the absence of inhibitor}} \times 100$$

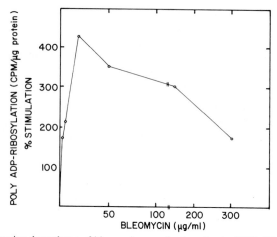

Fig. 2. Concentration dependence of bleomycin on activation of poly (ADP-rib) polymerase. Cells (6.7 × 10⁶/ml) were treated with various concentrations of bleomycin for 1 hr at 37°C with an appropriate control. Nuclei were isolated and assayed for poly (ADP-rib) polymerase activity. Enzyme activity in drug-treated cells is expressed as percentage of stimulation over control.

fragmentation of cellular DNA. Treatment of cells with this agent resulted in a similar stimulation of poly(ADP-rib) polymerase activity (Fig. 2). However, with increased degradation of DNA at higher bleomycin concentrations, a decrease in enzyme activity was noted.

B. Activation of Poly(ADP-Rib) Polymerase Activity by Nitrosoureas at the Nucleosomal Level of Chromatin

It was realized that before a precise understanding of the relationship between the cytotoxic events produced by nitrosoureas and poly ADP-ribosylation of nuclear proteins could be understood, a better characterized, and more simple nuclear subcomponent was required for study. Significant advances have been made recently in elucidation of the subunit structure of chromatin. It seemed important to be able to study the MNU stimulation at this level of chromatin and to exploit this information to learn more about both poly ADP-ribosylation and chromatin structure per se.

Briefly, chromatin consists of DNA packaged at successive intervals into small particles, nucleosomes. Approximately 140–160 base pairs of DNA are located around a globular disk-shaped particle, the nucleosome, which in turn is composed of an octamer of two asymmetric pairs of the four histones, H3, H4, H2a, and H2b. The internucleosomal region (linker) consists of 0–60 base pairs of DNA. This structure has been outlined in more detail in a recent review (Smulson, 1979). The polynucleosome is in turn thought to be coiled into its own superstructure within native chromatin. Our laboratory has been very active in determining in which regions within this chromatin framework various nitrosoureas are capable of interaction, and these data have been summarized in a review chapter in an earlier volume in this series (Smulson *et al.*, 1979).

Micrococcal nuclease specifically cleaves the internucleosomal DNA and thus is experimentally used to isolate a variety of subsets of nucleosomes (i.e., monomer, dimer, 3N, 4N, and so on). As shown in Fig. 3, these particles can be separated by their sedimentation properties in sucrose gradients. In this experiment, intact HeLa cells were incubated in the absence or presence of MNU under conditions where poly (ADP-rib) polymerase is stimulated. Nucleosomes were prepared and characterized for size of DNA (Fig. 3C), and the activity for the polymerase was assayed *in vitro*. The data indicate that a marked enhancement of activity was noted in preparations from MNU-treated cells, even at this subunit level of chromatin. Activity for the enzyme is found only in particles greater than a simple mononucleosome (1N) since we have determined previously that the enzyme is bound between nucleosomes (Mullins *et al.*, 1977; Giri *et al.*, 1978b). Before further analysis of the MNU effect at the nucleosomal level was

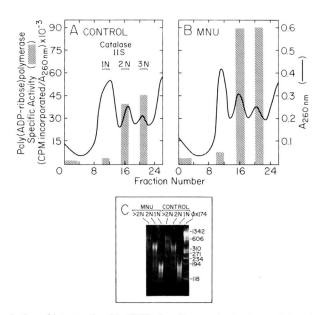

Fig. 3. Incubation of intact cells with MNU—its effect on the *in vitro* activity of poly (ADP-rib) polymerase at the nucleosomal level. HeLa cells (4×10^7) in 20 ml were incubated in the absence (A) or presence (B) of MNU (7.5 mM) for 1 hr at 37°C. Nuclei were prepared and digested with micrococcal nuclease to isolate nucleosomal subfractions by sucrose gradient centrifugation. The direction of sedimentation is from left to right. Catalase was utilized as an 11S marker. The optimal density was monitored at 260 nm (—). Approximately 0.2 A_{260} units of pooled fractions were assayed for poly (ADP-rib) polymerase activity as indicated by the shaded bars. 1N, 2N, and 3N refer to mononucleosomes, dinucleosomes, and trinucleosomes, respectively. (C) DNA was isolated from the pooled fractions corresponding to each peak and the samples were analyzed under nondenaturing conditions by 6% polyacrylamide gel electrophoresis as described. Hae III-digested ØX174 DNA fragments were used as base pair (from Sudhakar *et al.*, 1979).

possible, more detailed information on the interaction of the poly ADP-ribosylation system and chromatin substructure was required. The following data summarize some of our more recent findings in these areas.

C. Domains of Chromatin Involved in DNA Replication–Repair and ADP-Ribosylation

Because different functional regions of chromatin (i.e., transcriptional, DNA repair, heterochromatin, etc.) are more or less susceptible to digestion by micrococcal nuclease, a study was undertaken on the release of nucleosomes containing poly (ADP-rib) polymerase during digestion (Table II). The data show that the specific activity of the enzyme decreases in specific nucleosomal fragments as digest time increased (Jump *et al.*, 1979). Because of the stimulatory effect of alkylating agents such as nitrosoureas on activity, we wondered whether DNA replicating or repair regions of chromatin were likewise preferentially released by micrococcal nuclease. The rationale was that the polymerase would rearrange and specifically bind proximal to strand breaks and repair forks caused in nitresoureas. Accordingly, we examined the release of DNA replicating regions in chromatin (Fig. 4). In order to test this possibility, experimental conditions were developed using micrococcal nuclease to isolate nucleosomal fragments enriched in nascent DNA from the DNA replicative fork (Jump *et al.*, 1979). This experiment shows that like poly(ADP-rib) polymerase (Table II), the specific activity of the ^{3}H-pulse label thymidine (DNA replication) is highest in mono- and dinucleosomes cleaved preferentially by micrococcal nuclease, whereas the ^{14}C label (total DNA) was digested more slowly along with the bulk chromatin (A_{260}). Thus, DNA replicating and repair regions of chromatin and poly ADP-ribosylation regions might be closely coupled. More definitive experiments are

TABLE II. Poly(ADP-Rib) Polymerase Activity Associated with Mono-, Di-, and Trinucleosomal Fragments from Different Chromatin Digests[a]

Perchloric acid digest (%)	Poly(ADP-rib) polymerase sp. act. ($\times 10^{-5}$) (CPM incorp. A_{260} nm^{-1} (15 min)$^{-1}$		
	Mononucleosome	Dinucleosome	Trinucleosome
3	0.53	2.87	4.00
11	0.35	1.68	2.30
17	0.21	1.50	1.67

[a] Micrococcal nuclease digested chromatin was prepared and fractionated. The resultant mono-, di-, and trinucleosome regions of the gradients were analyzed for poly (ADP-rib) polymerase activity (Mullins *et al.*, 1977).

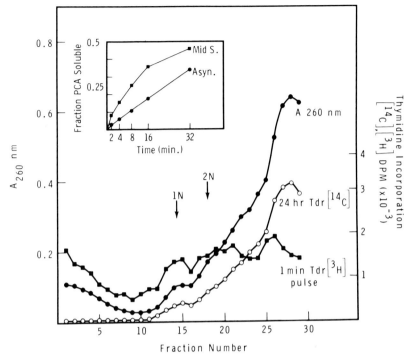

Fig. 4. Kinetics of micrococcal nuclease digestion of nuclei isolated from asynchronous and mid-S-phase cells. Asynchronous cells (3.5×10^5/ml) were uniformly labeled by two consecutive 24 hr treatments with 0.005 μCi/ml [^{14}C] thymidine followed by a 2 hr chase in label-free medium. Cells were pulse labeled with 40 μCi/ml [^{3}H] thymidine for 1 min. The pulse labeling was stopped by pouring the cell suspension over crushed ice. Nuclei were prepared and digested for 2 min with micrococcal nuclease (2 units/A_{260}). Chromatin fragments were separated in 10–31% sucrose velocity gradients: SW 40, 38K, 6 hr, 4°C. Direction of sedimentation is from left to right. Absorbance was at 260 nm (●—●), and Cl$_3$AcOH-insoluble radioactivity was determined for each fraction: ^{14}C(○—○); ^{3}H (■—■). 1N, mononucleosome; 2N, dinucleosome (from Jump *et al.*, 1979).

under way on this topic using new observations concerning the polymerase described at the end of this review.

One pivotal series of observations has been the relationship we have found *in vitro* between the order of complexity of chromatin (i.e., the length of nucleosomes) and activity for poly ADP-ribosylation (Butt *et al.*, 1978; 1979). The results might bear directly on our new observations on the effect of ADP-ribosylation on chromatin condensation and relaxation, discussed below, and how this might be involved in nitrosourea repair. These observations are summarized in Fig. 5. It has been observed *in vitro* that poly (ADP-rib) polymerase specific activity increases with nucleosome size, reaching a maximum on polynucleosomes of approximately 8–10 N. This is of potential importance since in the 300 A solenoid chromatin fiber, approximately 7 nucleosomes constitute one turn of the fiber. Based upon further analysis of enzyme molecules (Jump *et*

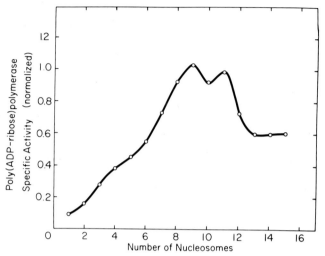

Fig. 5. The relationship between poly (ADP-rib) polymerase specific activity and the number of nucleosomes on a chromatin fragment. The specific activity of poly (ADP-rib) polymerase was normalized for three separate experiments. The average specific activity at normalized 1.00 was 40,000 CPM incorporated/OD_{260nm}/20 min (from Butt *et al.*, 1979).

al., 1980), we have tentatively concluded that the polymerase itself binds to DNA at a specified periodicity within chromatin. It is possible that this domain constitutes a "housekeeping" region for the polymerase. The enzyme would respond to DNA strand breaks or DNA replicating forks and generate crosslinked chains of poly ADP-ribose on histones linking and relaxing nucleosomes to allow for other enzymes (such as DNA polymerases) to interact with the DNA in this region. The experiments described below might explain one possible mechanism for the above hypothesis.

D. Role of Poly ADP-Ribosylation in the Condensation of Chromatin

One fascinating reaction of the poly ADP-ribosylation modification is the crosslinking of histone H1 to yield a H1 dimer complex connected by a chain of 15 poly (ADP-rib) units (Stone *et al.*, 1977). Data described below suggest that other crosslinked chromatin proteins might also result from the poly (ADP-rib) polymerase reaction. The crosslinking of histone H1 is particularly interesting since this histone is not part of the nucleosome core particle but rather helps to stabilize higher ordered forms of chromatin structure. A polynucleosome domain (i.e., DNA repair region) might be relaxed via poly (ADP-rib) glycohydrolase degrading of this crosslink and condensed by H1 dimer formation. Recently, we began an analysis of the biosynthesis of this H1 crosslink in purified oligonucleosomes (Nolan *et al.*, 1980). The electrophoretic analysis of H1 dimer is facilitated by the selective extraction of this histone in 5% perchloric acid. When histone H1 ADP-ribosylation is examined with ^{32}P-NAD, gel electrophoresis, and radioautograph, a series of partially ADP-ribosylated species is detected migrating from unmodified H1 to H1 dimer (Fig. 6, lane 1). These 15 partially ADP-ribosylated H1 molecules were shown to be intermediates in H1 dimer formation, since during a ''chase'' period with nonradioactive NAD, the partials were converted to H1 dimer (Fig. 6, lane 3).

To further explore findings on the relationship between oligonucleosome size and activity (Fig. 5), we consequently attempted to assess the influence of oligonucleosome complexity on *in vitro* poly (ADP-rib)-H1 dimer synthesis

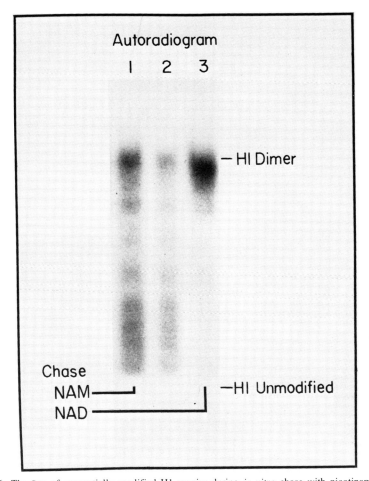

Fig. 6. The fate of sequentially modified H1 species during *in vitro* chase with nicotinamide or unlabeled NAD. Large nucleosomes were incubated *in vitro* with [^{32}P]-NAD for 5 min. Subsequently, either 5 mM (final concentration) nicotinamide (NAM) (slot 1) or a 10-fold excess of unlabeled NAD (slot 3) was added and the reactions allowed to proceed for an additional 10 min. Histone H1 was extracted as before; the radioactivity incorporated was essentially identical in the three samples. Aliquots were electrophoresed in acetic acid/urea gels and autoradiographed (from Nolan *et al.*, 1980).

(Fig. 7). Withstanding the complications of such a quantitation, the data show that (*a*) H1 dimer synthesis is roughly proportional to chromatin concentration; (*b*) H1 dimer *in vitro* synthesis is greater in large polynucleosomes (16N) than in smaller particles (8N). This latter observation might relate to a generalized crosslinking and/or folding of nucleosomes.

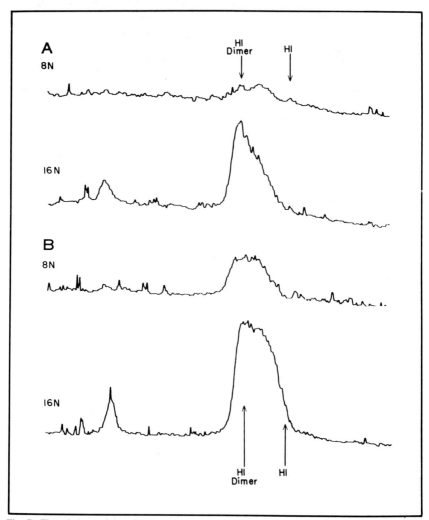

Fig. 7. The relative activity of polynucleosomes of 8N and 16N size in formation of the poly(ADP-rib)–H1 complex. Reactions were carried out with two limiting concentrations of purified 8N and 16N polynucleosomes in 1.5 ml of incubation mix containing 5.5 μCi of [^{32}P]-NAD. H1 complex was isolated and electrophoresed. The exposed X ray film was scanned in a densitomer (ORTEC Inc.). The abscissa represents arbitrary absorbance units at 550 nm. A: the concentration of 8N and 16N polynucleosomes in the reaction mixture was 0.2 A_{260} units/1 ml; B: the concentration of 8N and 16N polynucleosomes in the reaction mixture was 0.3 A_{260} units/1 ml (from Nolan *et al.*, 1980).

Further suggestion of a role for poly ADP-ribosylation in chromatin condensation was obtained when a study was conducted on the influence of NAD concentration on polynucleosome migration on *native* polyacrylamide gels. In such gel systems nucleosomes migrate as discrete nuclear protein particles. Small nucleosomes (i.e., mononucleosomes) migrate to the bottom of the gel while larger polynucleosomes migrate progressively higher with regard to repeat size. The particles can be detected by their staining with ethidium bromide (see Fig. 8A). We observed that when oligonucleosomes enriched in species 8–10 N are incubated *in vitro* with very low (25 nM) ^{32}P-NAD to generate very short chains of poly (ADP-rib), subsequently electrophoresed, stained with ethidium bromide, and exposed to x ray film (to detect ADP-ribosylation), the ^{32}P-labeled nucleosomes migrated identically with the bulk nucleosomes (compare lanes 1 in Fig. 8A and 8B). When nucleosomes were incubated with increasing concentrations of NAD, keeping ^{32}P-NAD constant, electrophoresed, and stained by ethidium bromide, there was no effect on the electrophoretic properties of the bulk of the chromatin (Fig. 8A, lanes 2–5) due to increasing ADP-ribosylation. Of great interest was the fact that with increasing NAD concentration, the subset of poly ADP-ribosylated polynucleosomes aggregated extensively (Fig. 8B, lanes 2–5) as evidenced by greatly reduced mobility in the electrophoretic system. This condensation of chromatin could equally well be demonstrated by sedimentation analysis on sucrose gradients (Butt and Smulson, 1980). Extensive analysis indicated that the aggregates probably contained chromatin actually organized in nucleosomal structure. These aggregated structures were very resistant to both micrococcal nuclease and phosphodiesterase digestion. Experiments indicated that poly ADP-ribosylated histones including histone H1 dimer and poly (ADP-rib) polymerase were associated with the aggregated forms of chromatin (Butt *et al.*, 1980).

Assuming that poly ADP-ribosylation is intimately associated with nitrosourea-induced repair regions of chromatin, one hypothesis that is currently being tested is whether we can utilize this aggregation phenomenon to purify polynucleosomes involved with DNA repair.

Poly (ADP-rib) chain lengths were determined in native and in highly complex samples of chromatin, generated by increasing NAD concentrations (Butt and Smulson, 1980). The *free* ^{32}P-labeled polymer was cleaved from the nuclear protein acceptors from four of the aggregated samples of Fig. 8B and subsequently electrophoresed. The data (Fig. 9) revealed a broad size range of poly (ADP-rib) chains ranging from one to more than fifty-ADP-rib chains in length. It was of considerable interest to note that the length of the most highly radioactive chains was also dependent on NAD concentration. Under conditions of maximal oligonucleosome condensation (100 μM NAD) the poly (ADP-rib) barely entered the gels. Based on the correlation between NAD concentration and both polymer chain lengths and extents of condensation, it is likely that poly (ADP-rib) association with nucleosomal proteins may be the major contribution via domain crosslinking to the aggregation phenomenon.

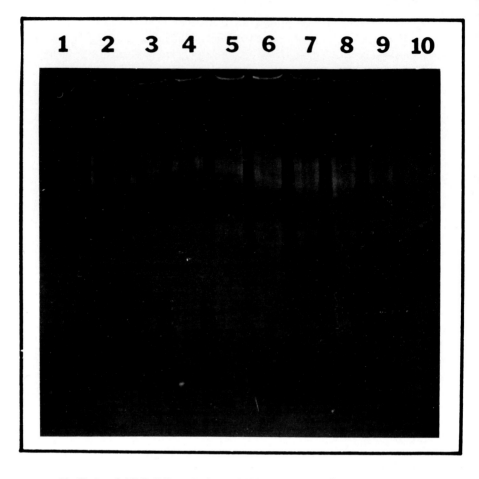

E. Role of ADP-Ribosylation of Histone H1 in Oligonucleosome Condensation

Histone H1 has been implicated in condensation of the spacer regions of DNA between mononucleosomes and in stabilizing higher ordered chromatin structures (Renz *et al.*, 1977; Worcel, 1978). Because of the marked influence of NAD concentration on *in vitro* chromatin condensations noted above (Figs. 8–9) it was of importance to study poly (ADP-rib) histone H1 crosslinking (see Figs. 6 and 7) under the same substrate ranges.

In the experiment shown in Fig. 10, intact nuclei were incubated with 2 nM, 0.1 μM, and 100 μM-NAD, subsequently selectively extracted for histone H1, and electrophoresed (Nolan *et al.*, 1980). At the two lower concentrations of NAD, where chromatin complex formation is not observed, no *in vitro* biosynthesis of histone H1 dimer was found (lane 5). In contrast, under conditions of chromatin condensation (100 uM NAD), considerable H1 dimer synthesis is

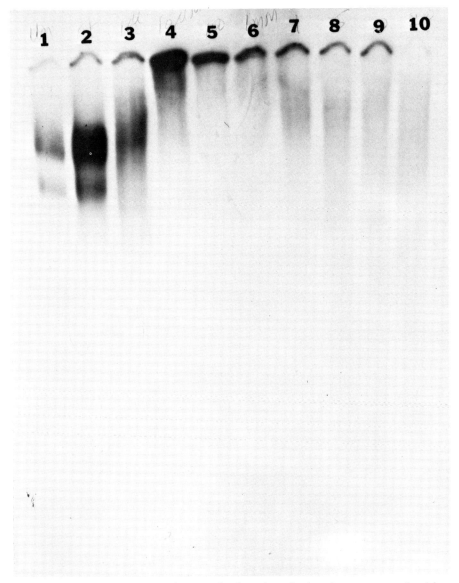

Fig. 8. NAD concentration-dependent complex formation of polynucleosomes, as analyzed by chromatin gel electrophoresis. An aliquot of chromatin (0.07 A_{260} units) from oligonucleosomes was incubated with various concentrations of NAD at $20°$ for 5 min while maintaining [^{32}P]-NAD constant at 0.25 μCi per assay. The reaction was terminated by placing the tubes on ice and by the addition of nicotinamide to final concentration of 5 mM. The samples were subjected to electrophoresis on 3–8% native chromatin gradient polyacrylamide gels. A: ethidium bromide stain; B: autoradiograph. Lane 1, chromatin incubated with 25 nM NAD for 5 min at $5°$; lanes 2–6, chromatin incubated at $20°$ with 25 nM, 1 μM, 10 μM, and 1 mM NAD, respectively. In lanes 7–10, the samples were incubated with 100 μM NAD for 5 min, terminated with nicotinamide, and 10 μg of proteinase K was added; the samples were incubated for an additional 5, 10, and 20 min at $20°$, respectively (from Butt and Smulson, 1980).

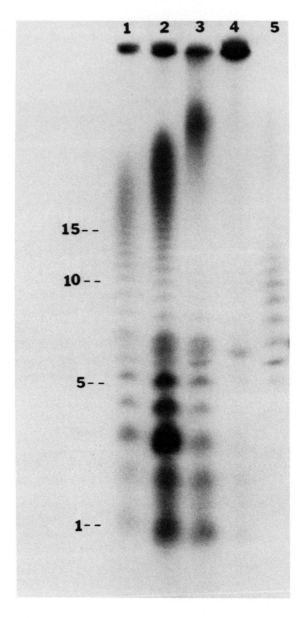

Fig. 9. Correlation between poly (ADP-rib) chain length, nucleosome aggregation, and NAD concentration. Samples enriched in octanucleosomes were incubated with 1μCi [^{32}P]-NAD in the presence of various concentrations of unlabeled NAD as described in Fig. 8. The reaction was terminated and polymer was released by incubation in 100 mM Tris pH 9.5 and further purified. Lanes 1–4 represent polymer from chromatin incubated with 25 nM, 1 μM, 10 μM and 100 μM NAD, respectively, and lane 5 represents marker polymer with average chain length of 10N (from Butt and Smulson, 1980).

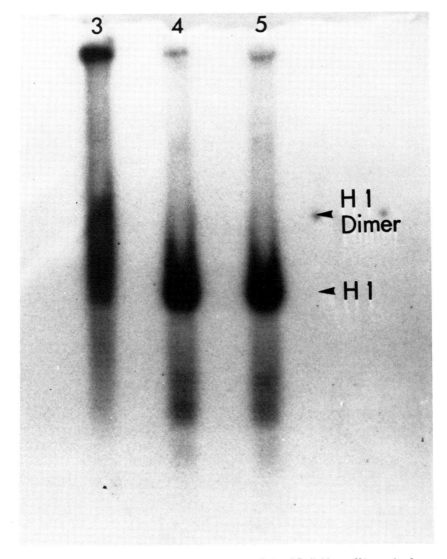

Fig. 10. The Effect of substrate (NAD) concentration on Poly ADP-rib-histone H1 complex formation. Nuclei (3×10^7) were incubated under optimal conditions for poly (ADP-rib) polymerase with 0.5 μCi of [^{32}P]-NAD in a volume of 0.5 ml at NAD concentrations of 2 nM (lane 5); 0.1 μM (lane 4); 100 μM (lane 3). H1 histone was selectively extracted from each sample as described by Stone *et al.*, (1977) and electrophoresed in acetic acid/urea polyacrylamide slab gels exposed for radioautography (From Nolan *et al.*, 1980).

observed. Pulse-chase (with high nonradioactive NAD) also demonstrated this concentration effect (Fig. 6).

As an initial approach to studying the potential influence of the poly (ADP-rib) polymerase catalyzed crosslinking of histone H1 on chromatin structure, the aggregation phenomenon described above was studied using purified oligonucleosomes containing a normal complement of H1 and in the same particles specifically depleted of histone H1 (Butt *et al.*, 1980). This approach depends on selectively removing histone H1 from polynucleosomes under conditions whereby nucleosome stability is maintained and little of the nonhistone proteins, especially poly (ADP-rib) polymerase, are removed. Nelson *et al.* (1979) have described a mild procedure (using Dowex 50W-X2) for selective removal of histone H1. Octonucleosomes were prepared and incubated identically in the presence and absence of Dowex. Total nuclear proteins remaining in the two samples were analyzed by polyacrylamide electrophoresis. Under these conditions greater than 95% of H1 was removed from the chromatin; the nonhistone protein complement of H1-depleted chromatin was the same in both samples. A 40% loss in poly (ADP-rib) polymerase activity was noted; however, it is not clear whether this was reflective of the removal of H1 per se or removal of polymerase. Withstanding these complications, it was of interest that the H1-depleted nucleosomes were found to be incapable of complex formation (Fig. 11), while nucleosomes treated under identical conditions but containing histone H1 clearly responded to the elevated NAD levels by forming complex material. We have recently reported on the successful reconstitution of purified poly (ADP-rib) polymerase with polymerase depleted polynucleosomes (Jump *et al.*, 1980). However, attempts to restore NAD-promoted aggregation of chromatin by reconstitution of depleted chromatin with either stoichiometric levels of H1 and/or exogenous purified polymerase (Jump and Smulson, 1980) have been unsuccessful thus far.

These results, coupled with the earlier data, do, however, suggest that ADP-ribosylation of histone H1 plays an important role in the chromatin condensation reactions. The question remains whether nitrosourea stimulation of ADP-ribosylation is related to this condensation ability of poly (ADP-rib). It might be anticipated *in vivo* that nuclear proteins existing in domains of chromatin that are extensively condensed due to poly ADP-ribosylation might have differing susceptibility to carbamoylation induced by nitrosoureas. As a first approach toward investigating the influence of (ADP-rib)-induced chromatin aggregation, nuclear protein carbomoylation due to MNU and CCNU was compared in nuclei preincubated in the presence or absence of 2mM NAD (Fig. 12). Nuclei from log-phase cells were treated with (^{14}C)-carbonyl-labeled nitrosoureas in the absence of NAD modification or after poly ADP-ribosylation of nuclear proteins was allowed to take place. Acid-soluble proteins were extracted and analyzed on Triton acid-urea gel electrophoresis. MNU specifically carbamoylated histone H2B more than H1 (lane C). The condensation of nucleosomes by NAD did not affect the above-noted differential modification of histones by MNU, but consid-

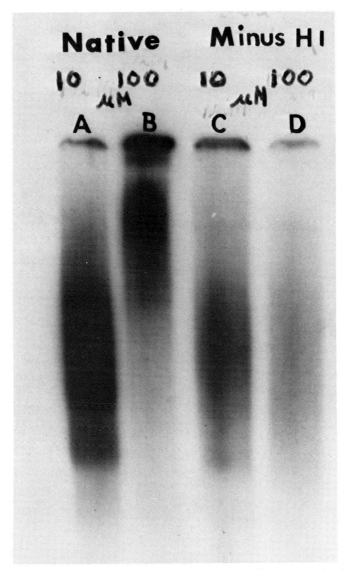

Fig. 11. The effect of histone H1 depletion on ADP-rib-induced chromatin complex formation. Polynucleosomes (8–10N) were prepared and treated with Dowex 50W-X2 to selectively remove histone H1 (Nelson *et al.*, 1980). An aliquot of polynucleosomes (0.07 A_{260} units) with H1 (lanes 1 and 2) or without H1 (lanes 3 and 4) were incubated with 1 μM [^{32}P]-NAD (lanes 1 and 3) or 100 μM [^{32}P]-NAD (lanes 2 and 4). The reactions were terminated by placing the tubes on ice, and the addition of nicotinamide to 5 mM final concentration. The samples were electrophoresed on 3–8% gradient polyacrylamide gels and exposed for radioautography (from Butt *et al.*, 1980).

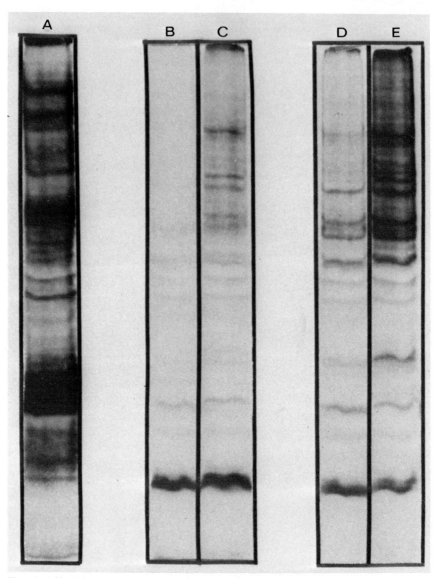

Fig. 12. Effect of poly (ADP-ribose) modification on the covalent interaction of nitrosoureas with nuclear proteins. Proteins were extracted (0.4 N H_2SO_4) from nuclei modified with either (^{3}H)-NAD (50 μCi), (^{14}C)-carbonyl-MNU (25 μCi) or (^{14}C)-cyclohexyl-CCNU (25 μCi) after preincubation either in the presence or absence of 2 mM NAD. The extracted proteins were analyzed on Triton-acetic acid-urea acrylamide gels. Modified proteins in the gel were identified following fluorography and exposure of the dried gel to a preflashed SB-5 X ray film for an appropriate time. Lane A represents ADP-ribosylated proteins; lanes C and E, proteins modified by MNU and CCNU, and lanes B and D, proteins modified by ^{14}C-nitrosoureas, after ADP-ribosylation was first allowed to take place.

erably reduced the interaction of nitrosoureas with nonhistone proteins migrating slowly in the gel (lane B). Great variation was noted with CCNU modification of histones depending on whether or not nuclei were pretreated with NAD. Differences noted included the preferential modification of histone H2B as opposed to H1 during incubation of nuclei with CCNU (Fig. 12). NAD condensation of chromatin similarly reduced the binding of CCNU to nonhistone proteins (Fig. 12). H1 was the major acceptor for modification by poly ADP-ribosylation (lane A), whereas H2B was preferentially modified by the drugs.

III. CONCLUSION

In summary, the data from our laboratory supports a role for poly ADP-ribosylation of nuclear proteins in nitrosourea-induced DNA repair, and suggests a putative mechanism by which this might occur. Poly (ADP-rib) polymerase requires DNA for activity. It appears to be localized in periodic domains of folded polynucleosomes, perhaps every turn of 8–10 nucleosomes, although it might be capable of rearrangement. Its role might involve a "housekeeping" function for this domain of chromatin. The enzyme seems to be activated by single-strand DNA break generation of alkylating agents. Activation would result in a crosslinking of polynucleosomes in this or adjacent regions of chromatin. This condensation could be "relaxed" via the poly (ADP-rib) glycohydrolase enzyme. This change in chromatin structure in some way allows DNA repair enzyme to function in the damaged region. The condensation of nucleosomes seems to involve crosslinking of nuclear proteins. The crosslinking of histone H1 via poly ADP-ribosylation to a dimer (and perhaps higher orders of protein crosslinking) seems to be particularly significant in this process.

ACKNOWLEDGMENT

The expert typing of Helen Dookhan is greatly appreciated.

REFERENCES

Berger, N. A., Sikorski, G. W., Petzold, S. O., and Kurohara, K. K. (1980). *Biochem. 19,* 289–293.
Butt, T. R., Brothers, J., Giri, C., and Smulson, M. E. (1978). *Nucleic Acid Res. 5,* 2775–2788.
Butt, T. R., Jump, D., and Smulson, M. E. (1979). *Proc. Nat. Acad. Sci. 76,* 1628–1632.
Butt, T. R., and Smulson, M. E. (1980). *Biochem.* In press.
Butt, T. R., DeCoste, B., Jump, D. B., Nolan, N., and Smulson, M. E. (1980). *Biochem.* In press.
Giri, C. P., West, M. H. P., Ramirez, M. L., and Smulson, M. E. (1978a). *Biochem. 17,* 3495–3500.
Giri, C. P., West, M. H. P., Ramirez, M. L., and Smulson, M. E. (1978b). *Biochem. 17,* 3501–3504.

Juarez-Salinas, Sims, J. L., and Jacobson, M. K. (1979). *Nature 282,* 740–741.

Jump, D. E., and Smulson, M. E. (1980). *Biochem. 19,* 1024–1031.

Jump, D. E., Butt, T. R., and Smulson, M. E. (1979). *Biochem. 18,* 983–990.

Jump, D. E., Butt, T. R., and Smulson, M. E. (1980). *Biochem. 19,* 1031–1037.

Mullins, D. W., Giri, C., Smulson, M. E. (1977). *Biochem. 16,* 506–513.

Nelson, P. P., Albright, S. C., Wiseman, J. M., and Garrard, W. T. (1979). *J. Biol. Chem. 254,* 11751–11760.

Nolan, N. L., Butt, T. R., Wong, M., Lambrianidou, A., and M. E. Smulson. (1980). *Eur. J. Biochem.* In press.

Renz, M., Nehls, P., and Hozier, J. (1977). *Proc. Nat. Acad. Sci. 74,* 1879–1883.

Smulson, M. E. (1979). *Trends in Biochem. Sci. 4,* 4225–4228.

Smulson, M. E., Schein, P., Mullins, D. W., and Sudhakar, S. (1977). *Cancer Res. 37,* 3006.

Smulson, M. E., Sudhakar, S., Tew, K. D., Butt, T. R., and Jump, D. B. (1979). *In* "Effects of Drugs in the Cell Nucleus" (H. Busch, S. T. Crooke, and Y. Daskal, eds.). Academic Press, New York.

Stone, T. R., Lorimar, W. S., and Kiedwell, W. R. (1977). *Eur. J. Biochem. 81,* 9–18.

Sudhakar, S., Tew, K. D., and Smulson, M. E. (1979). *Cancer Res. 39,* 1405–1410.

Worcel, A. (1978). *Cold Spring Harbor Symp. on Quant. Biol. 42,* 313–324.

SECTION II
CLINICAL STUDIES

Chapter 11

SUBACUTE AND CHRONIC TOXICITIES ASSOCIATED WITH NITROSOUREA THERAPY

John S. Macdonald
Raymond B. Weiss
Donald Poster
Luz Hammershaimb

I. INTRODUCTION

Nitrosourea compounds are important antitumor drugs with antineoplastic activity in a variety of human cancers (Carter *et al.*, 1972). These drugs produce two major types of dose-limiting acute toxicities. One is cumulative myelosuppression (Carter *et al.*, 1972; DeVita *et al.*, 1965), commonly seen with the chloroethyl nitrosoureas—BCNU, CCNU, and methyl CCNU[a]; the other is renal tubular toxicity associated with the naturally occurring methyl nitrosourea, streptozotocin (Schein *et al.*, 1974). Recently, other less common toxicities, including pulmonary, renal, and central nervous system adverse effects, have become apparent with the prolonged and intensive use of nitrosoureas. This chapter will review the information available on the incidence, etiology, prognosis, and clinical presentation of these toxic manifestations, and it will also discuss the possible role of nitrosoureas in producing second malignant tumors in man.

[a] The abbreviations used are BCNU: 1,3-bis(2-chloroethyl)-1-nitrosourea; CCNU: 1-(2-chloroethyl)-3-cyclohexyl-1-nitrosourea; chlorozotocin: 1,3-(2-chloroethyl)-3-nitrosoureido-D-glucopyranose; methyl CCNU: 1-(2-chloroethyl)-3-(4-methyl cyclohexyl)-1-nitrosourea.

II. PULMONARY TOXICITY

Pulmonary adverse reactions and toxicity have become increasingly common results of cytotoxic drug therapy (Weiss and Muggia, 1980; Weiss *et al.*, 1980; Collis, 1980; Durant *et al.*, 1979). It would appear that the lung is an organ that is particularly sensitive to injury by cytotoxic agents. It is also apparent that there may be several types of clinical pulmonary side effects induced by chemotherapy agents, although the most common syndrome is pulmonary fibrosis (Weiss and Muggia, 1980; Collis, 1980). Table I lists a number of drugs that have been reported to cause pulmonary fibrosis and shows that the drugs most commonly involved in pulmonary toxicity are the classical alkylating agents. Since pulmonary adverse effects are common with alkylating drugs, it is not surprising that nitrosoureas, which are also alkylating compounds, would be proven to produce pulmonary toxicity. It should be clear also that in most instances pulmonary fibrosis is not an acute toxicity of antineoplastic agents, but only becomes apparent after the drugs have been in clinical trial for several years. This is true for all of the alkylating agents, but not for bleomycin, whose pulmonary toxicity was noted in early clinical trials and is the major dose-limiting toxicity (Collis, 1980).

The toxic lung damage associated with nitrosoureas is a fibrosing alveolitis and interstitial pneumonitis, which can produce pulmonary fibrosis, restrictive pulmonary failure, and death (Weiss *et al.*, 1980; Collis, 1980). Pulmonary toxicity has been reported with three nitrosoureas in clinical use: BCNU (Weiss *et al.*, 1980; Aronin *et al.*, 1980; Durant *et al.*, 1979; Selker *et al.*, 1980; Schreml *et al.*, 1978), methyl CCNU (Lee *et al.*, 1978), and chlorozotocin (Ahlgren *et al.*, 1980). Pulmonary toxicity has not been reported with CCNU or streptozotocin. The nitrosourea most often associated with pulmonary toxicity is BCNU. This drug was also the earliest drug of this class to enter clinical trial.

TABLE I. Pulmonary Toxicity from Cancer Chemotherapy Agents[a]

Drug	Date lung toxicity reported	Date of first use
Alkylators		
Busulphan	1961	1953
Cyclophosphamide	1967	1958
Melphalan	1972	1958
Chlorambucil	1972	1955
Mitomycin C	1978	1957
Nitrosoureas		
BCNU	1976	1962
Methyl CCNU	1978	1971
Other		
Bleomycin	1969	1969

[a] Adapted from Collis (1980).

The incidence of BCNU pulmonary toxicity among 935 patients is examined in Table II. As can be seen, the incidence of this toxicity varied in three series between 1.3 and 30%, for an overall incidence of 4.6%. The differences in pulmonary toxicity reported in these series may relate to several factors. Nine of the patients with pulmonary toxicity in the series of Durant *et al.* (1979) had non-Hodgkin's lymphomas, whereas the patients in the series of Selker *et al.* (1980) and Aronin *et al.* (1980) all had malignant brain tumors. Durant's lymphoma patients were treated with multiple drug therapy, but for shorter durations and to lesser total nitrosourea dosages than were the patients of Selker *et al.* and Aronin *et al.* Since the patients in the latter two series had no exposure to other pulmonary toxic drugs and had no pulmonary irradiation, it would seem that a rate of 20% to 30% may approximate the incidence of BCNU pulmonary toxicity in patients treated with large drug doses.

Total BCNU dose and duration of therapy have been implicated as risk factors for BCNU pulmonary toxicity (Weiss *et al.*, 1980; Collis, 1980). The series of Aronin *et al.* and Selker *et al.* support the suggestion that large cumulative doses and/or prolonged administration of BCNU are associated with increased pulmonary toxicity. Aronin *et al.* reported that pulmonary toxicity occurred in patients receiving an average total dose of 1146 mg/m^2, which was significantly ($p = 0.005$) greater than the mean dose of 777 mg/m^2 in patients who did not show any evidence of pulmonary toxicity. The possibility of developing drug toxicity increased rapidly when total doses above 1500 mg/m^2 were given. At this level, 50% of patients had pulmonary toxicity. In the study of Selker *et al.* (1980), the total BCNU doses at which pulmonary toxicity was seen varied between 240 and 2340 mg/m^2. However, 50% of patients had received doses of greater than 1200 to 1500 mg/m^2 when toxicity occurred.

In the study of Durant *et al.* (1979), no correlation was seen between total doses of BCNU and pulmonary toxicity. The total doses given in this series varied between 400 and 2075 mg/m^2, but only 2/10 patients had received total doses in excess of 100 mg/m^2. It should be remembered that 9/10 patients described by Durant *et al.* (1979) had received polychemotherapy and that 4/10 (40%) also received thoracic irradiation. These therapeutic manipulations may have enhanced the potential of BCNU to produce pulmonary toxicity at low total doses (Weiss *et al.*, 1980).

It is difficult to clearly define preexisting conditions that may lead to BCNU-

TABLE II. Incidence of BCNU-Associated Pulmonary Toxicity

Total no. pts.	No. with pulmonary toxicity	Pulmonary toxicity (%)	Investigator
795	10	1.3	Durant *et al.* (1979)
93	19	20	Aronin *et al.* (1980)
47	14	30	Selker *et al.* (1980)

induced pulmonary toxicity. The most convincing risk factor appears to be the presence of lung disease. In the study of Aronin *et al.* (1980), patients developing toxicity had a significantly higher ($p = 0.004$) incidence of some form of pretreatment pulmonary pathology. These workers noted that preexisting lung disease not only increases the risk of BCNU pulmonary toxicity, but also permits pulmonary toxicity to occur at lower total BCNU doses. In patients with pulmonary disease, a single BCNU dose has produced lung damage. Because of significant risk of pulmonary toxicity, Aronin *et al.* (1980) have recommended that patients should not be treated with BCNU if baseline pulmonary function studies show either a forced vital capacity or a diffusing capacity less than 70% of the predicted value.

It has been suggested that both prior therapy with cyclophosphamide and thoracic irradiation increase the risk of pulmonary toxicity with BCNU (Weiss *et al.*, 1980). In the series of trials by Durant *et al.* (1979) the patients showing pulmonary toxicity had received prior cyclophosphamide therapy. As noted previously, these patients frequently evidenced pulmonary toxicity at total BCNU doses of less than 1000 mg/m². In similar fashion, of the four patients in this series receiving thoracic irradiation, BCNU-associated lung damage occurred at total BCNU doses between 400 and 600 mg/m². Although the association between nitrosourea pulmonary toxicity and therapy with thoracic irradiation and cyclophosphamide is suggestive, it has not been clearly proven.

It should be noted that increased age does not appear to be a risk factor for nitrosourea pulmonary toxicity. In the study of Aronin *et al.* (1980), patients exhibiting pulmonary toxicity were on the average younger than patients without toxicity. However, the difference was not statistically significant. These investigators also noted that younger patients have greater bone marrow tolerance to BCNU therapy and, thus, receive more courses of BCNU at higher mean doses than older patients. Since the total dose of nitrosourea is the most significant risk factor for pulmonary toxicity, younger patients would be at the greatest risk.

The symptoms and physical signs reported in all large series on BCNU pulmonary toxicity have been similar (Aronin *et al.*, 1980; Selker *et al.*, 1980; Durant *et al.*, 1979). Symptoms were dyspnea, tachypnea, and nonproductive cough. Physical findings were either normal or confined to bibasilar crepitant rales on auscultation. The roentgenographic findings associated with nitrosourea pulmonary toxicity are similar to those found with other pulmonary toxic drugs such as bleomycin (Weiss and Muggia, 1980). These usually include bilateral interstitial infiltrates in a micronodular, finely reticulated pattern. Although chest x rays are most commonly abnormal, both Durant *et al.* (1979) and Selker *et al.* (1980) report patients with documented BCNU pulmonary toxicity and normal chest films. When pulmonary function studies are performed, three abnormalities are found. There is resting hypoxemia, a restrictive defect on ventilation studies, and decreased carbon monoxide diffusing capacity. The clinical presentation of pulmonary toxicity induced by nitrosoureas other than BCNU is similar to the findings described above for BCNU.

TABLE III. Chlorozotocin + Methyl CCNU Induced Pulmonary Toxicity
Three Case Reports

Drug	Total dose	Duration of therapy	Pathology	Author
Chlorozotocin	(1) 480 mg/m^2	4 mo.	—	Ahlgren *et al.* (1980)
	(2) 670 mg/m^2	7 mo.	Pulmonary fibrosis	
Methyl CCNU	4100 mg	25 mo.	Pulmonary fibrosis	Weiss *et al.* (1980)

Although corticosteroids have been used in the treatment of nitrosourea-induced pulmonary toxicity, there is no evidence that these drugs are effective (Weiss and Muggia, 1980; Weiss *et al.*, 1980). The prognosis of pulmonary toxicity induced by nitrosoureas is generally grave. In the review by Weiss *et al.* (1980) of 16 case reports of patients with nitrosourea pulmonary toxicity, there were 13 deaths as a direct result of pulmonary dysfunction. In the experience of Durant *et al.* (1979), 60% (6/10) of patients with pulmonary toxicity died of this complication. In the series of Aronin *et al.* (1980) and Selker *et al.* (1980), patients with pulmonary toxicity faired somewhat better. Aronin *et al.* reported three deaths among 19 patients, and Selker *et al.* reported five deaths in 14 affected patients. The reasons for the improved survival in these two series are not clear but may relate to increased awareness of the potential pulmonary toxicity of nitrosoureas, with consequent earlier diagnosis and cessation of therapy. Further studies will be needed to confirm the potential benefits of early diagnosis.

Table III summarizes reports of pulmonary toxicity with nitrosoureas other than BCNU. As can be noted, only three cases have been reported. These include two patients treated with chlorozotocin (Ahlgren *et al.*, 1980) and one patient (Weiss *et al.*, 1980) who received methyl CCNU. These cases had a clinical pathologic picture identical to that seen with BCNU pulmonary damage. It is of interest to note that the two patients treated with chlorozotocin developed pulmonary toxicity at relatively modest doses of 480 and 670 mg/m^2. This may correlate with the fact that chlorozotocin is a more potent alkylating agent than the other chloroethyl nitrosoureas (Hoth *et al.*, 1978). Since it has been hypothesized that the alkylating activity of nitrosoureas may cause the pulmonary toxicity associated with these drugs, it is possible that the most potent alkylator among the nitrosoureas would be the most potent producer of pulmonary toxicity. The low incidence of pulmonary toxicity associated with methyl CCNU may result from the fact that this drug is most commonly used in patients with advanced gastrointestinal cancer who are likely to die before there has been prolonged nitrosourea therapy.

III. RENAL TOXICITY

With the exception of the dose-limiting renal toxicity secondary to therapy with the naturally occurring methyl nitrosourea, streptozotocin (Schein *et al.*,

1974), renal toxicity with clinically used nitrosoureas is rare. Although preclinical toxicology studies with chloroethyl nitrosoureas showed potential renal toxicity with these drugs (Carter *et al.*, 1972), clinical toxicity has been unusual. Table IV details some of the information available. Renal insufficiency has occurred in association with methyl CCNU (Nichols and Moertal, 1979; Harmon *et al.*, 1979), CCNU (Silver and Morton, 1979), BCNU (Schacht and Baldwin, 1978), and chlorozotocin (Baker *et al.*, 1979). As can be determined from Table IV, renal toxicity is uncommon, with only 26 patients reported as having this complication. The major factor that all of these patients had in common was prolonged therapy with chloroethyl nitrosoureas, resulting in high cumulative doses. Most patients in this compilation of series received total nitrosourea doses of greater than 1000 mg/m^2, and the duration of therapy exceeded 1 year in all cases.

The well-described patients in the series of Harmon *et al.* (1979) indicate the typical clinical pathologic findings in chloroethyl-nitrosourea- induced renal insufficiently. These workers reported renal involvement in 6 of 17 pediatric brain tumor patients who were treated with methyl CCNU at 120 mg/m^2 every 6 weeks for a projected course of 2 years. Although the renal function studies of the six patients developing renal failure remained normal during nitrosourea treatment, a progressive decrease in renal size was noted. This decrease continued after cessation of therapy, and the patients developed irreversible renal failure. All affected patients had received a total dose greater than 1500 mg/m^2 of methyl CCNU. In two patients who received smaller doses for shorter periods of time, a decrease in renal size was seen but did not result in renal failure before they died of their malignancies. The pathology of all affected patients in this series was similar. Glomerulosclerosis, thickening and wrinkling of glomerular basement membranes, and tubular atrophy and loss were observed in all patients with renal failure. These histopathologic findings are characteristic of the observations made in the findings of all reported patients with chlorocthyl nitrosourea renal toxicity (Table IV).

It should be emphasized that renal toxicity with chloroethyl nitrosoureas is an uncommon finding. Nichols and Moertel (1979) (Table IV), in reviewing the extensive series of patients treated with methyl CCNU at the Mayo Clinic, point

TABLE IV. Nephrotoxicity with Chloroethyl Nitrosoureas

Investigators	Drug	No. pts./ total no. pts.	Dose drug in pts. toxicity	Mean duration of therapy (months)	Reversible
Nichols and Moertel (1979)	Methyl CCNU	4/857	880–3740 mg/m^2	15 → 36	no
Schacht *et al.* (1978)	BCNU, CCNU	14/160	1200–2000 mg/m^2	12 → 24	no
Harmon *et al.* (1979)	Methyl CCNU	6/17	>1500 mg/m^2	>17	no
Silver *et al.* (1979)	CCNU	1/1	2300 mg	24	no
Baker *et al.* (1979)	Chlorozotocin	1/1	2100 mg	18	no

out that only 4 (0.5%) of 857 patients were felt to have drug-associated nephrotoxicity. Three of these four patients had received greater than 1000 mg/m^2 of
methyl CCNU, and duration of therapy was greater than 15 months in all affected
patients. The strong dose relationship between chloroethyl nitrosourea total dose
and renal toxicity is indicated in the Mayo series by the fact that 3 (14%) of the 22 patients who received greater than 1000 mg/m^2 of drug developed nephrotoxicity
thought to be nitrosourea related.

In summary, it would appear that chloroethyl nitrosoureas have little potential
for renal toxicity when given at modest total dose. However, when therapy is to
be continued for a prolonged period of time and total doses greater than 1000
mg/m^2 are attained, it is important to be alert to the possibility of renal toxicity.
The most valuable aid in predicting clinically significant toxicity is kidney size.
The experience of Harmon et al. (1979) would suggest that nephrotomograms
should be periodically obtained, and if progressive renal size decrease is seen,
nitrosourea therapy should be stopped. Although these workers did report decreased renal size in two patients who had methyl CCNU therapy stopped at total
dosage less than 1500 mg/m^2 and did not develop renal toxicity, it should be
pointed out that these patients died of malignancy. It is possible that patients with
any degree of decreased renal size secondary to nitrosourea administration who
survive for prolonged periods of time will eventually develop renal failure. Since
the renal failure associated with nitrosourea therapy is irreversible (Table IV), the
most prudent clinical approach is to attempt to abort this syndrome by stopping
nitrosoureas if renal size decreases.

IV. OTHER ADVERSE EFFECTS

A. Second Tumors

The association between alkylating agent therapy and second tumors is well
established (Rosner et al., 1975). Acute nonlymphocytic leukemia has been
reported in four patients after therapy with chloroethyl nitrosoureas (Table V).
The two cases of Cohen et al. (1976) and the one case of Osband et al. (1977)
were patients who had primary brain tumors. The patient of Wiggins et al.
(1978) had a primary colon cancer with pulmonary metastases. As can be seen
from Table V, patients were treated with BCNU, CCNU, and methyl CCNU.
The patient of Wiggins et al. (1978) also had minimal exposure (3 weeks) to
streptozotocin. All of these patients had prolonged myelosuppression after nitrosourea therapy which was subsequently followed by the development of acute
leukemia. It should also be emphasized that three of the four patients had received radiation therapy in addition to nitrosourea therapy, and it has been
previously reported that radiation in combination with alkylating agents may
induce acute leukemias (Rosner et al., 1975). All of the patients described in
Table V had survived for at least 2 years with their primary malignant diseases

TABLE V. Acute Nonlymphocytic Leukemia after Therapy with Chloroethyl Nitrosoureas

Investigators	Original diagnosis	Radiation therapy	Drug	Total dose	Duration therapy	Time between cessation of nitrosourea therapy and diagnosis of acute leukemia (months)
Cohen *et al.* (1976)	brain tumor	yes	(1) BCNU	720 mg/m²	36 mo	44
			CCNU	700 mg/m²		
			VM 26	10,000 mg/m²		
	brain tumor	no	(2) methyl CCNU	2700 mg/m²	36 mo	18
Osband *et al.* (1977)	brain tumor	yes	methyl CCNU	1200 mg/m²	10 mo	?
Wiggins *et al.* (1978)	colon cancer	yes	methyl CCNU	400 mg	4 mo	13
			streptozotocin	2.5 gm	3 wk	

and had either received extensive combined modality therapy with chemotherapy and irradiation or high cumulative doses of nitrosoureas, as is indicated by the total methyl CCNU dose of 2700 mg/m² received by the second patient of Cohen *et al.* (1976). Thus, although second tumors are possible after nitrosourea treatment, this complication is exceedingly rare and is only possible in the patient who has prolonged survival after his primary malignant disease and receives high cumulative doses of nitrosourea or combined modality therapy with nitrosourea and irradiation.

B. Other Toxicities

Two other unusual toxicities have recently been reported as possibly associated with nitrosourea therapy. These are encephalopathy (Burger *et al.*, 1980) and optic neuritis (Louie *et al.*, 1978; McLennan *et al.*, 1978). Burger and colleagues (1980) reported encephalopathy associated with cerebral white matter necrosis in four patients treated with BCNU. These patients were treated with extraordinarily high doses (1200 mg/m²) in preparation for bone marrow transplantation. Within 25 to 47 days after BCNU therapy, 3/4 patients had developed symptoms ranging from diplopia to quadriparesis. All patients were dead within 87 days of treatment and exhibited two patterns of neuropathologic findings. Some patients showed randomly distributed discreet foci of necrosis in the white matter with occasional evidence of fibrinoid necrosis in vessels associated with the lesions. Other patients exhibited large bilaterally symmetrical areas of white matter necrosis with edema, vascular thrombi, and fibrinoid necrosis. The neuropathologic findings were similar to the leukoencephalopathy occasionally seen with methotrexate therapy (Liu *et al.*, 1978). This highly lethal syndrome of leukoencephalopathy with BCNU therapy has only been associated with the extraordinarily high drug doses administered by Burger *et al.* (1980). It is unlikely that such a high nitrosourea dose would ever be used except in preparation for bone marrow transplant.

The second unusual complication of nitrosourea therapy is optic neuroretinitis. McLennan *et al.* (1978) reported the occurrence of neuroretinitis in a patient with multiple myeloma who had been treated with BCNU and procarbazine. Subsequently, Louie *et al.* (1978) described another five patients with optic neuritis after chloroethyl nitrosourea therapy. Although this association is something that clinicians caring for patients receiving nitrosoureas should be aware of, optic neuritis has not been clearly demonstrated to be due to nitrosoureas. Of the six patients reviewed by Louie *et al.* (1978), five had been treated with combination chemotherapy and one had received radiation therapy to the brain plus CCNU without other chemotherapy. Four of these six patients had either brain metastases (2/4) or primary brain tumors (2/4). Both of these conditions may be associated with optic neuritis. Thus, it appears that there is no unambiguous evidence from these patients to suggest that optic neuritis was due to chloroethyl nitrosoureas alone. Confirmation of this potential association will require that physicians treating patients with nitrosoureas carefully observe their patients for evidence of possible visual disturbances.

V. CONCLUSIONS

Nitrosoureas have been shown to be useful drugs in cancer treatment (Carter *et al.*, 1972). With continued and more widespread use of these agents, it has become apparent that there are a number of adverse toxic affects that can occur with nitrosourea therapy. The most important nonacute toxicity of chloroethyl nitrosoureas appears to be pulmonary fibrosis. As reviewed in this chapter, this toxicity appears to be dose related, and the histopathology is similar to pulmonary fibrosis associated with other alkylating agents and with bleomycin. The other adverse effect of nitrosourea therapy that has been widely reported is renal insufficiency. This is uncommon but may occur in patients treated to high cumulative doses with a variety of chloroethyl nitrosoureas. Less common adverse effects including second tumors, optic neuritis, and leukoencephalopathy have been observed in association with chloroethyl nitrosoureas but further experience is required to clearly define the exact clinical pathologic relationship between nitrosoureas and these unusual adverse effects.

ACKNOWLEDGMENT

The authors appreciate the excellent editorial and technical assistance of Mr. William Soper and Ms. Anne Gooding in preparing this manuscript.

REFERENCES

Ahlgren, J. D., Smith, F. P., Kerwin, D. M., Sikic, B., Weiner, J. H., and Schein, P. S. (1980). *Cancer Treat. Rep.* In press.

Aronin, P. A. Mahaley, M. S., Rudnick, S. A., Dudka, L., Donohue, J. F., Selker, R. G., and Moore, P. (1980). *N. Eng. J. Med. 303,* 183–188.

Baker, J. J., Lokey, J. L., Price, N. A. (1979). *Letter, N. Eng. J. Med. 662,* 301.

Burger, P. C., Kamenar, E., Schold, S. C., Fay, J. W., Phillips, G. L., Herzig, G. P. (1980). *Cancer.* In press.

Carter, S. K., Schabel, E. M., Broder, L. G., and Johnston, T. P. (1972). *Adv. Cancer Res. 16,* 273–332.

Cohen, R. J., Wiernik, P. H., Walker, M. D. (1976). *Cancer Treat. Rep. 60,* 1257–1261.

Collis, C. H. (1980). *Cancer Chemother. Pharmacol. 4,* 17–27.

DeVita, V. T., Carbone, P. P., Owens, A. H., Gold, G. L., Krant, M. J., and Edmondson, J. (1965). *Cancer Res. 25,* 1876–1881.

Durant, J. R., Norgard, M. J., Murad, T. M., Bartolucci, A. A., Langford, K. H. (1979). *Annals Intern. Med. 90,* 191–194.

Harmon, W. E., Cohen, H. S., Schneeberger, E. E., and Grupe, W. G. (1979). *N. Eng. J. Med. 300,* 1200–1203.

Hoth, D., Woolley, P., Green, D., Macdonald, J., and Schein, P. (1978). *Clin. Pharmacol. Therapeut. 23,* 712–722.

Lee, W., Moore, R. P., and Wampler, G. L. (1978). *Cancer Treat. Rep. 62,* 1355–1358.

Liu, H. M., Maurer, H. S., Vongsivut, S., and Conway, J. J. (1978). *Human Pathol. 9,* 635–648.

Louie, A. C., Turrisi, A. T., Muggia, F. M., and Bono, V. H. (1978). *Med. Ped. Oncol. 5,* 245–247.

McLennan, R., and Taylor, H. R. (1978). *Med. Ped. Oncol. 4,* 43–48.

Nichols, W. C., and Moertel, C. G. (1979). *Letter, N. Eng. J. Med. 301,* 1181.

Osband, M., Cohen, H., Cassady, J. R., and Joffe, N. (1977). *Proc. ASCO & AACR 18,* 303.

Rosner, F., and Grunwald, H. (1975). *Am. J. Med. 58,* 339–353.

Schacht, R. G., Baldwin, D. S. (1978). *Kidney Int. 14,* 661.

Schein, P. S., O'Connell, M. J., Blom, J., Hubbard, S., Magrath, I. T., Bergevin, P., Wiernik, P. H., Ziegler, J. L., and DeVita, V. T. (1974). *Cancer 34,* 993–1000.

Schreml, W., Bargon, G., and Angen, B. (1978). *Blut 36,* 353–358.

Selker, R. G., Jacobs, S. A., Moore, P. B., Wald, M., Fisher, E. R., Cohen, M., Bellot, P. (1980). *J. Neurosurg. 7,* 560–565.

Silver, H. K. B., Morton, D. L. (1979). *Letter, Cancer Treat. Rep. 13,* 226–227.

Weiss, R. B., and Muggia, F. M. (1980). *Am. J. Med. 68,* 259–266.

Weiss, R. B., Poster, D. S., and Penta, J. S. (1981). *Cancer Treat. Rev.* In press.

Wiggins, R. G., Jackson, R. J., Fialkow, P. J., Woolley, P. V III, Macdonald, J. S., Schein, P. S. (1978). *Blood 52,* 659–663.

Chapter 12

THE TOXICITY OF HIGH-DOSE BCNU WITH AUTOLOGOUS MARROW SUPPORT[1]

Tak Takvorian
Fred Hochberg
George Canellos
Leroy Parker
Nicholas Zervas
Emil Frei

I. INTRODUCTION

The nitrosourea compounds demonstrate antitumor activity in a wide variety of animal tumor model systems and human solid malignancies. However, the nitrosoureas are unique among chemotherapeutic agents because of the delayed and cumulative myelosuppression that limits both the frequency of administration and total dose. This may account in part for the limited success of the nitrosoureas in cancer treatment. Delayed and prolonged myelosuppression are uniquely suited for "rescue" with stored autologous bone marrow. Preclinical studies in the dog (Abb *et al.*, 1978) have demonstrated that the lethal hematologic toxicity of BCNU can be prevented by infusion of bone marrow cells stored at 4°C., 24 hr after drug infusion. A pilot study was undertaken in patients with solid tumors to explore the antitumor potential of BCNU doses in

[1]Supported in part by HEW and National Cancer Institute Grant 1-CM-67037 and The Friends of the Farber Center.

the range of 600–1400 mg/m² with noncryopreserved autologous marrow rescue. Dose-response considerations and the visceral toxicity of the escalated doses are the subject of this report.

II. MATERIALS AND METHODS

Twenty-three patients were entered on study: 17 patients with biopsy-proven grade III or IV astrocytoma, 3 patients with melanoma, and 3 patients with advanced colorectal adenocarcinoma. There were 8 females and 15 males, with a median age of 38. Six brain tumor patients and one patient with melanoma had received adjuvant CCNU (140 mg to 2840 mg) prior to entry onto study, and one of these had additionally received high-dose methotrexate. The three patients with colorectal carcinoma demonstrated refractoriness to to 5-FU before entry onto study. All patients had clinical and radiographic evidence of tumor progression. Those patients receiving corticosteroids because of their underlying disease were maintained on a fixed dosage for 2 weeks prior to entry onto protocol and for a minimum of 2 months posttreatment with high dosage BCNU. The brain tumor patients were also on a variety of antiseizure medications including phenytoin, phenobarbital, valproic acid, primidone, and carbamazepine. Informed consent was obtained prior to initiation of the protocol in all cases.

Bone marrow was aspirated from the pelvis (anterior and posterior iliac crests and spines) while the patient was under general anesthesia, using 14-gauge Rosenthal needles. The technique for aspiration and removal of particulate matter were a modification of that described by Thomas and Storb (1970). A mean of 2 × 10¹⁰ nucleated cells were obtained during this 2 hr procedure. Clotting of the marrow was prevented by aspirating the cells into anticoagulant citrate dextrose solution, USP, Formula A containing 50 units/ml heparin sodium (preservative-free); the ratio of marrow to anticoagulant was 8:1. Marrow was refrigerated at 4°C. for 48–72 hr before infusion and had a viability > 90% by colony-forming assay by modification of the technique of Pike and Robinson (1970).

The protocol is outlined in Table I. On return from the recovery room, the patients were treated with one-half of the total dose of BCNU to be administered. It was given intravenously through a peripheral intravenous catheter or a Hickman line inserted into the internal jugular vein, with the second half of the drug dose administered after an interval of 12 hr. Each dose of drug was adminis-

TABLE I. High-Dose BCNU with Autologous Marrow Support

Time 0:	Bone marrow harvest under general anesthesia centrifuged to remove fat and particulate matter, refrigerated at 4°C
Time 6 hr:	BCNU infusion over 1 hr
Time 18 hr:	BCNU infusion over 1 hr
Time 48 hr:	Bone marrow reinfused

tered over 60–90 min. The refrigerated bone marrow was infused intravenously approximately 24 hr (range 24–48 hr) after the last BCNU infusion or 48 hours after marrow procurement.

There were two treatment groups: All patients who had previously received nitrosourea chemotherapy were treated at 800 mg/m² total dose. Patients not previously treated with a nitrosourea prior to entry onto this protocol were treated at doses ranging from 600 mg/m² to 1400 mg/m² Twenty-one patients received only one course of treatment and were observed for tumor response and toxicity; two patients were re-treated at their same BCNU dose at a time of tumor recurrence.

BCNU was dissolved in Absolute Ethanol, USP, and was diluted with normal saline or 5% dextrose in water for a final ethanol concentration of 5–10% (v/v). Patients were pretreated with pentobarbital and perphenazine; and this was continued for 20 hr to control nausea and vomiting. Patients treated at ≥ 1200 mg/m² received an oral antibiotic prophylaxis regimen to help prevent infection during the period of aplasia, consisting of nystatin 10⁵ units/cc, 5–10cc po qid, starting at the time of treatment with chemotherapy, and followed in 72 hr with the addition of trimethoprim (160 mg)-sulfamethoxazole (800 mg) po every 12 hr. Both antibiotics were discontinued with the appearance of fever or documented infection. Red blood cells were transfused for a hematocrit less than 30%, and platelets were transfused prophylactically when the platelet count fell below 20,000/mm³.

Marrow engraftment was studied by serial bone marrow biopsy and aspiration for cellularity and the ability to form *in vitro* colonies. Quantitative evaluation of tumor response included serial computerized axial tomographic studies; thallium, technetium gluco-heptonate, and gallium brain scans; and neurologic examination at frequent intervals.

Weekly evaluations of renal and liver function were performed. Chest radiographs were followed at 2–4 week intervals, and pulmonary function was monitored with spirometry and diffusing capacity of CO_2, arterial blood gases, and gallium scanning of the lung.

III. RESULTS

A. Acute Toxicity

During the immediate chemotherapy-treatment period, fluid balance was the major problem. Mild hypotension was seen in most patients (a drop in diastolic blood pressure of 10–20 mm/hg); it was multifactorial in etiology and responded to fluid replacement. All patients were relatively dehydrated secondary to surgical preparation and bone marrow aspiration, but emesis, antiemetic and analgesic medications, BCNU, and its ethanol vehicle all contributed to transient hypotension. However, patients with evidence of increased intracranial pressure were particularly vulnerable to rapid shifts in fluid volume. Two patients with glioma had CNS herniation during the peritreatment period (one during the infusion of drug), which reversed with mannitol administration. One patient with melanoma, age 62, suffered a subendocardial myocardial infarction secondary to transient hypotension during the treatment period.

BCNU behaves more as an irritant rather than a vesicant in causing local pain at injection sites, and this is probably related more to the ethanol vehicle of the drug rather than the drug per se. One patient developed a slow-healing (2 months) ulcer at a site of drug infiltration, and one patient treated at high dosage (1200 mg/m² BCNU) developed a severe sterile phlebitis along the peripheral vein route of administration, with resultant fever and flexure contraction secondary to the sclerosed vein. Conjunctival suffusion occurred in all patients acutely with drug delivery, as did flushing of the skin with peripheral vasodilatation. In a patient with traumatic extubation postsurgery, suffusion of mucous membranes with the infusion of chemotherapy necessitated reintubation.

Nausea and vomiting were controlled by concomitant perphenazine and sedation with pentobarbital, although these drugs contributed to hypotension. Vomiting occurred, if at all, 4–6 hr after the first infusion of BCNU only. There were no acute sequelae due to the infusion of refrigerated, fresh bone marrow.

B. Hematopoietic Toxicity

Myelosuppression was the major and expected toxicity. At the lowest dose studied in this protocol, 600 mg/m², the pattern of delayed myclosuppression was not dissimilar to that of conventional dosage, occurring 4 weeks after treatment, despite autologous marrow infusion. However, at higher doses of BCNU, the nadir of both WBC and platelet counts was both more profound and occurred earlier than at lower doses, shifting the nadir of myelosuppression toward the left (see Table II). That is, at 600 mg/m² BCNU the WBC nadir of 2800/mm³ occurred on day 28 and a platelet nadir of 41,000/mm³ also occurred on day 28

TABLE II. Myelosuppression and the Days to Nadir of White Blood Cells and Platelets

BCNU dose	Days to WBC nadir	Days to platelet nadir
600 mg/m²	d. 28 (2800 WBCs)	d. 28 (41,000 Plts.)
800 mg/m²	d. 15 (1100)	d. 17 (26,000)
800 mg/m² (S/P former CCNU)	d. 13 (275)	d. 17 (20,000)
1000 mg/m²	d. 14 (450)	d. 14 (22,000)
1200 mg/m²	d. 9 (150)	d. 15 (10,000)
1400 mg/m²	d. 10 (<100)	d. 12 (15,000)

TABLE III. Relationship between Mononuclear Cells and CFU-C Infused and Myelosuppression at 1200 mg/m² BCNU

Patient	Cell no. reinfused × 10¹⁰	Cell no. infused/kg × 10⁸	Total CFU-C × 10⁶	Total CFU-C/ kg × 10⁴	Day with WBC ≤500	Day with WBC ≤1000	Day with PMN ≤500	Day with PMN ≤1000
K.R.	0.9	1.5	0.46	0.78	12	15	14	16
L.G.	2.6	3.0	1.3	1.5	9	10	10	10
K.R.	1.6	3.4	2.34	4.9	7	12	12	14
W.H.	2.1	2.5	4.69	5.5	10	10	10	13
J.F.	1.2	3.2	6.6	9.0	5	9	5	12
J.S.	3.1	3.7	13.6	1.6	4	5	4	12

versus a WBC nadir of $<100/mm^3$ on day 10 and a platelet nadir of 15,000/mm³ on day 12 posttreatment at 1400 mg/m² BCNU. Although the time to nadir appeared to be inversely related to the dose of drug absolutely, the recovery of counts was directly influenced by the quantity (number of cells infused/kg) and "quality" (total CFU-C infused/kg) of the marrow infused at any dosage level (see Table III). The time to recovery from nadir of white blood cells and platelets is tabulated in Tables IV and V respectively.

Patients who had formerly received nitrosourea before entry onto this protocol, however, were all treated at 800 mg/m² and had a more profound and prolonged myelsuppression than would have been expected at this dose level in a previously untreated patient, occurring at days 13 and 17 for white blood cells and platelets respectively. In this group five of seven patients never recovered their platelet counts greater than 100,000/mm³ and three of these remained less than 50,000/mm³. All patients without former nitrosourea ultimately achieved a stable platelet count greater than 100,000/mm³.

In both groups of patients, the platelet nadir occurred after the WBC nadir, was more prolonged, and recovered more slowly and in a step-wise incremental manner. However, the latter two features were more pronounced in the group of patients who had prior nitrosourea. In no patient was a second nadir seen in white blood cells or platelets.

TABLE IV. Myelosuppression Recovery

Dose	No. pts.	Median days with total WBC ≤100	Median days with total WBC ≤500	Median days with total WBC ≤1000
600 mg/m²	(2)	0	0	0
800 mg/m²	(2)	2(0–3)	3(0–6)	4(0–8)
1000 mg/m²	(2)	0	2(1–3)	6(1–10)
1200 mg/m²	(6)	3(0–6)	8(4–12)	10(5–15)
1400 mg/m²	(2)	5(4–6)	8(6–10)	9(7–11)
800 mg/m² (S/P former CCNU)	(6)	1(0–5)	8(1–13)	12(6–19)

TABLE V. Myelosuppression Recovery

Dose	No. pts.	Median days with total PMN			No. of plt. transfusions
		≤100	≤500	≤1000	
600 mg/m²	(2)	0	0	0	0
800 mg/m²	(2)	2(0–4)	5(0–10)	5(0–10)	1
1000 mg/m²	(2)				1
1200 mg/m²	(6)	5(2–8)	10(4–14)	13(10–16)	5(1–31+)
1400 mg/m²	(2)	7(6–8)	9(7–10)	14(13–14)	9(1–17+)
800 mg/m² (S/P former CCNU)	(6)	2(1–7)	11(3–20)	15(10–25)	6(1–26+)

C. Infection

The morbidity of myelosuppression was limited to infection. Eleven patients developed fever posttreatment. As noted above, one melanoma patient treated at 1200 mg/m² BCNU developed a severe sterile phlebitis acutely with the infusion of chemotherapy and concomitant fever, which resolved as the phlebitis healed, prior to the period of aplasia; also, one patient developed culture-negative fever acutely with the infusion of his cryopreserved bone marrow, which rapidly resolved prior to the period of aplasia.

In seven of the remaining nine patients, fever developed with myelosuppression, but an infectious etiology was not found. Three of these seven patients rapidly defervesced with the initiation of broad-spectrum antibiotics, and in the remaining four patients defervescence occurred only when the total polymorphonuclear count exceeded 500/mm³. Two patients died during aplasia, while febrile: one patient with glioma developed *Clostridia* sepsis and died 9 days after treatment with BCNU at 1400 mg/m². A second patient had a fever of multifactorial origin including mucositis, a urinary tract infection, and dermatomal *H. Zoster,* but subsequently died of neurological complications.

Two patients had a *Staph aureus* infection of their Hickman line necessitating antibiotics and removal of the line, without additional sequelae. Despite the high dosage of steroids in many of the glioma patients (dexamethasone 8mg–64mg/day), steroids did not appear to mask infection, and no occult or opportunistic infection was seen.

No hemorrhage or bleeding complication was noted in any of the patients. Platelets were transfused prophylactically for a platelet count of less than 20,000/mm³. However, nine patients required prolonged platelet support (greater than five platelet transfusions), often complicated by the development of antiplatelet antibodies and requiring single-donor, leukocyte-poor, or HLA-matched platelets. In the group without former nitrosourea, this was seen in five of the patients treated at doses of 1200 mg/m² BCNU; but this was also seen in four of the seven patients with prior nitrosourea therapy, treated at 800 mg/m² BCNU.

In eight patients, recovery of peripheral granulocyte counts was preceded by a transient, but impressive, atypical lymphocytosis. This presented as a rapidly rising white blood cell count from the time of nadir, in one patient reaching a WBC count of 20,000/mm³ with 90+% lymphocytes, more than half of which were atypical. The range in other patients was 500-4000 WBCs/mm³ with 6-71% atypical lymphocytes. This was in contrast to the more gradual rising pattern of polymorphs, bands, and monocytes that characterized the recovery of peripheral counts in all other patients. The atypical lymphocytes were medium-to-large in size with markedly increased, deeply basophilic cytoplasm, many with azurophilic granules. Nuclei were multilobed or deeply fissured with very prominent nucleoli. The lymphocytosis lasted 5-7 days and was replaced with a greater percentage of "normal"-appearing lymphocytes and a monocytosis, transitioning gradually to a normal differential.

Fever was seen in three of these patients but was not a necessary concomitant. Two patients had a concomitant focal pneumonitis, which appeared to be interstitial on chest x ray and became diffuse in one patient, suggestive of a viral pneumonia or pneumocystis. However, pulmonary function did not deteriorate, and the changes were self-limited, resolving as the lymphocytosis resolved without treatment. Mild hepatitis, which slowly reversed in time, developed in two patients after the resolution of the atypical lymphocytosis. Cerebritis was suspected in one patient, but this also resolved.

Viral cultures, including CMV and E-B virus, and titres were negative in all patients, and no etiology could be discerned for this constellation of abnormalities.

D. Hepatic Toxicity

Liver toxicity was assessed by serial determination of SGOT, LDH, alkaline phosphatase, and bilirubin in all patients. No abnormalities in hepatic enzyme tests of any significance were observed for the first 10 weeks following treatment at BCNU doses of less than or equal to 1000 mg/m². At total doses of 1200-1400 mg/m² BCNU abnormalities were detected in 1 of 2 evaluable patients. This one patient with melanoma treated at 1200 mg/m² BCNU developed liver necrosis 1 month after treatment and died 2 weeks later in hepatic coma. At autopsy, the liver demonstrated centrilobular necrosis, consistent with a toxic drug insult to the liver, with no tumor or evidence of infection seen.

Long-term follow-up of hepatic function was available in six patients (range 10 weeks to 15 months), and abnormal liver function was seen in two patients with glioma. In both patients this developed insidiously at 4 months after treatment at doses of 800 mg/m² and 1200 mg/m² BCNU and was asymptomatic. Hyperbilirubinemia never developed; there was mild elevation in SGOT and alkaline phosphatase to five times the upper normal limit. The elevation in SGOT has stabilized and slowly improved, but the alkaline phosphatase continues to climb in both patients. Liver biopsy has not been done, and they have been maintained on stable dose steroids.

E. Renal Toxicity

No abnormality of renal function has been noted as monitored by serial BUN and creatinine, and, in some cases, creatinine clearance. Length of follow-up is similar to that of liver function testing noted above.

F. Pulmonary Toxicity

Symptomatic pulmonary toxicity developed in two patients. One patient with recurrent glioma developed the rapid onset of shortness of breath and deterioration of pulmonary function 4 months after treatment at 1200 mg/m^2 BCNU. An open-lung biopsy revealed prominent hypertrophy and hyperplasia of the type II pneumoncytes, active mononuclear cell infiltration, and inflammation and interstitial edema. No infecting organism was cultured or seen on special stains of the pathologic tissue. The interstitial pneumonitis is attributed to BCNU. On stable dosage steroids, the patient's chest radiograph improved and pulmonary function improved and stabilized at a functionally tolerable, asymptomatic level, although the DLCO remained at 60% expected.

One patient with malignant melanoma also developed a biopsy-proven interstitial pneumonitis 5 weeks after treatment at 800 mg/m^2 BCNU which was rapidly progressive, leading to pulmonary insufficiency and death within 2 weeks. On autopsy, some tumor was found in the lungs, but the pulmonary insult was again attributed to BCNU pneumonitis, in the absence of any other cause.

G. Miscellaneous Toxicity

1. Gastrointestinal

No definite GI toxicity has been noted except for the mild nausea and vomiting, occasionally seen acutely, with the infusion of drug. Two patients developed mucositis during the period of aplasia which healed as the blood counts recovered, but one additional patient treated at 1400 mg/m^2 BCNU died of *Clostridia* sepsis, secondary to seeding from diffuse sloughing of the entire gastrointestinal tract during the period of aplasia. Mild, self-limited diarrhea has been noted in several patients at variable time intervals posttreatment.

2. CNS

No CNS toxicity unique to BCNU has been seen. One patient with glioma developed seizures 48 hr after treatment at 1400 mg/m^2 BCNU and became comatose, dying several weeks later, but on postmortem examination no abnormality other than viable tumor, tumor necrosis, and edema could be discerned. Autopsy examination of 9 additional patients at 10 days to 1 year posttreatment with a variety of doses of BCNU likewise revealed no specific central nervous system pathology. The development of encephalopathy in glioma patients was presumed to be secondary to tumor rather than to drug toxicity, but the number of patients evaluable at the highest dosage levels (1400 mg/m² BCNU) is limited to one of three patients at 3 months. Patients with nonglial tumors did not develop encephalopathy.

Two patients, however, did develop bilateral neurofibrillary retinal infarcts antemortem 1 to 2 months after treatment with 800 mg/m² BCNU. Histopathologic examination including electron microscopy revealed no tumor emboli, infection, or other etiology for the gliosis, and drug toxicity is presumed.

No cardiac abnormality or dysfunction has been noted in this patient population. Postmortem examination has been done in 10 patients and has revealed no specific CNS or peripheral end-organ toxicity of BCNU except as noted above.

H. Antitumor Effect

All 17 glioma patients had relapsed by clinical and CT scan evidence at a median of 10 months (range 0–18 months) following initial surgery, radiation ± adjuvant nitrosourea before entry onto this protocol. Two patients had a significant decrease in the two-dimensional area of tumor, as demarcated on CT scan by the boundary of contrast enhancement of greater than 25% within 2 months of treatment on this protocol with a single dose of BCNU while on constant or reduced steroid dose. However, two-thirds of patients demonstrated other CT scan findings of potential significance including (a) a decrease and change in the pattern of contrast enhancement, for example, leaving a rim of contrast-enhancing tissue in the place of a formerly large homogenous mass of contrast-enhancing tumor (see Fig. 1); (b) a corresponding reduction in the area of low-density mass surrounding the tumor (mass effect) and edema with consequent improvement in the degree of shift of midline structures (see Fig. 2). The Karnovsky status of patients did not change significantly.

The median survival postrelapse from conventional therapy for patients who had not received former nitrosourea is 6 months, with the longest remission being 12 months after a single course of treatment with BCNU and 18 months in a patient who was retreated after 9 months when he developed evidence of tumor recurrence. Only one patient with former nitrosourea therapy is alive at the time of this writing without evidence of tumor progression at 9 months.

Two of the three patients with colorectal carcinoma had marked reduction in palpable lymphadenopathy and a fall of the elevated CEA to the normal range, but the effect was transient, lasting only 2 months. One of the three patients with melanoma, who died with hepatic necrosis 6 weeks after treatment, was found to have no residual tumor on postmortem examination after treatment at 1200 mg/m² BCNU; and one other melanoma patient who died with interstitial pneumonitis 5 weeks after treatment at 800 mg/m² BCNU had evidence of partial tumor necrosis on postmortem examination.

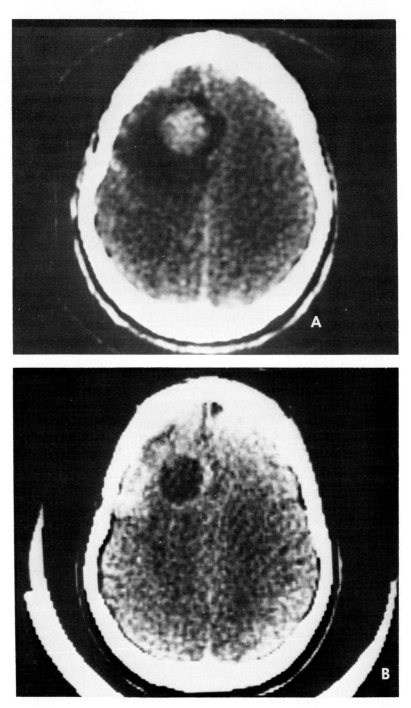

Fig. 1. A: Baseline Ct scan prior to treatment with BCNU at a time of tumor recurrence. B: CT scan of same patient 2 months after treatment at 600mg/m² BCNU. Note change in character of enhancement.

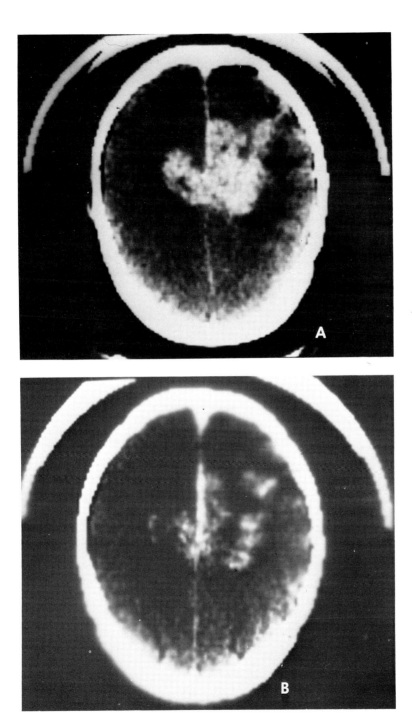

Fig. 2. A: Baseline CT scan prior to treatment with BCNU at a time of tumor recurrence. B: CT scan of same patient 1 month after treatment at 800mg/m² BCNU. Note improvement in mass effect, edema, and decreased contrast enhancement.

At the doses employed in this study there was no clear-cut evidence of dose response in terms of antitumor activity.

IV. DISCUSSION

The major end-organs of toxicity of nitrosoureas include the bone marrow, lungs, liver, and kidney. A single course of BCNU (240 mg/m^2) produces a characteristic delayed myelosuppression at 3–4 weeks that lasts an additional 1–2 weeks. Thrombocytopenia appears earlier, lasts a shorter period, and subsides earlier than leukopenia, which often has a biphasic nadir. Repetitive administration of drug is cumulative in terms of more profound myelosuppression.

In the original phase I study of BCNU by DeVita *et al.* (1965), 26% of 134 patients studied developed an abnormality in one or more liver function tests at a mean onset of 34 days posttreatment, which was reversible. Ten percent of patients developed unexplained elevations in BUN and creatinine, although more recent studies relate total cumulative dose with an increasing incidence of irreversible renal damage. Pulmonary toxicity is a newly recognized consequence of nitrosourea therapy.

Myelosuppression in this study was severe but manageable and not dose-limiting with the use of autologous marrow support. At the lower doses, the pattern of late myelosuppression was not eradicated, although it was perhaps ameliorated. At high doses of BCNU, the myelosuppression effect was similar to that of other alkylating agents, without a second "late" nadir, once the marrow engrafting and recovery had taken place. It would appear that at lower doses of BCNU, a more primitive, noncycling hematopoietic stem cell is affected, which is manifested several weeks later in suppression of the differentiating line. Infusion of marrow did not eradicate the late myelosuppression of BCNU, perhaps because the lower dose of drug did not adequately condition the marrow space to accept and engraft the transplanted marrow. However, at higher doses of BCNU, proliferating, partially differentiated, committed cells are selectively affected, more rapidly creating a "space" in which the transplanted marrow may engraft, obviating any late effect of chemotherapy. Additionally, the elimination of the partially differentiated cells may accelerate the recruitment and proliferation of endogenous, nonharvested stem cells which would also tend to counter any later myelosuppression (Parker *et al.*, 1980).

There is now experimental evidence in the rodent that the treatment of the bone marrow stroma with BCNU affects the success of subsequent hematopoietic cell recovery after transplantation (Cohen, 1980). In our series bone marrow whose stem cell population, and perhaps stroma also, had been "affected" by prior nitrosourea did not engraft as well as "virgin" marrow, particularly in the megakaryocytic series. Presumably, poorer total mononuclear cell yields at bone marrow harvest in this group of patients and less CFU-C/kg of body weight infused account for the severe myelosuppression. Pretreatment peripheral counts and bone marrow cellularity did not predict this problem. However, despite prolonged myelosuppression, infection and bleeding were not a major problem in either treatment group.

The absolute necessity of autologous marrow support cannot be determined without nonmarrow support controls at each dose level, but it is our belief that the transplanted marrow ameliorates the period of myelosuppression. This is inferred from the fact that the rate of recovery of peripheral blood counts appears to be directly related both to the total number of mononucleated cells and CFU-C/kg of body weight reinfused. Recent experience by others also supports the contention that autologous marrow infusion ameliorates chemotherapy-induced myelosuppression (Appelbaum, 1978; Abrams, 1980).

Pulmonary toxicity remains the most dangerous visceral toxicity of nitrosourea chemotherapy. Its development is currently not accurately predictable, and it often progresses to a fatal conclusion. The accumulated reports in the literature suggest a "final common pathway" for alkylation damage to the lung. The detoxification response by the lung is hypertrophy and proliferation of type II pneumatocytes with consequent scarring and interstitial fibrosis. Although the incidence in our small series is 8.7% of 23 patients, with a 4.5% fatality, the literature suggests a range of 1.3% (Durant, 1979) to 15% (Selker, 1978) for cumulative nitrosourea in larger series. Although there appears to be no simple maximum beyond which all patients will develop pulmonary toxicity, it has recently been suggested by Aronin (1980) that >1500 mg/m² cumulative BCNU is associated with a greater than 50% probability of developing restrictive lung disease, albeit dose and schedule may be important covariables with total accumulated dosage. At present there is no predictive test for pulmonary toxicity, and attention must be given to monitoring these patients for clues to the denouement of pulmonary toxicity and the development of animal model systems so that selective rescue can be developed.

Although renal toxicity has not yet been seen in this study, it is possible that the length of follow-up is not sufficient for this to be manifest. Renal toxicity with the chronic administration of nitrosourea has only recently been described: Schacht and Baldwin (1978) noted progressive chronic renal failure in 14 of 160 patients with brain tumor who received chronic nitrosourea (14/17 patients received >1200 mg/m² and 9/9 patients received >1800 mg/m²), and a similar experience has been reported in children (Harmon et al., 1979) and at the Mayo Clinic (Nichols and Moertel, 1979). As with pulmonary toxicity, the dose per course and schedule of drug may be important covariables along with total accumulated dose.

The results of this study of escalating doses of BCNU are similar to those reported by Phillips et al. (1980), who also explored the toxicity of BCNU between doses of 600 mg/m² and 2850 mg/m², administered in divided dosage over 3 days. Cryopreserved autologous marrow was utilized, with liver necrosis being a dose-limiting toxicity at greater than 2 gm/m²; a dose of 1200 mg/m² BCNU was defined as the upper limit of reasonable safety for a single course of

BCNU. In our study severe visceral toxicity did not appear to be strictly dose-related, but rather appeared to be idiosyncratic within the range studied. Toxicity appeared to be more significant in patients who had former nitrosourea and consequently a higher total dose of BCNU.

No definite statement regarding the antitumor effect of high-dose BCNU with autologous marrow support can be made at this time. Median survival, after initial relapse in gliomas, may be somewhat increased in patients who have not formerly been treated with a nitrosourea; but this is a small and selected series, and a randomized study with a carefully matched control arm is needed to confirm or extend these preliminary observations. A more accurate means to determine tumor status is required since the CT scan is not sufficiently sensitive to detect occult progression. A more definitive correlation between the changes seen on CT scan and pathologic change *in vivo* is also required. In this regard, we are exploring radioisotope scans using thallium to better identify viable as opposed to necrotic tumor. In addition, a greater antitumor effect of this approach might be expected in circumstances of smaller tumor volume and when given prior to radiation therapy.

V. CONCLUSION

In summary, we conclude that (*a*) toxicity with high-dose BCNU is marked, but may be acceptable in certain circumstances of uniformly fatal tumors for which better and safer treatment is not currently available; (*b*) a maximal "safe" single dosage has not been defined but will be less than or equal to 1200 mg/m^2 BCNU; (*c*) myelosuppression is ameliorated by the infusion of noncryopreserved autologous marrow; (*d*) significant nonmyeloid toxicity is dose-limiting, primarily to the lung and liver; (*e*) patients with prior nitrosourea therapy had no benefit and did have excessive myelotoxicity from this treatment; and (*f*) a greater antitumor potential of this approach might be expected in circumstances of smaller tumor volume and when treatment is given prior to radiation therapy in patients with glioma.

REFERENCES

Abb, J., Netzel, B., Rodt, H. V., and Thierfelder, S. (1978). *Cancer Res. 38,* 2157.

Abrams, R. A., Glubiger, D., Simon, R., Lichter, A., and Deisseroth, A. B. (1980). *Lancet 1,* 385.

Appelbaum, F. R., Herzig, G. P., Ziegler, J., Graw, R. G., Levine, A. S., and Deisseroth, A. B. (1978). *Blood 52,* 85.

Aronin, P. A., Mahley, M. S. Jr., Rudnick, S. A., Dudka, L., Donahue, J. F., Selker, R. G., and Morre, P. (1980). *N. Eng. J. Med. 303,* 183.

Cohen, G., Canellos, G., and Greenberger, J. (1980). *In* "Biology of Bone Marrow Transplantation: ICN-UCLA Symposia." In press.

DeVita, V. T., Carbone, P. P., Owens, A. H. Jr., Gold, L., Krant, M. J., and Edmonson, J. (1965). *Cancer Res. 25,* 1876.

Durant, J. R., Norgard, M. J., Murad, T. M., Bartolucci, A. A., and Langford, K. H. (1979). *Ann. Inter. Med. 90,* 191.

Harmon, W. E., Cohen, H. J., Schneeberger, E. E., and Grupe, W. E. (1979). *The New Engl. J. Med. 300,* 1200.

Nichols, W. C., and Moertel, C. G. (1979). *New Eng. J. Med. 301,* 1181.

Parker, L. M., Takvorian, T., Hochberg, F. H., Canellos, G. P., and Frei, E. III. (1980). *Proc. Am. Assoc. Cancer Res. 21,* #564.

Phillips, G. L., Fay, J. W., Wolff, S. N., Karanes, C., Herzig, R. H., and Herzig, G. P. (1980). *Proc. Am. Assoc. Cancer Res. 21,* #721.

Pike, B. L., and Robinson, W. A. (1970). *J. Cell Physiol. 76,* 77.

Schacht, R. G., and Baldwin, D. C. (1978). *Kidney Internat. 14,* 661.

Selker, R. G., Jacobs, S. A., and Moore, P. (1978). *Proc. Am. Soc. Clin. Oncol. 19,* 333.

Thomas, E. D., and Storb, R. (1970). *Blood 36,* 507.

Chapter 13

CLINICAL PHARMACOLOGY OF THE NITROSOUREAS[1]

Victor A. Levin

I. INTRODUCTION

The cloroethylnitrosoureas (CENUs) are clinically important anticancer agents with activity against a wide variety of human cancers. All clinically active CENUs are prodrugs whose cytotoxicity is related to their ability to chemically transform to chloroethylating species near the intracellular site of action (Kramer et al., 1974; Colvin et al., 1976; Kohn, 1977; Weinkam and Lin, 1979).

In vitro and in vivo activity appears to be a function of the pharmacokinetics of the individual CENU. The total amount of alkylating product in the tumor cell determines cell kill (Weinkam and Deen, 1980) and is related to both the integrated tumor-cell drug exposure (C_T or concentration $\times$ time) and the intracellular half-life ($t_{1/2}$) of the parent CENU.

Most CENUs are rapidly cleared from plasma. The rate of clearance is related to tissue distribution and chemical reaction rates: More lipophilic CENUs distribute more extensively than less lipophilic ones. For BCNU, protein-binding catalyzed degradation accelerates breakdown (Weinkam et al., 1980b); for lipophilic CENUs, serum lipid partitioning slows chemical and protein-binding reactions (Weinkam et al., 1980a). Hepatic metabolism can increase biotransformation and reduce antitumor activity (Levin et al., 1979), or alter the

[1]Supported in part by National Cancer Institute Project Program Grant CA-13525 and American Cancer Society Faculty Research Award 155.

TABLE I. Determinants of CENU Plasma Drug Levels

I. Compartmental distribution, volumes and rate constants
    A. Protein binding
    B. Lipophilicity
    C. Ionization
    D. Tissue blood flow

II. Extent and rate of elimination
    A. Excretion
        1. Renal reabsorption and tubular secretion
        2. Fecal
    B. Degradation
        1. Chemical transformation
            a. Protein interactions
            b. Lipid interactions
        2. Biotransformation
            a. Soluble enzymes
            b. Tissue enzymes

prodrug by hydroxylation of the cyclohexyl ring (Heal *et al.*, 1978). Thus, while CENUs share common modes of activation and expression of antitumor activity, because so many factors contribute to the rates of activation and the extent of biodistribution, their individual pharmacokinetics would be expected to differ significantly, and, for a given CENU, vary among patients.

Table I lists the multiple factors that influence CENU pharmacokinetics. In the sections that follow, I will discuss some of these factors for BCNU, CCNU, MeCCNU, PCNU, ACNU, and chlorozotocin (CLZ), and examine briefly the potential therapeutic advantage of increasing tumor capillary drug levels and permeability.

II. GENERAL METHODS

Because of the development by Weinkam of sensitive mass spectrometric techniques to measure parent species and some intermediates, the pharmacokinetics of BCNU and PCNU in humans are fairly well defined. Pharmacokinetic parameters for the other nitrosoureas have been measured using radiolabeled drug, with or without chromatographic verification of parent species and intermediates, or by using modifications of the Bratton-Marshall reaction for nitroso groups (Forist, 1964; Loo *et al.*, 1966; Sponzo *et al.*, 1973; Woolley *et al.*, 1980).

Before the development of sensitive mass spectrometric methods, CENUs and their intermediates and metabolites were characterized by normal phase high-pressure liquid chromatography (HPLC) and thin layer chromatography. However, because of the low wavelength and the low molar absorptivity of the nitroso chromophore, this technique is not sensitive. For instance, using normal or reverse phase HPLC, an assay limit of 5 μg/ml was found for CCNU (Weinkam *et al.*, 1980b). With newer detectors and columns, I. Kroll has measured BCNU to less than 0.5 μg/ml using reverse phase techniques (personal communication). Because ACNU has a chromophoric pyrimidinyl moiety, HPLC techniques may be satisfactory (Nakamura *et al.*, 1977; Mori *et al.*, 1979).

The colorometric assay of Loo *et al.* (1966) can detect lower levels of CENUs ($\sim$1 μg/ml) in HPLC effluent than is possible using conventional ultraviolet detection methods. However, the sensitivity is not sufficient to determine patient plasma levels of CENU, and the method lacks the specificity necessary to distinguish between parent species and biologically active intermediates that retain the nitroso moiety.

We measured BCNU levels in various media using direct sample insertion selected ion monitoring (SIM) chemical ionization mass spectrometry (CIMS) with $^2H_8$-BCNU as an internal standard (Weinkam *et al.*, 1978). The sensitivity of this technique is approximately 50–100 ng.

PCNU was measured using either gas chromatography–mass spectroscopy (GC–MS) or GC using a nitrous detector. After extraction from plasma, drug is converted to a stable O-methylcarbamate by reaction in anhydrous basic methanol (Weinkam and Liu, 1981). Derivativization allows measurement of as little as 2–5 ng of PCNU. This analytical technique can be used for BCNU, CCNU, MeCCNU, and ACNU with similar sensitivity.

III. PHARMACOKINETIC DATA

Before comparing the human pharmacokinetics of some of the CENUs, we should first review factors affecting chemical reaction rates that could influence and account for variability in plasma pharmacokinetics.

Table II summarizes the available information on the $t_{1/2}$ in phosphate buffer and human serum at 37°C and pH 7.4. The $t_{1/2}$ for CLZ in serum is not known, and the $t_{1/2}$ for ACNU was measured in our laboratory using GC/CIMS techniques previously described for BCNU and PCNU (Weinkam and Liu, 1981).

TABLE II. CENU Half-Lives in pH 7.4 Phosphate Buffer (37°C) and Human Plasma, pH 7.4 (37°C)

Drug	Buffer $t_{1/2}$ (min⁻¹)	Plasma $t_{1/2}$ (min⁻¹)	Reference
CCNU	53	30	Weinkam *et al.* (1980b)
BCNU	49	14	Levin *et al.* (1978a)
			Weinkam *et al.* (1980b)
PCNU	25	25	Weinkam *et al.* (1980b)
ACNU	33	29	(Unpublished results)
CLZ	48		Heal *et al.* (1978)

Decomposition of BCNU, CCNU, and MeCCNU but not PCNU or ACNU are affected by protein. Weinkam *et al.* (1980b) found that the human-albumin-catalyzed decomposition of BCNU, CCNU, and MeCCNU occurs at a maximum velocity at normal plasma albumin concentrations, and that because of the excess of albumin compared to plasma levels of CENU under normal physiologic conditions, the rate of decomposition is independent of albumin concentration. Thus CENUs can decompose both by general base-catalyzed reactions in aqueous media and by protein-catalyzed mechanisms that require the formation of a CENU-protein complex before decomposition. Because of the high reactivity of decomposition products, a large percentage of them react with, and are covalently bound to, albumin.

Table III summarizes the effect of total hydrophobic lipids on *in vitro* serum $t_{1/2}$ for MeCCNU, CCNU, and BCNU (Weinkam *et al.*, 1980a). These data demonstrate that the more lipophilic the CENU, the more it is stabilized against hydrolysis with increasing serum lipid concentration. Between total hydrophobic lipid concentrations of 6 mg/ml and 22 mg/ml, the BCNU $t_{1/2}$ increased 1.6-fold, the CCNU $t_{1/2}$ increased 1.8-fold, and the MeCCNU $t_{1/2}$ increased 2.5-fold.

In addition to these factors that influence chemical transformation rates, CENUs can be enzymatically deactivated by denitrosation reactions (Levin *et al.*, 1979) or modified by hydroxylation (Heal *et al.*, 1978). Because these reactions depend on liver microsomal enzymes, the rates of biotransformation would be expected to vary as a function of factors such as patient age, nutritional status, and other drugs being taken.

Pharmacokinetics were computed using a two-compartment open model (Benet and Ronfeld, 1969) because available human CENU plasma curves indicated a biexponential plasma decay for parent BCNU, PCNU, ACNU, and CLZ. The schema for the two-compartment open model is shown in Fig. 1.

Table IV summarizes the available pharmacokinetics in humans of selected CENUs that were computed using a computer program based on Benet's equations (Benet and Ronfeld, 1969). Only data obtained for BCNU, PCNU, and ACNU accurately reflect the pharmacokinetics of parent CENU; data for CLZ are specific for the nitroso group, and data for CCNU are specific only for ^{14}C

TABLE III. Effect of Added Total Hydrophobic Serum Lipids on the In Vitro Half-Life of CENU

		$t_{1/2}$ (min^{-1})		
Drug	Log P	6 mg/ml[a]	12 mg/ml	22 mg/ml
MeCCNU	3.3	69	116	173
CCNU	2.8	43	58	77
BCNU	1.5	29	38	46

[a] Normal patient range 3.5 to 8.5 mg/ml (adapted from Weinkam *et al.*, 1980a).

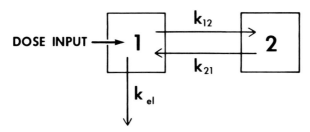

Fig. 1. Two-compartment open model for drug elimination. V_1 is the volume of the central compartment, 1, and V_2 is the volume of the peripheral compartment, 2.

radioactivity in the cyclohexyl and carbonyl moieties; data for MeCCNU, based on ^{14}C radioactivity, is less satisfactory than the CCNU data and are not included in Table IV. ACNU kinetics reported by Mori, *et al.* (1979) were used to calculate the pharmacokinetic parameters. Based on Benet's equations, the volumes of distribution at steady-state (V_{DSS}) were 3.25, 2.36, 0.31, and 0.23 l/kg for BCNU, PCNU, ACNU, and CLZ, respectively. With increasing hydrophilicity it would have been expected that V_{DSS} would decrease, particularly in view of the similarity in molecular weight and chemical transformation $t_{1/2}$ among the four CENUs. Paradoxically, ACNU (log P = 0.9) had a smaller V_{DSS} than PCNU (log P = 0.4). The V_{DSS} for ACNU and CLZ are only slightly larger than the volume of body extracellular water. Further work is needed to explain the ACNU pharmacokinetic data because it does not fit our well-established model of the relationship of biodistribution to log P values. Conceivably the difference in aqueous solubility between PCNU and ACNU accounts for the fact that the V_{DSS} for PCNU is larger than that for ACNU.

The V_{DSS} for CCNU is not reliable because parent species were not quantitated. However, because the hydroxylated products are less lipophilic than the parent and are more stable (Heal *et al.*, 1978), the V_{DSS} for total CCNU and hydroxylated products may be a reasonable value.

The elimination constant, k_{el}, is similar for the two water-soluble nitrosoureas, ACNU and CLZ (4 and 8 min respectively), but faster than for the lipophilic nitrosoureas, BCNU and PCNU (22 and 24 min respectively). However, the k_{el} for CCNU is unrealistic because it reflects ^{14}C radioactivity distributions. As expected, these trends hold for clearance values.

TABLE IV. Pharmocokinetics of CENUs in Humans

Drug	Mol wt	Log P	V_1 (l/kg)	V_2 (l/kg)	V_{DSS} (l/kg)	k_{el} (min^{-1})	k_{12} (min^{-1})	k_{21} (min^{-1})	Clearance (ml/min/kg)	Reference
CCNU	234	2.8	0.71	0.99	1.70	0.00028	0.0027	0.0020	1	Sponzo *et al.* (1973)
BCNU	214	1.5	1.14	2.11	3.25	0.032	0.114	0.063	56	Levin *et al.* (1978a)
PCNU	263	0.4	1.25	1.10	2.36	0.029	0.063	0.085	37	Unpublished results
ACNU	309	0.9	0.10	0.22	0.31	0.164	0.212	0.088	16	Mori *et al.* (1979)
CLZ	313	−1.0	0.14	0.08	0.23	0.087	0.102	0.179	12	Kovach *et al.* (1979)

IV. DISCUSSION

To be useful, clinical pharmacokinetics should contribute insight into the selection of agents or the best method for the administration of drugs that can improve safety or efficacy in the clinical setting. In this section, computer simulation of regional tumor cell integrated drug exposure (steady-state C_T) and intracellular drug products is of value and will be discussed.

A. Pharmacokinetics

Because specific assays were used, the determination of relevant pharmacokinetic data is satisfactory for BCNU, ACNU, and PCNU. Data are approximate for CLZ and unsatisfactory for CCNU and MeCCNU.

Studies of CENUs have shown that the interpatient similarities and predictability of pharmacokinetics are best for the water-soluble CENUs, ACNU and CLZ, and worst for the more lipophilic CENUs, BCNU and PCNU. This reflects differences in protein-catalyzed reactions, body fat distribution, serum lipids, and other factors (Levin *et al.*, 1978a; Weinkam *et al.*, 1980a).

B. Computer Modeling

From *in vitro* studies, we know that CENU cell kill is related to the integrated drug exposure dose (Wheeler *et al.*, 1975; Weinkam and Deen, 1980), and that the mode of CENU ethylation is the same for all CENUs. Weinkam and Deen (1980) proposed that the total intracellular product, P_i, formed by chemical transformation of CENUs would be directly related to cell kill, and substantiated this for BCNU, CCNU, MeCCNU, and PCNU.

In vivo this situation is complicated by the many factors that affect CENU plasma pharmacokinetics (Table I), as well as factors that affect CENU diffusion from a tumor capillary to cells at varying distances from the capillary. We have approached this problem mathematically (Levin *et al.*, 1980b) by considering

TABLE V. Factors Influencing Drug Diffusion from Capillary to Tumor Cells at Varying Distances from Capillary

1. Capillary drug level
2. Blood flow
3. Capillary surface area and permeability
4. Cellular permeability
5. Cellular–extracellular drug ratio
6. Extracellular drug half-life
7. Intracellular drug half-life
8. Drug diffusion coefficient

factors that might slow extracellular diffusion within the tumor (Table V). While many of these parameters cannot be known precisely for human tumors, some estimates from *in vitro* and animal experiments can be applied as approximations. Because production of CENU active species reflects total cellular exposure to parent CENU as well as the rate of the chemical activation intracellularly, the clinical efficacy of one CENU compared to another should be predictable from an analysis of some of the physical and biological parameters listed in Table V.

The area under the plasma disappearance curve (AUC) at maximum clinical dose, tumor capillary permeability coefficient, extracellular $t_{1/2}$, intracellular $t_{1/2}$, and the cell-extracellular partition ratio are sufficient to approximate the total cellular exposure to active species for most CENU drugs currently used (Levin *et al.*, 1980b). Most of these parameters are known for BCNU and PCNU; for ACNU and CLZ, they can be approximated from experimental studies, assuming they are similar to BCNU and PCNU and differ mainly in capillary and cellular permeability and cellular-extracellular ratios. Table VI summarizes the constants used and the source or method of obtaining each parameter. These data were used to computer simulate cellular drug levels for different capillary-to-cell distances; equations can be found in Appendix I of Levin *et al.* (1980b). Fig. 2 shows the computer-simulated plots for BCNU, PCNU, ACNU, and CLZ.

At low blood-flow rates, which are synonomous with low capillary density (large intercapillary distances) in this model, BCNU appears consistently to form more intracellular cytotoxic product than PCNU, ACNU, and CLZ. From a clinical point of view, the results of computer simulation appear to be reasonably accurate: BCNU is more active than PCNU and CLZ. ACNU shows activity, but there are insufficient data to compare it with BCNU. Another prediction from Fig. 2 is that nitrosourea-sensitive, high blood-flow tumors such as lymphoma would be expected to have a better clinical response than low blood-flow tumors such as squamous cell carcinoma, adenocarcinoma, or brain tumors.

TABLE VI. Parameters used to Generate Computer-Simulated Plots for Figure 1

Drug	Dose (mg/m²)	AUC (μM min)	ECF $t_{1/2}$ (min⁻¹)	ICF $t_{1/2}$ (min⁻¹)	P_c (cm/sec × 10⁴)[a]	Cell ECF	References
BCNU	220	2130[b]	49	49	0.1–1	2[c]	Levin *et al.* (1978a; 1978c)
PCNU	100	700[b]	25	25	0.01–0.1	1.7[c]	Levin *et al.* (unpublished observations) Weinkam *et al.* (in press)
ACNU	100	1003	33[d]	29[d]	0.01–0.1	1[c]	Mori *et al.* (1979)
CLZ	200	1311	48	48	0.001–0.01	1[c]	Heal *et al.* (1978)

[a] Range of permeability expected from studies with subcutaneous and intracerebral 9L tumors and normal brain. BCNU and PCNU P_c have been determined (Levin, 1980; Levin *et al.*, 1980a; 1980b; unpublished observations), while ACNU and CLZ values were extrapolated.

[b] Plasma AUC after peak plasma level attained the following drug administration multiplied by 2 to account for AUC of ascending portion of curve.

[c] Cell/media ratios from *in vitro* cell uptake studies (unpublished observations).

[d] Jane Liu, unpublished observations; method of Weinkam and Liu, 1981.

[e] Default value since literature not found; probably an overestimate based on V_{DSS} value from Table 1.

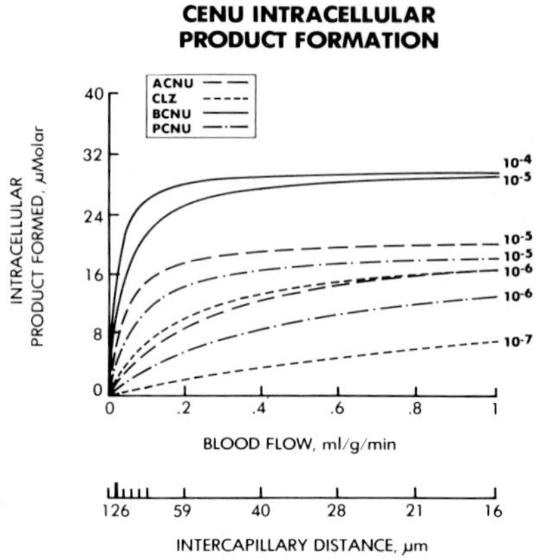

**CENU INTRACELLULAR
PRODUCT FORMATION**

Fig. 2. Computer simulation of human tumor levels based on pharmacokinetic data and *in vitro* approximations cited in Tables I, II, and VI. The model assumes that the rate of intracellular product formed is proportional to the integrated drug exposure × rate of chemical breakdown in aqueous solution (assumed to be equivalent to cytosol). The drug dose upon which this modeling was done represents the usual maximum tolerated patient dose for each CENU shown.

While this type of computer modeling is considered "soft" because it is based on approximations, it can be of considerable value. If an analog group of drugs with similar modes of activation and antitumor activity are compared, this approach can be of enormous benefit. Differences in clinical activity predictably can be found to relate to differences in body distribution and chemical transformation, as we found for the CENUs. In the future it would be advantageous to determine the necessary pharmacokinetic parameters for a group of analogs simultaneously in early phase I studies and choose the best analog for more deliberate phase II testing.

In addition to choosing the CENU with the best physical–chemical structure for optimal tumor entry, there are two additional ways to maximize tumor levels. One would be to reduce myelotoxicity so that more drugs could be given; the other would be to increase capillary drug level or increase tumor capillary permeability. The former approach has been taken (see Herzig *et al.*, this volume); the latter requires discussion.

C. Methods to Increase Tumor Drug Delivery

To achieve higher capillary drug levels without increasing systemic toxicity, intraarterial chemotherapy can be used, particularly when a tumor is within the region supplied by one artery that can be assessed by femoral angiography. Some of the theoretical assumptions necessary in the design of intraarterial therapies have been discussed by Eckman *et al.* (1974).

Based on the favorable characteristics of lipophilicity, molecular size, and rate of spontaneous hydrolysis, BCNU is an ideal drug for intraarterial therapy. Intrahepatic arterial infusion in patients harboring nitrosourea-sensitive hepatic tumors was encouraging. Ensminger *et al.* (1978) found that after hepatic artery infusion, the AUC in the hepatic vein was 2.3-fold higher than in the peripheral vein, which is also the approximate advantage found in monkey brain tissue when intracarotid artery BCNU was compared to intravenous BCNU (Levin *et al.*, 1978b).

Among tumors and within regions of individual tumors, blood volume and blood flow can vary considerably (Levin *et al.*, 1980a; 1980b). Because of this, collective tumor permeability can be reduced, even though individual tumor capillaries are quite "leaky"; reduced capillary density and shunting of blood can artificially reduce the capillary surface area available for drug diffusion. Because many CENUs are blood-flow-limited in their transcapillary tumor transport and have rapid rates of chemical transformation, they will be sensitive to tumor blood flow and capillary density (Levin *et al.*, 1980a). Therefore, anything that will increase tumor perfusion, such as osmotic agents that will passively open capillaries, will increase tumor drug levels and the cell kill of sensitive tumors. Aside from published reports of the intraarterial administration of hypertonic solutions to "open" the blood–brain barrier (Rapoport, 1971), little attention has been given to the use of intravenous hyperosmotic agents to passively increase tumor blood flow and allow better access for drugs such as BCNU and CCNU. Clearly, this approach should be investigated as a "benign" means to increase the therapeutic index of CENUs.

V. CONCLUSION

The large number of CENUs that have entered clinical trials throughout the world offer the opportunity to conduct large-scale comparative clinical pharmacokinetic studies. Surprisingly, little effort has been made in this way to better understand the clinical relevance of each analog and to guide future anticancer drug analog development.

The many factors that impact on CENU pharmacokinetics have been reasonably well established; they include pH-dependent hydrolysis, lipid stabilization, protein catalysis, enzymatic denitrosation, enzymatic hydroxylation, and tissue distribution dependence on lipophilicity. It is readily apparent from Table VI and Fig. 2 that the AUC for plasma clearance of parent species together with $t_{1/2}$ determines the amount of active species produced in cancer cells at varying distances from capillaries. As a generalization, for any analog series of anticancer drug whose activity is dependent on chemically transformed products, prediction of efficacy may well be possible using techniques similar to those outlined in this chapter.

ACKNOWLEDGMENTS

The author thanks Jane T.-Y. Liu for measuring ACNU chemical transformation, Dr. Herbert Landahl for help with the computer simulations of CENU data, Dr. Robert Weinkam for helpful discussions over the past several years, Beverly J. Hunter for manuscript preparation, and Neil Buckley for editorial assistance.

REFERENCES

Benet, L. Z., and Ronfeld, R. A. (1969). *J. Pharm. Sci. 58,* 639-641.

Colvin, M., Brundrett, R. B., Cowens, W., Jardine, L., and Ludlum, D. B. (1976). *Biochem. Pharmacol. 25,* 695-699.

Eckman, W. W., Patlak, C. S., and Fenstermacher, J. D. (1974). *J. Pharmacokinet. Biopharm. 2,* 257-285.

Ensminger, W. D., Thompson, M., Come, S., and Egan, E. M. (1978). *Cancer Treat. Rep. 62,* 1509-1512.

Forist, A. A. (1964). *Anal. Chem. 36,* 1338-1339.

Heal, J. M., Fox, P. A., Doukas, D., and Schein, P. S. (1978). *Cancer Res. 38,* 1070-1074.

Kohn, K. (1977). *Cancer Res. 37,* 1450-1454.

Kovach, J. S., Moertel, C. G., Shutt, A. J., Frytak, S., O'Connell, M. J., Rubin, J., and Ingle, J. N. (1979). *Cancer 43,* 2189-2196.

Kramer, B. S., Fenselau, C. C., and Ludlum, D. B. (1974). *Biochem. Biophys. Res. Commun. 56,* 783-788.

Levin, V. A. (1980). *J. Med. Chem. 23,* 682-684.

Levin, V. A., Hoffman, W., and Weinkam, R. J. (1978a). *Cancer Treat. Rep. 62,* 1305-1312.

Levin, V. A., Kabra, P. M., Freeman-Dove, M. A. (1978b). *J. Neurosurg. 48,* 587-593.

Levin, V. A., Kabra, P. A., and Freeman-Dove, M. A. (1978c). *Cancer Chemother. Pharmacol. 1,* 233-242.

Levin, V. A., Stearns, J., Byrd, A., Finn, A., and Weinkam, R. J. (1979). *J. Pharmacol. Exp. Ther. 208,* 1-6.

Levin, V. A., Wright, D. C., Landahl, H. D., Patlak, C. S., and Csejtey, J. (1980a). *Br. J. Cancer 41 (Suppl. IV),* 74-78.

Levin, V. A., Patlak, C. S., and Landahl, H. D. (1980b). *J. Pharmacokinet. Biopharm. 8,* 257-295.

Loo, T. L., Dion, R. L., Dixon, R. L., and Rall, D. P. (1966). *J. Pharm. Sci. 55,* 492-497.

Mori, T., Katsuyoshi, M., and Katakura, R. (1979). *Nerve Brain (No To Shinkei) 31,* 601-606.

Nakamura, K., Asami, M., Kawada, K., and Sasahara, K. (1977). *Ann. Rep. Sankyo Res. Lab. 29,* 66-74.

Rapoport, S. I., Hori, M., and Klatzo, I. (1971). *Science 173,* 1026-1028.

Sponzo, R. W., DeVita, V. T., and Oliverio, V. T. (1973). *Cancer 31,* 1154-1158.

Weinkam, R. J., and Lin, H. S. (1979). *J. Med. Chem. 22,* 1193-1201.

Weinkam, R. J., and Deen, D. F. (1980). *Proc. Am. Assoc. Cancer Res. 21,* 1 (abstract).

Weinkam, R. J., Wen, J. H. C., Furst, D. E., and Levin, V. A. (1978). *Clin. Chem. 24,* 45-49.

Weinkam, R. J., Finn, A., Levin, V. A., and Kane, J. P. (1980a). *J. Pharmacol. Exp. Ther. 214,* 318-323.

Weinkam, R. J., Liu, T.-Y. J., and Lin, H. S. (1980b). *Chem.-Biol. Interact. 31,* 167-178.

Weinkam, R. J., and Liu, T.-Y. J. (1981). *Cancer Treat. Reps.* In press.

Wheeler, K. T., Tel, N., Williams, M., Sheppard, S., Levin, V. A., and Kabra, P. (1975). *Cancer Res. 35,* 1464-1469.

Woolley, P., Luc, V., Smythe, T., Rahman, A., Hoth, D., Smith, F., and Schein, P. (1980). *Proc. Am. Soc. Clin. Oncol. 21.* 336 (abstract).

Chapter 14

NITROSOUREAS IN THE THERAPY OF HODGKIN'S DISEASE

Arlan J. Gottlieb
Clara D. Bloomfield
Arvin S. Glicksman
Nis I. Nissen
M. Robert Cooper
Thomas F. Pajak
James F. Holland

I. INTRODUCTION

The Cancer and Leukemia Group B (CALGB) has had extensive experience in the use of various nitrosoureas in the therapy of Hodgkin's disease. BCNU (1,3 bis[2 chlorethyl]-1-nitroso-), CCNU (1-[2 chlorethyl]-4-cyclohexyl-1-nitroso-), methyl CCNU (1-[2 chlorethyl]-4-methylcyclohexyl-1-nitroso-), streptozotocin (NSC 85998), and, most recently, chlorozotocin (2-[3-(2 chlorethyl)]-3-nitroso-) have been employed. These studies were initiated in 1969 and continue to the present. They encompass an evaluation of the toxicity and antineoplastic efficacy of nitrosoureas used as single agents and in drug combinations used both as the sole modality of therapy and in combination with radiotherapy.

The use of a nitrosourea as an alkylating agent in multidrug treatment programs of advanced Hodgkin's disease provides a number of extremely effective drug combinations for primary treatment as well as for the therapy of relapsed patients. Resultant toxicities do not appear to be more severe than those encoun-

Copyright © 1981 by Academic Press, Inc.
All rights of reproduction in any form reserved.
ISBN 0-12-565060-4

tered with other combination drug programs incorporating other alkylating agents for therapy of Hodgkin's disease.

II. MATERIALS AND METHODS

Unless otherwise indicated, the data reported were accrued from prospective randomized clinical trials conducted at the multiple member institutions of CALGB.

In randomized studies, patients were assigned to treatment programs by the

TABLE I. CALGB Programs Employing Nitrosoureas in the Therapy of Hodgkin's Disease and Non-Hodgkin's Lymphoma

CALGB protocol number	Studies closed to new patient entry Description
6953	Comparison of CCNU and BCNU in the treatment of advanced Hodgkin's disease
6951	A comparative study of a BCNU containing 4-drug program versus MOPP versus 3-drug combinations in advanced Hodgkin's disease
7252	A comparison of Methyl CCNU and CCNU in patients with advanced forms of Hodgkin's disease, lymphoma and "reticulum cell sarcoma"
7251	A comparison of two four-drug combinations containing CCNU (COPP, CVPP) with two four-drug combinations containing mechlorethamine (MOPP, MVPP) for the treatment of advanced Hodgkin's disease
CALGB protocol number	**Studies open to new patient entry** **Description**
7451	Radiotherapy versus chemotherapy with BOPP versus combined modality therapy with BOPP and radiotherapy in the treatment of Stage III Hodgkin's disease
7551	Combination chemotherapy with CVPP with and without radiotherapy for the treatment of advanced Hodgkin's disease
7552	Combination chemotherapy with CVPP and BAVS alone and in sequential combination for previously treated Stages III$_B$ and IV Hodgkin's disease
7751	Involved field radiotherapy plus chemotherapy with CVPP versus chemotherapy with CVPP in the treatment of poor risk patients with Stage I or II Hodgkin's disease
7772	Chlorozotocin in previously treated Hodgkin's disease and non-Hodgkin's lymphoma

TABLE II. Combination Drug Programs[a]

Nitrosourea-containing combinations		
CVPP	CCNU	75 mg/m² po day 1
	Vinblastine	4 mg/m² iv day 1,8
	Procarbazine	100 mg/m² po day 1–14
	Prednisone[c]	40 mg/m² po day 1–14
BOPP[b]	BCNU	80 mg/m² iv day 1
	Vincristine	1.4 mg/m² iv day 1,8
	Procarbazine	100 mg/m² po day 1–14
	Prednisone[c]	40 mg/m² po day 1–14
BAVS	Bleomycin	5 μ/m² iv day 1,8
	Adriamycin	50 mg/m² iv day 1
	Vinblastine	4 mg/m² iv day 1,8
	Streptozotocin	1500 mg/m² iv day 1,8
COPP	CCNU	75 mg/m² po day 1
	Vincristine	1.4 mg/m² iv day 1,8
	Procarbazine	100 mg/m² po day 1–14
	Prednisone[c]	40 mg/m² po day 1–14
Others		
MOPP[b]	Mechlorethamine	6 mg/m² iv day 1,8
	Vincristine	1.4 mg/m² iv day 1,8
	Procarbazine	100 mg/m² po day 1–14
	Prednisone[c]	40 mg/m² po day 1–14
MVPP	Mechlorethamine	6 mg/m² iv day 1,8
	Vinblastine	4 mg/m² iv day 1,8
	Procarbazine	100 mg/m² po day 1–14
	Prednisone[c]	40 mg/m² po day 1–14

[a] Programs are repeated monthly for a minimum of 6 months of therapy.

[b] The BOP program omits the procarbazine; the OPP program omits the alkylating agent.

[c] Prednisone is given during courses 1 and 4 for 6 cycle programs and months 1, 5, and 9 for 12 cycle programs.

sealed envelope technique. The Latin square design was employed to balance treatment assignments by pretreatment characteristics. Data were provided from each participating institution and analyzed by the study chairman, the CALGB statistician, and, where appropriate, by a study chairman of a modality other than chemotherapy (e.g., radiotherapy, immunotherapy). Stratification of cases prior to randomization to therapy was done prospectively in some studies (e.g., splenectomy and prior therapy in CALGB 7251). A description of the various CALGB investigations is provided in Table I. In the CALGB studies summarized, the initial two digits indicate the year of activation. Definition of the acronyms used for the combination drug programs and dose schedules are provided in Table II.

III. RESULTS

A. Nitrosoureas as Single Agents

A randomized comparison of BCNU (200 mg/m² iv q6w) and CCNU (100 mg/m² po q6w) in advanced, previously treated Hodgkin's disease was initiated in 1969 (CALGB 6953) (Selawry and Hansen, 1972; Hansen *et al.*, 1981). The pretreatment characteristics of the patient groups were comparable. In the BCNU-treated group there was 1 complete and 7 partial responses in 29 patients (overall 28%). In 31 patients treated with CCNU, 5 complete and 14 partial responses were observed (overall 60%). The percentage of stipulated drug dose given in the first three treatment courses ranged from 85–90% for both drugs. Toxicities were similar.

In a subsequent study, the efficacy and toxicity of CCNU (100 mg/m² po) and methyl CCNU (150 mg/m² po) given every 6 weeks were compared in 109 patients with advanced, previously treated Hodgkin's disease and non-Hodgkin's lymphoma (CALGB 7252) (Maurice *et al.*, 1978). Patient groups and toxicity were comparable. The overall response rates were 42% for CCNU versus 15% for methyl CCNU in Hodgkin's disease; 21% for both CCNU and methyl CCNU in lymphocytic lymphoma; and 15% and 27% for CCNU and methyl CCNU respectively in reticulum cell sarcoma.

From these studies it was concluded that as a single agent CCNU, rather than BCNU or methyl CCNU, displayed the highest antitumor activity in the therapy of Hodgkin's disease. The activity displayed by the nitrosoureas as single agents has not as yet been tested in combination with other agents by CALGB in the primary therapy of non-Hodgkin's lymphoma. The activity of chlorozotocin is currently under study in both Hodgkin's disease and non-Hodgkin's lymphoma (CALGB 7772). Thus far, the drug shows little activity in patients for whom therapy with other nitrosoureas had previously failed.

B. Nitrosoureas in Combination Chemotherapy

A significant improvement in the therapy of advanced Hodgkin's disease resulted from the introduction of therapy with MOPP in the late 1960s (DeVita *et al.*, 1970). Whether the same level of response might be achieved by the use of a three-drug program (i.e., OPP or BOP) or with the substitution of BCNU for mechlorethamine in MOPP (i.e. BOPP) was investigated in a study initiated in 1969 and closed to new patients in 1972 (CALGB 6951) (Nissen *et al.*, 1979). Six monthly courses of induction therapy were employed. Previously untreated patients and patients failing radiotherapy and/or chemotherapy were eligible for study providing the latter group had not received three or more of the drugs to be employed in the study in the previous 2 months.

Four hundred and sixty patients with Stage IIIB or IV Hodgkin's disease were

TABLE III. Comparability of Induction Treatment Groups CALGB 6951[a]

Induction program	BOP	OPP	BOPP	MOPP
Number entered	112	117	114	117
Number evaluable	107	112	103	104
Males (%)	61	68	61	68
Median age	37	32	33	38
Mean age	40	36	36	38
% 40 yrs. or over	47	38	37	47
Institutional histology (%)				
Lymphocyte predominant	7	9	11	13
Nodular sclerosis	25	22	33	22
Mixed cellularity	33	38	28	37
Lymphocyte depleted	8	4	7	9
Prior therapy (%)				
Previously untreated	33	34	43	34
Radiotherapy only	18	28	23	26
Chemotherapy only	9	6	7	4
Chemotherapy and radiotherapy	40	32	27	36
Prior splenectomy (%)	10	11	14	12
Clinical stage III (%)	12	19	23	14
Involved site (%)				
Liver	51	50	42	43
Lung	24	21	27	28
Bone marrow	5	6	7	8

[a] From Nissen et al. (1979).

initially randomized to the four treatment programs (OPP, BOP, BOPP, MOPP); 426 were eligible for evaluation. Two years after activation of the study, it became apparent that the frequency of complete remission induction with the three-drug combinations was inferior to that accomplished by employing 4 drugs. Accordingly, randomization to OPP and BOP was discontinued.

The comparability of the induction treatment groups for the initial phase of the study is shown in Table III. The frequency of complete response for the group as a whole and for previously untreated patients is provided in Table IV. The efficacy of induction with OPP and BOP was statistically inferior to that resulting from therapy with BOPP or MOPP. The median time for complete response was 3 months in all programs. Hematologic toxicity was greatest in the MOPP group where 27% of patients treated experienced severe or life-threatening toxicity (BOPP versus MOPP, $p = 0.05$). Vincristine neuropathy was similar in all full-dose 4-drug programs. The severity of nausea and vomiting were similar for MOPP and BOPP, although these side effects were half as frequent in the BOPP program, wherein the most potent emetic BCNU was given only once monthly as compared to twice monthly for mechlorethamine in MOPP. An average of 80%

TABLE IV. Frequency of Complete Response in CALGB
6951[a]

Induction program	All patients		Previously untreated	
BOP	43/107	(40%)	17/35	(49%)
OPP	47/112	(42%)	15/37	(41%)
BOPP	69/103	(67%)	32/43	(74%)
MOPP	64/104	(62%)	23/35	(66%)
Total	426		150	

[a] From Nissen *et al.* (1979). For the entire patient group, statistical analysis indicated the difference in frequency of complete response to be BOP versus BOPP, $p < 0.001$; OPP versus BOPP, $p < 0.001$; OPP versus MOPP $p < 0.004$; BOP versus MOPP $p < 0.002$, BOPP versus MOPP, not significant; 3 = drug versus 4 = drug programs, $p < 0.01$.

of the BCNU dose was administered in cycle 6 of BOPP as compared to 60% of the mechlorethamine in cycle 6 of MOPP. The procarbazine dose was reduced to 75% and 55% during the sixth cycle of BOPP and MOPP, respectively.

Previous analysis has indicated that the duration of remission for patients in complete remission following OPP, MOPP, and BOPP induction is equivalent and superior to that achieved with BOP ($p = 0.002$) in the group as a whole, in previously untreated patients, and in those under the age of 40 (Nissen *et al.*, 1979). An update of the duration of the initial complete response is given in Fig. 1. Therapy with BOPP was found to be as effective as therapy with MOPP with

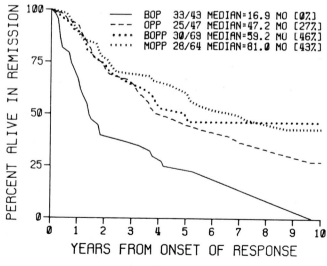

Fig. 1.

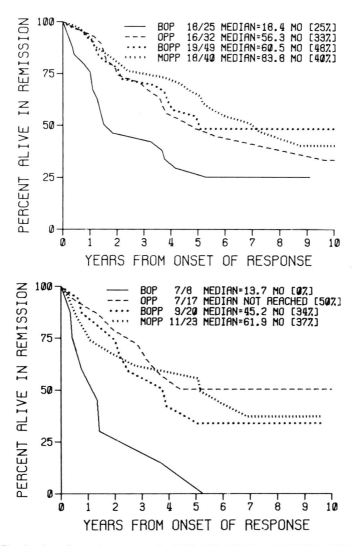

Fig. 1. The duration of complete response for BOP, OPP, BOPP, and MOPP in CALGB 6951 (Nissen *et al.*, 1979). Relapsed patients, the total number of patients initially at risk, and the median response for each program are shown with each figure. The study was closed to new patient entry in February 1972. Data are updated to October 1979. All evaluable patients (A), patients with no prior chemotherapy (B), and patients with prior radiotherapy only (C) are shown. At this time, no statistically significant differences were observed between the duration of response following therapy with BOPP, MOPP, or OPP. The BOP program was statistically inferior in each group.

regard to the duration of complete remission when patients were analyzed as a group as well as if they had been previously untreated or had relapsed from prior radiotherapy. While the duration of response to OPP has not to date proven statistically inferior to that seen with BOPP or MOPP, the number of patients lost to observation has curtailed the power of statistically detecting differences.

It is noteworthy that this study also included a postinduction randomization of patients in complete or partial response to maintenance therapy consisting of either chlorambucil, chlorambucil with monthly vincristine and prednisone, or with weekly vinblastine. The value of single-agent maintenance therapy in patients in complete remission following MOPP therapy in the study was assessed by comparison with an historic group of unmaintained MOPP-treated patients in complete remission (CALGB 6712). The effects of single-agent maintenance were analyzed for patients as cohorts and as matched pairs. No value for maintenance therapy with either chlorambucil or vinblastine could be shown.

A direct comparison of CCNU and mechlorethamine in the drug therapy of advanced Hodgkin's disease was next to be initiated. Therapy of stage III and IV Hodgkin's disease with two four-drug combinations containing CCNU (CVPP and COPP) was compared to two combinations that were otherwise identical (MVPP and MOPP) except that mechlorethamine was employed as the alkylating agent (CALGB 7251) (Cooper *et al.*, 1980). Previously untreated patients, and patients failing preceding radiotherapy and/or chemotherapy were studied provided they had not received chemotherapy with three or more of the drugs possibly to be employed in their treatment. Of the 566 patients entered, 532 were evaluable. The comparability of the patient groups for pretreatment variables is given in Table V (Cooper *et al.*, 1980). The complete response rates observed are given in Table VI.

After the initial course of induction chemotherapy patients treated with CVPP received a somewhat higher percentage of the stipulated dose of alkylating agent and vinca alkaloid during each cycle. For MOPP, the percentage of stipulated dose of mechlorethamine delivered in the second to sixth course was 77%, 74%, 67%, 61%, and 64% as compared to 80%, 85%, 80%, 71%, and 70% for CCNU in the CVPP program. The vincristine given in MOPP was 84%, 74%, 65%, 61%, and 62% in the second through sixth treatment cycle, while the corresponding percentages for vinblastine in CVPP were 78%, 85%, 81%, 73%, and 75%. These differences in "optimal dosing" were also observed when the subsets of patients without prior therapy and those receiving only prior radiotherapy were analyzed.

The occurrence of severe or life-threatening leukopenia and thrombocytopenia in the entire patient group, in previously untreated patients, and in those previously treated was similar among each of the four induction programs. Gastrointestinal side effects were less evident when CCNU was used. Severe nausea and vomiting was seen in 11% of the patients treated with MOPP, 4% of those treated with MVPP, 6% of those treated with COPP, and 5% of those treated with CVPP (CVPP versus MOPP, $p = 0.01$). Moreover, these effects occurred half as

TABLE V. Comparability of Induction Treatment Groups CALGB 7251[a]

Induction program	MOPP	MVPP	COPP	CVPP
Number entered	141	139	139	147
Number evaluable	138	124	133	137
Males (%)	59	67	57	64
Median age	30	36	36	27
Mean age	37	38	41	35
% 40 yrs. or over	40	39	45	36
Institutional histology (%)				
Lymphocyte predominant	7	12	6	8
Nodular sclerosis	41	32	33	32
Mixed cellularity	39	46	52	45
Lymphocyte depleted	13	10	9	15
Prior therapy (%)				
Previously untreated	52	55	54	54
Radiotherapy only	25	26	28	29
Chemotherapy only	7	4	5	2
Chemotherapy and radiotherapy	16	15	13	15
Prior splenectomy (%)	36	44	34	38
Clinical stage III (%)	29	33	39	32
Involved site (%)				
Liver	30	39	36	38
Lung	27	19	18	26
Bone marrow	16	16	9	12

[a] From Cooper et al. (1980).

frequently with the nitrosourea combinations as compared to those containing mechlorethamine. Less neurotoxicity was also observed in the vinblastine programs compared to those containing vincristine (CVPP and MVPP versus MOPP and COPP, $p = 0.01$).

The difference between complete response rates observed with MOPP (57%)

TABLE VI. Frequency of Complete Remissions (CR) in Advanced Hodgkin's Disease According to Induction Therapy and Prior Treatment Status in CALGB 7251[a]

Induction program	Total patients	All patients CR (%)	No prior therapy No.	No prior therapy CR (%)	Prior radiotherapy only No.	Prior radiotherapy only CR (%)	Prior chemotherapy No.	Prior chemotherapy CR (%)
MOPP	138	57	70	54	33	64	32	63
MVPP	125	66	67	73	32	72	23	39
COPP	132	65	71	56	37	86	23	61
CVPP	137	69	74	62	39	85	24	63

[a] From Cooper et al. (1980).

and CVPP (69%) was significant ($p = 0.05$) and mainly attributable to the response attained in previously untreated patients (54% versus 62%) and patients who had relapsed from prior radiotherapy without chemotherapy (64% versus 85%). The vinblastine-containing programs (i.e., CVPP and MVPP) proved to

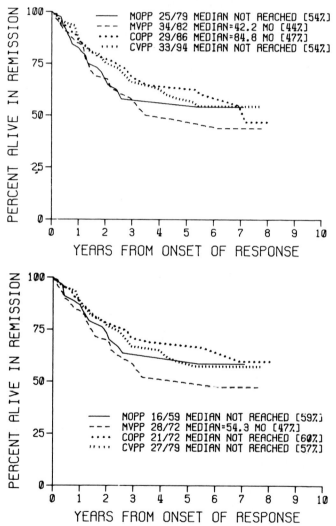

Fig. 2. The duration of complete response for MOPP, MVPP, COPP, and CVPP in CALGB 7251 (Cooper *et al.*, 1980). Relapsed patients, the total number of patients initially at risk and the median response for each program are shown below the figure. The study was closed to new patient entry in August 1975. Data are updated to October 1979. The duration of complete response is shown for all patients (A), patients with no prior chemotherapy (B), and patients with prior radiotherapy only (C). No statistically significant differences in duration of response were observed for any subset of patients for any of the treatment programs.

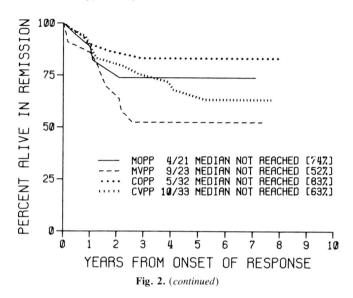

Fig. 2. (*continued*)

have significantly higher complete response frequencies than the vincristine combinations (MOPP and COPP) in previously untreated patients (65% versus 55%, $p = 0.04$). Whereas MVPP proved to have the highest induction rate in untreated patients, it was not statistically superior to CVPP ($p = 0.17$). In those patients who had failed prior radiotherapy but had not received chemotherapy, the CCNU-containing programs induced more frequent complete remissions than those containing mechlorethamine (85% versus 68%, $p = 0.007$).

Comparison of the duration of the initial complete response for each treatment program for patients with no prior chemotherapy or those failing only prior radiotherapy is shown in Fig. 2. The duration of complete response for CCNU-treated patients compared to mechlorethamine-treated patients is shown in Fig. 3 for all evaluable patients, previously untreated patients, and patients failing only prior radiotherapy.

A statistically significant difference in the duration of complete response when CCNU-containing programs were compared to those containing mechlorethamine was previously noted ($p = 0.025$) (Cooper *et al.*, 1980). Most recent evaluation of these data show these differences to be no longer significant whether analyzed for all patients ($p = 0.08$) or for patients relapsing after prior radiotherapy only ($p = 0.13$).

Although patients were not eligible for study if three or more of the drugs to be used in their therapy had been previously given, the comparative data have not as yet been analyzed for the nature of the prior chemotherapy in this cohort of complete responders.

It must be noted that both complete and partial responders were randomized to maintenance therapy consisting of either biweekly vinblastine or biweekly vinblastine with reinforcement (consisting of a single course of induction chemotherapy every 2 months) for 2 years. The 8-week cycle of therapy was repeated for 2 years (Cooper *et al.*, 1980). There was also a randomization to immunotherapy with the methanol extracted residue of BCG (MER) for patients in complete remission for 2½ years. No salutary effect was observed from the administration of MER.

C. Studies Open to New Patient Entry

Various randomized studies employing nitrosoureas in combination therapy with or without radiotherapy continue to accrue patients and remain coded. The

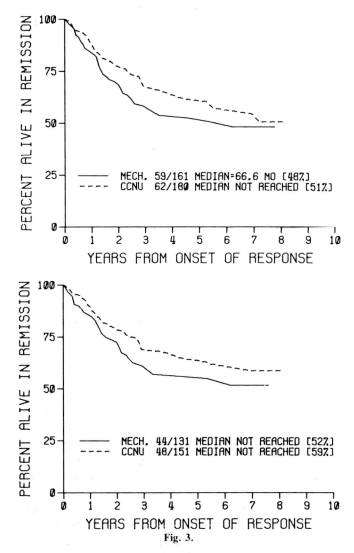

Fig. 3.

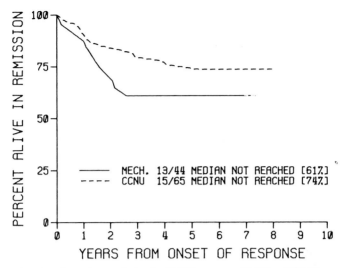

Fig. 3. The duration of complete remission for CCNU programs (CVPP and COPP) as compared to mechlorethamine (Mech.) programs (MVPP and MOPP) in CALGB 7251. Data are updated to October 1979 and are provided for all evaluable patients (A), patients with no prior chemotherapy (B) and patients with prior radiotherapy only (C) (Cooper *et al.*, 1980).

treatment of previously untreated III_A and III_B Hodgkin's disease is addressed in CALGB 7451, which compares (*a*) total nodal radiotherapy; (*b*) chemotherapy with BOPP (6 cycles); (*c*) total nodal radiotherapy followed by BOPP (6 cycles); (*d*) chemotherapy with BOPP (6 cycles) followed by total nodal radiotherapy (Stutzman *et al.*, 1979). Maintenance therapy with chlorambucil for patients treated with BOPP was discontinued when our analysis failed to show a significant impact of such therapy. Randomization to chemotherapy followed by radiotherapy was stopped after entry of 51 patients as a result of unacceptable toxicity during the irradiation phase. In addition, interim analysis indicated that fewer patients with III_B disease finished the prescribed course of therapy when they were initially treated with radiotherapy. Accordingly, patients with III_B disease are not included in CALGB 7551 (see below).

Over 200 patients have been entered on study 7451. Complete response frequencies range from 75% to 85% of evaluable asymptomatic patients. For symptomatic patients, complete responses are seen in 60% to 80% of evaluable patients. The results are being analyzed by various pretreatment characteristics including differences that may exist between the responses observed in patients with disease involving the spleen and/or splenic, celiac, or portal nodes (III_1) as compared to "lower abdominal nodes" (i.e., lower para-aortic, iliac, or mesenteric nodes) (III_2).

Combined chemotherapy and radiotherapy is also being tested in stage III_B and IV previously untreated Hodgkin's disease in CALGB 7551 (Rafla, 1979; Coleman *et al.*, 1979). Two chemotherapy programs with either 6 courses of CVPP or 12 courses of monthly CVPP therapy are compared to combined modality therapy with irradiation and drugs. The two combined modality programs consist of six courses of treatment with CVPP, an 8-week rest period followed by involved field radiotherapy; or three courses of therapy with CVPP, a 4-week rest, involved field radiotherapy, an 8-week rest period, and a concluding three courses of CVPP. Because of hematologic toxicity the initial rest period was, of necessity, lengthened to 8 weeks.

Complete responses have been observed in 51–68% of patients with predominantly stage IV disease. Approximately 50 evaluable patients have concluded therapy in each group. Between 60% and 75% of these patients continue in complete remissions at 3 years. No significant differences among these treatment programs has as yet emerged.

The response to therapy of patients with stage III$_B$ or IV Hodgkin's disease failing prior radiotherapy or chemotherapy are being evaluated on CALGB 7552 (Rafla, 1979 ; Coleman *et al.*, 1979). The study compared 12 monthly cycles of therapy with CVPP to 12 monthly cycles of therapy with BAVS. A third program alternates therapy with CVPP and BAVS for a total of 12 cycles. Initially, half the patients were randomized to receive immunotherapy with MER. Immunotherapy has been discontinued since, at best, it provided no augmentation in therapeusis (Vinciguerra *et al.*, 1980).

Complete response rates in patients failing prior radiotherapy currently range from 70 to 82%. In the patients receiving any prior chemotherapy with or without radiotherapy, complete response rates vary from 20 to 40%. The durability of the complete response in all treatment programs is clearly superior in the patients failing only prior radiotherapy. Statistically significant differences in duration of complete remission have not yet emerged between the treatment programs.

Combined radiotherapy and chemotherapy with CVPP is being compared to therapy with CVPP in a group of fully staged patients with stage I and II Hodgkin's disease who are considered less likely to achieve long-term survival when treated by radiation alone. Included in the poor risk group are patients with mixed cellularity or lymphocyte-depleted histology, B symptoms, age 40 or older, or a mediastinal mass greater than 35% of the transverse thoracic diameter (at T$_5$–T$_6$). Involved field radiotherapy followed in 4 weeks by six cycles of therapy with CVPP is being compared to chemotherapy alone with six cycles of CVPP (CALGB 7751). Approximately 50 patients have been entered to date. It is too early to report results.

IV. DISCUSSION

A variety of questions relating to the use of nitrosoureas in the therapy of Hodgkin's disease have been addressed by the studies of CALGB. Efficacy trials employing nitrosoureas as the sole antineoplastic agent (i.e., CALGB 6953, 7252) have indicated the superiority of CCNU over methyl CCNU and BCNU as single agents in the therapy of patients with previously treated Hodgkin's disease. The design of a second group of studies compared the activity of CCNU and BCNU in combination with a vinca alkaloid, a corticosteroid, and procarbazine to the latter group of agents in combination with mechlorethamine (MOPP versus BOPP in CALGB 6951 and MOPP versus CCVP versus COPP versus MVPP in CALGB 7251). The symmetric design of the latter study allows assessment of the activity of the nitrosourea as compared to mechlorethamine *in combination* with a vinca alkaloid, corticosteroid, and procarbazine. Similar comparisons may be made of the activity of the vinca alkaloids in combination. No direct, randomized comparison has been made between BCNU (BOPP) and the CCNU combinations (CVPP and COPP).

These studies differ in concept from CALGB 7552 wherein combination chemotherapy with CVPP is compared to therapy with BAVS or with alternating combinations of CVPP and BAVS. Assessment of the contribution of streptozotocin or any of the components of the alternating combination to the activity of BAVS cannot be made nor can it be taken as having been previously established for the combination when used in sequence. However, the exclusion of an alkylating agent from a combination including vincristine, prednisone, and procarbazine (OPP in CALGB 6951) is associated with a significant fall in the rate of complete remission, and is independent of whether mechlorethamine or a nitrosourea is employed. Similarly, the poor activity displayed by BOP establishes the important therapeutic combination of procarbazine in these combinations.

The curative potential of therapy with MOPP in advanced Hodgkin's disease has been amply demonstrated (DeVita *et al.,* 1980; Coltman, 1980). The complete remission rates obtained have ranged from just over 50 to 80% of patients treated (DeVita *et al.,* 1980; Coltman, 1980). A continuing complete remission is observed in approximately two-thirds of patients with well-scrutinized and reevaluated complete responses to therapy. Relapse is uncommon after approximately 4–5 years of complete response. These data have been shown to be affected by a variety of demographic factors, pretreatment variables, and symptom status, as well as physician and patient compliance to the rigors of dosing and scheduling. Due to the success of therapy with MOPP over the past 10 years, the evaluation of new or alternate treatment programs demands that the durability of the complete response produced must be evaluated together with the frequency of complete response. This can best be accomplished by randomized controlled trials employing comparable treatment groups. The results of the testing of the BOPP program satisfy these criteria. A comparably high complete remission rate and continuing complete response rate was achieved with BOPP as with MOPP in a study where the minimum follow-up time for patients in complete response was 6½ years. Similarly, the CCNU-containing programs, CVPP and COPP (CALGB 7251 closed to new patient entry in 1975), give every indication of being as efficacious as similar programs containing mechlorethamine.

In patients with Hodgkin's disease who have failed prior radiotherapy but who have had no prior chemotherapy, a 63% complete remission rate was achieved by BOPP and a 68% remission rate by MOPP in CALGB 6951. However, in CALGB 7251, CCNU-containing combinations (i.e., COPP, CVPP) were significantly more efficacious in inducing complete remission after failure from radiation therapy than were mustard-containing combinations (MOPP, MVPP). There appears to be little difference in the hematologic toxicities, but there is a reduction in nausea and vomiting when CCNU is substituted for mustard. Thus the possibility that nitrosourea combinations may offer more effective salvage therapy after radiotherapeutic failure bears additional investigation. It is noteworthy that patients failing prior radiotherapy may represent an even more favorable prognostic group than previously untreated patients. A similar suggestion has been made in patients treated with MOPP (DeVita *et al.*, 1980).

A number of active studies assess the relative efficacy of nitrosourea combinations combined with radiotherapy as compared to therapy with radiation or with drugs alone. The timing of radiation and combined nitrosourea therapy was designed as a consequence of the rather extensive experience with MOPP and radiation therapy. In CALGB 7451, the administration of six cycles of BOPP before total nodal irradiation resulted in severe and life-threatening hematologic toxicity that required cessation of radiotherapy. The alternate program of radiotherapy followed by chemotherapy is moderately toxic but manageable.

In CALGB 7551 radiotherapy is sandwiched between courses of chemotherapy with CVPP (Rafla, 1979). It was found that the initiation of radiotherapy in less than 7 weeks following the conclusion of chemotherapy resulted in 8% life-threatening leukopenia and 16% life-threatening thrombocytopenia. When the rest prior to radiotherapy was extended to greater than 7 weeks, no life-threatening toxicity resulted.

Nonlethal toxicities in potentially curative programs for the treatment of fatal diseases take on major importance when they affect the frequency of complete response, and ultimately the rate of cure. These effects may be mediated in part by physician and patient compliance and attenuation of the drug schedule. In the programs described, the nitrosoureas are administered once monthly and produce less nausea and vomiting than programs employing twice monthly mechlorethamine. The better gastrointestinal tolerance of the nitrosourea combinations at equal therapeutic gain may thus aid in their acceptance.

A quantum leap was made in the therapy of advanced Hodgkin's disease with the introduction of MOPP therapy. Yet the curability of the disease with some 6–8 monthly courses of therapy remains between 30–50% of patients treated. In addition, for patients relapsing from MOPP-induced complete remissions, the continued sensitivity to reinduction with MOPP (Fisher *et al.*, 1979) argues against 6–8 months of therapy as being adequate treatment for those patients who appear to be quite sensitive to the drug combination. The failure of more prolonged therapy with MOPP to improve the percentage of patients continuing in long-term unmaintained complete response has been used to suggest an acquired kinetic resistance of the neoplastic cell. A late intensification of therapy has been suggested as a means of improving the durability of the complete response (Norton and Simon, 1977). In addition, alternation of drug combinations that are designed in a manner thought to maximize the metabolic attack on the tumor cell and minimize the development of shared drug resistance represents a promising approach (Santoro *et al.*, 1980). Such a strategy is being tested in CALGB 7552. Moreover, a therapeutic advantage may be gained for some patients by combining drugs and radiation therapy. Thus, whether as the sole modality of therapy or in combination with other treatment modalities, the highly active drug combinations containing nitrosoureas provide additional flexibility in chemotherapy, and will play a significant role in the continued approach to the maximal curability of advanced Hodgkin's disease.

ACKNOWLEDGMENTS

The irreplaceable participation of the member institutions, study chairman, and individual investigators of CALGB is herewith acknowledged. Special thanks to Ms. L. P. Glowienka for statistical help.

REFERENCES

Coleman, M., Vinciguerra, V. P., Rafla, S., and Stutzman, L. (1979). *Proc. Am. Soc. Clin. Oncol.* 20, 412, 428.

Coltman, C. A. Jr. (1980). *Semin. Oncol.* 7, 155–173.

Cooper, M. R., Pajak, T. F., Nissen, N. I., Stutzman, L., Brunner, K., Cuttner, J., Falkson, G., Grunwald, H., Bank, A., Leone, L., Seligman, B. R., Silver, R. T., Weiss, R. B., Haurani, F., Blom, J., Spurr, C. L., Glidewell, O. J., Gottlieb, A. J., and Holland, J. F. (1980). *Cancer 46*, 654–662.

DeVita, V. T., Serpick, A. A., and Carbone, P. P. (1970). *Ann. Intern. Med. 73*, 881–895.

DeVita, V. T., Simon, R. M., Hubbard, S. M., Young, R. C., Berard, C. W., Moxley, J. H., Frei, E., Carbone, P. P., and Canellos, G. (1980). *Ann. Intern. Med. 92*, 587–595.

Fisher, R. I., DeVita, V. T., Hubbard, S. B., Simon, R., and Young, R. C. (1979). *Ann. Intern. Med. 90*, 761–763.

Hansen, H. H., Selawry, O. S., Pajak, T. F., Spurr, C. L., Falkson, G., Brunner, K., Cuttner, J., Nissen, N. I., and Holland, J. F. (1981). *Cancer 47*, 14–18.

Maurice, P., Glidewell, O., Jacquillat, C., Silver, R. T., Carey, R., TenPas, A., Cornell, C. J., Burningham, R. A., Nissen, N. I., and Holland, J. F. (1978). *Cancer 41*, 1658–1683.

Nissen, N. I., Pajak, T. F., Glidewell, O., Pedersen-Bjergaard, J., Stutzman, L., Falkson, G., Cuttner, J., Blom J., Leone, L., Sawitsky, A., Coleman, M., Haurani, F., Spurr, C. L., Harley, J. B., Seligman, B., Cornell, C., Henry, P., Senn, H., Brunner, K., Martz, G., Maurice, P., Bank, A., Shapiro, L., James, G. W., and Holland, J. F. (1979). *Cancer 43*, 31–40.

Norton, L., and Simon, R. (1977). *Cancer Treat. Rep. 61*, 1307–1317.

Rafla, S. (1979). *Int. J. Radiat. Biol. 140*, Supplement 2.

Santoro, A., Bonadonna, Bonfante, V., and Valgussa, P. (1980). *Proc. Amer. Soc. Clin. Oncol. 21*, 470.

Selawry, O. S., and Hansen, H. H. (1972). *Proc. Am. Assoc. Cancer Res. 13*, 46.

Stutzman, L., Nisce, L., and Friedman, M. (1979). *Proc. Am. Assoc. Clin. Oncol. 20*, 407.

Vinciguerra, V., Coleman, M., Pajak, T. F., Rafla, S., Stutzman, L., and Nissen, N. I. (1980). *Cancer Clin. Trials.* In press.

Chapter 15

HODGKIN'S DISEASE AND MALIGNANT LYMPHOMAS

Richard A. Gams
John R. Durant

I. INTRODUCTION

It is nearly fifteen years since the Southeastern Cancer Study Group (SECSG) started evaluating nitrosoureas in the management of patients with Hodgkin's disease and non-Hodgkin's lymphoma. These studies were stimulated by reports that these agents showed a broad spectrum of activity in many animal tumor systems as well as by the encouraging results from early clinical trials in man.

Although BCNU was initially used as a single agent (Lessner, 1968) the SECSG soon became interested in utilizing this agent as part of combination regimens. Evidence in animal tumor systems suggested that the nitrosoureas were synergistic with cyclophosphamide (Valeriote *et al.*, 1974). Consequently, the cooperative group developed a sequential series of studies employing the combination of BCNU and cyclophosphamide with other active agents to determine their effectiveness in Hodgkin's disease and non-Hodgkin's lymphomas.

II. MATERIALS AND METHODS

All member institutions of SECSG participated in the various studies described in this chapter. In general, approximately 5% of patients proved to be ineligible because of incorrect histologic diagnosis or inability of the pathology referee to classify the lymphoma. Of the remaining eligible patients, approximately 10% have been considered inevaluable primarily because of major protocol violations or because patients were lost to follow-up. Patients were considered eligible for the various studies if they had a biopsy diagnosis of Hodgkin's disease or non-Hodgkin's lymphoma confirmed by the pathology referees for the SECSG and by members of the pathology repository committee of the lymphoma task force. Excluded from all studies were patients with localized disease (stage I or in some instances stages I and II) and patients without measurable lesions. In the earlier studies patients were admitted regardless of having had major prior therapy. In the later studies, however, patients who had received major prior chemotherapy or radiation therapy in excess of 3000 rads within three or more anatomic areas were considered ineligible.

Remission status was determined according to SECSG criteria as either CR (disappearance of all signs and symptoms of disease) or PR (disappearance of all symptoms and a decrease of 50% or greater in all measurable disease without the appearance of new lesions). All studies abnormal at the start of therapy were completed at the time of final evaluation.

Toxicity was evaluated according to standard SECSG criteria. Survival curves were calculated using the method of Kaplan and Meier (1958), and the method of Gehan (1965) was used for the comparison of the survival curves. Chi-square statistics were used to test the differences in remission rates.

III. RESULTS AND DISCUSSION

A. Hodgkin's Disease

In 1966 the SECSG began a broad phase II trial of BCNU in patients with advanced Hodgkin's disease resistant to other forms of management (Lessner, 1968). A total of 66 evaluable patients with refractory stage IIIB and IV Hodgkin's disease were treated with BCNU given intravenously at doses of 100 mg/m^2 on 2 or 3 consecutive days every 4 weeks. Thirty patients had a useful response with four patients (7%) achieving complete remission. Nausea and vomiting occurred in approximately 25% of patients. Hematologic toxicity (primarily delayed thrombocytopenia and granulocytopenia) was similar to that reported by others. Slightly more than half the patients had significant reductions in the platelet count, and in 68% of these it fell to 50,000 cells/μliter or less.

The characteristics of the patients responding to therapy are shown in Table I. Virtually all of the responders were resistant to nitrogen mustard or cyclophosphamide, clearly indicating that BCNU represented a compound with major new activity in advanced Hodgkin's disease.

The combination of the above-mentioned toxicity and the inconvenience of intravenous administration prompted the SECSG in 1968 to undertake a study of BCNU given orally in various doses and schedules (Lessner, 1970). BCNU

TABLE I. Patients with Hodgkin's Disease Responding to BCNU

Patient	Age	Sex	Disease stage	Resistant to mustard compounds	Response
1	44	M	IVB	Yes	PR
2	33	M	IVB	Yes	PR
3	33	M	IVB	Yes	PR
4	20	F	IVB	Yes	PR
5	33	M	IIIB	–	CR
6	63	M	IIIB	Yes	PR
7	35	F	IVB	Yes	PR
8	21	M	IVB	Yes	PR
9	30	M	IVB	Yes	CR
10	37	M	IVB	Yes	PR
11	46	M	IVB	Yes	CR
12	49	M	IVB	–	CR
13	16	F	IIIB	Yes	PR
14	36	M	IVB	Yes	PR
15	49	M	IVB	Yes	PR
16	21	F	IVB	Yes	PR
17	22	F	IVB	Yes	NR

given by this route produced dose-limiting gastrointestinal toxicity with very little effect on the bone marrow. In 26 patients no CRs and only five PRs were obtained. The SECSG concluded that BCNU given orally was impractical and abandoned further studies with the oral form of this agent.

Based on the evidence that BCNU was synergistic with cyclophosphamide in animal tumor systems (Valeriote et al., 1974), the SECSG then undertook a series of studies exploring BCNU in various combinations for the management of advanced Hodgkin's disease (Durant and Lessner, 1973). After preliminary dose-ranging studies were performed, the three drug combinations shown in

TABLE II. Four-Drug BCNU Combinations in Advanced Hodgkin's Disease

Drugs	Dose mg/m^2	Number	CR	PR
BCNU (iv)	100 q 28 d	31	9	14
CTX (iv)	600 q 28 d			
VCR (iv)	1 q 14 d			
Procarb (po)	100 × 10 d			
BCNU (iv)	100 q 28 d	8	3	2
VCR (iv)	1 q 14 d			
VLB (iv)	5 q 28 d			
Procarb (po)	100 × 10 d			
BCNU (iv)	100 q 28 d	22	9	8
CTX (iv)	600 q 28 d			
VLB (iv)	5 q 28 d			
Procarb (po)	100 × 10 d			

Table II were investigated. Each of the three four-drug combinations produced useful responses in the majority of patients who were refactory to other forms of chemotherapy. Nevertheless, in the regimen in which vincristine was substituted for cyclophosphamide, a neuromyalgic syndrome that was often prolonged and severe occurred in several patients. Since this regimen did not produce a superior response rate and did not contain cyclophosphamide, further testing of this regimen was discontinued after only 8 patients were treated. Otherwise, significant but tolerable hematologic and gastrointestinal toxicity was obtained with each regimen, and all had comparable response rates.

These results encourage the SECSG to develop a regimen for advanced Hodgkin's disease that might prove superior to the standard MOPP regimen by producing a larger number of complete responses or reducing the toxicity. To this end the group selected the regimen of BCNU, cyclophosphamide, vinblastine, and procarbazine because it differed from MOPP regimens including vincristine and required only one visit per month. In addition, this regimen produced very little peripheral neuropathy. To these drugs was added prednisone at a dose of 60 mg/m^2 per day for 10 days, to be administered in six monthly cycles.

The broad phase III study of the five-drug regimen (BVCPP) was undertaken to determine the complete response rate and toxicity and to evaluate the need for subsequent maintenance chemotherapy. All patients achieving a complete response at the end of six monthly cycles of BVCPP were randomized to receive (*a*) six further cycles of BVCPP; (*b*) six monthly cycles of MOPP; (*c*) no further

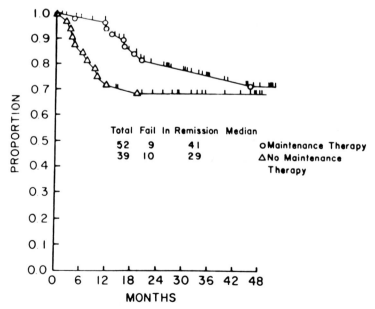

Fig. 1. Disease-free survival of patients with advanced Hodgkin's Disease who achieved CR with BVCPP (maintenance therapy versus none $p = 0.028$).

therapy. The results of this study (Durant *et al.*, 1978) confirmed its value as a viable alternative to the MOPP regimen.

From July 1971 to November 1975, 324 eligible and evaluable patients with advanced Hodgkin's disease (stages IIIB and IV) were admitted for treatment with BVCPP. Of the 180 patients (56%) who had had no prior therapy 122 (68%) achieved complete remission and were randomized to one of the two maintenance arms or to no further therapy.

Figure 1 shows a superior disease-free survival for those patients who completed maintenance therapy according to protocol compared to those randomized to no further therapy. Figure 2 shows a superior survival for the maintained patients. Since these results were in conflict with those reported by other groups, multivariate analysis was performed to assure the comparability of the maintained and nonmaintained patients. When this analysis was performed, maintenance treatment was not identified as an important determination of either duration of response or survival. Instead it was found that a higher proportion of patients who received maintenance therapy were under age 40 (49% maintained versus 35% unmaintained) and a greater percentage had favorable histology (67% maintained versus 50% unmaintained). Though neither of these differences were statistically significant, multivariate analysis indicated that they were very important and resulted in the differences in remission duration and survival that would otherwise have falsely been attributed to maintenance therapy. Consequently, it was concluded that maintenance therapy beyond the achievement of

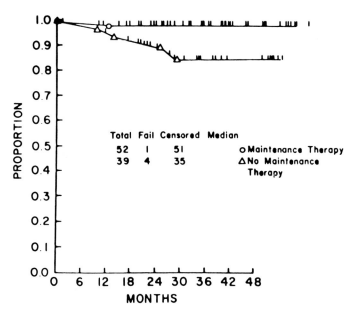

Fig. 2. Survival of patients with advanced Hodgkin's Disease who achieved CR with BVCPP (maintenance therapy versus none $p = 0.037$).

complete remission added no additional benefit to disease-free survival or survival.

In patients who had received major prior chemotherapy alone or in addition to radiation therapy, only 19 of 67 (28%) achieved complete remission. In these patients as well, there were no differences evident in the duration of remission or survival when those randomized to one or the other of maintenance arms were compared to those receiving no further therapy.

These results clearly indicated that BCVPP was an effective alternative to MOPP in the management of advanced Hodgkin's disease. Complete response rates and duration of response were virtually identical to those seen with MOPP. BCVPP required fewer outpatient visits, and gastrointestinal side effects were felt to be less severe.

Encouraged by these results, we undertook a study to explore the possibility that the addition of more non–cross-resistant agents to the BVCPP regimen might improve response rates and survival (Gams et al., 1979). Based on reports that low-dose bleomycin improved results with the MOPP regimen, bleomycin was added to the BVCPP regimen at a dosage of 2 mg/m^2 on days 1 and 8. Patients were randomized between this six-drug combination (BVCPP–Bleo) given monthly for 6 months and the same combination of agents alternating with adriamycin (60 mg/m^2 on day 1, DTIC 625 mg/m^2 on days 1 and 8, and bleomycin 2 mg/m^2 on days 1 and 8). Thus, patients randomized to the alternating arm would receive BVCPP–Bleo on odd numbered months and adriamycin, DTIC, and bleomycin in even numbered months for a total of 6 months of treatment.

Multivariate analysis indicated that patients entered on this latter regimen were strictly comparable to those patients with advanced Hodgkin's disease treated with the five-drug combination. The toxicity of the various regimens is shown in Table III. It can be seen that hematologic toxicity was not significantly

TABLE III. Hematologic Toxicity of BVCPP, BVCPP-Bleo (Cont.) or BVCPP Alternating with Adria, DTIC-Bleo (Alt.)

	SECSG 340 BVCPP	75 HD 301 Alt.	75 HD 301 Cont.
Hemoglobin			
Mild-mod.	41 (21%)	12 (21%)	7 (13%)
Severe	7 (2%)	0	3 (5%)
Life threat.	0	0	0
Granulocyte			
Mild-mod.	89 (26%)	19 (34%)	8 (15%)
Severe	69 (20%)	12 (21%)	14 (27%)
Life threat.	25 (7%)	1 (1%)	3 (5%)
Platelet			
Mild-mod.	100 (29%)	20 (36%)	8 (15%)
Severe	30 (9%)	3 (5%)	4 (7%)
Life threat.	7 (7%)	2 (3%)	2 (3%)

TABLE IV. Response to BVCPP, BVCPP-Bleo (Cont.) and
BVCPP-Bleo Alternating with Adria-DTIC-Bleo (Alt.) in Patients with
Advanced Hodgkin's Disease

| | SECSG 340 | 75 HD 301 | |
	BVCPP	Alt.	Cont.
CR	127 (68%)	31 (65%)	32 (78%)
PR	19 (10%)	7 (15%)	4 (10%)
No prog.	6 (3%)	6 (12%)	0 (0%)
Worse	19 (10%)	4 (8%)	3 (7%)
Dead	17 (9%)	0 (0%)	2 (5%)
Total	188	48	41

different among the five-drug regimen, the six-drug regimen, or the alternating combinations.

Table IV shows the response rates to the various treatment arms. Although there may be some advantage to adding bleomycin to the five-drug regimen, the results are not significantly different. Furthermore, the addition of alternating non-cross-resistant combinations at least in this study conferred no advantage.

This lack of difference in response rates among the various regimens is reflected in the survival curve shown in Fig. 3. Consequently, it must be concluded that despite their potential value as salvage regimens, the addition of new non-cross-resistant agents to the basic five-drug BCVPP combination offers no advantage.

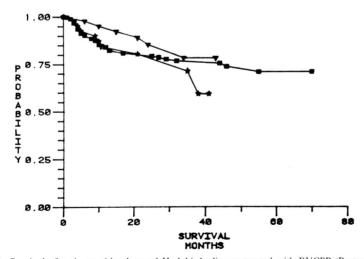

Fig. 3. Survival of patients with advanced Hodgkin's disease treated with BVCPP (Protocol 340, ■——■) or BVCPP-Bleo (protocol 366 Cont., ★——★) or BVCPP-Bleo/Adria-DTIC-Bleo (protocol 366 Alto, ▼——▼).

B. Non-Hodgkin's Lymphomas

During the initial studies with BCNU as a single agent (Lessner, 1968), there were 20 patients entered with refractory non-Hodgkin's lymphomas. Of these, 5 were found to have useful responses. Based on this evidence of activity and the considerations indicated above in the section on Hodgkin's disease, the SECSG explored combination chemotherapy including BCNU in non-Hodgkin's lymphomas. In the first of these studies BCNU was combined with cyclophosphamide, vincristine, and prednisone in the management of patients with diffuse "histiocytic" lymphomas (DHL) (Durant *et al.*, 1975). Of 28 previously untreated patients with DHL, 14 (50%) achieved a complete remission, and an additional 7 (25%) had a good partial response for an overall remission rate of 75%. The median survival for previously untreated patients with diffuse histiocytic lymphoma who achieved a complete remission was nearly 2 years. The median duration of unmaintained remission was 34.9 weeks. These results

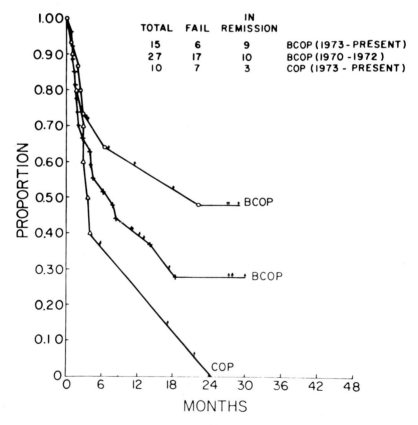

Fig. 4. Disease-free survival of patients with diffuse histiocytic lymphoma who achieved CR with BCOP or COP.

suggested that BCNU in combination with other agents, especially cyclophosphamide, could produce a major impact in non-Hodgkin's lymphomas.

The SECSG then undertook a phase III study which in part compared BCNU,
cyclophosphamide, vincristine, and prednisone (BCOP) with cyclophosphamide,
vincristine, and predisone (COP) (Durant *et al.*, 1977). It was found that patients
could be divided into good (nodular) and poor (diffuse) prognostic categories,
but within each of these categories BCOP and COP gave virtually identical
response, durations of response, and survival rates. The sole exception to this
was the category of diffuse histiocytic lymphoma.

In this latter category, 10 of 29 patients (34%) initially started on BCOP and
either maintained on it or dropped because of failure to respond achieved a CR at
6 months whereas only 3 of 23 similar patients (13%) treated with COP did so.
This difference is reflected in the remission duration curve shown in Fig. 4 and
survival curve shown in Fig. 5. There were too few long-term survivors treated
with COP to be included in Fig. 5. These results once again demonstrated the
utility of BCOP in diffuse histiocytic lymphoma.

In the current SECSG study, BCOP is being compared directly with the

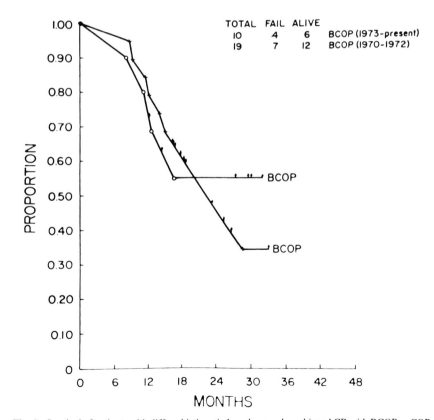

Fig. 5. Survival of patients with diffuse histiocytic lymphoma who achieved CR with BCOP or COP.

combination of cytoxan, adriamycin, vincristine, and prednisone (CHOP) in non-Hodgkin lymphomas with poor histology. Insufficient numbers of patients have been accrued in each histologic category to make any reliable statement about the comparability of the two regimens.

IV. CONCLUSION

These studies reflect the ongoing commitment of the SECSG to exploring the utility of nitrosoureas in Hodgkin's disease and non-Hodgkin lymphomas. BCNU in combination with other agents has proven effective in these circumstances and represents a useful alternative to other active drug combinations.

REFERENCES

Durant, J. R., and Lessner, H. E. (1973). *Cancer 32*, 277–285.
Durant, J. R., Loeb, V., Dorfman, R., Chan, Y-K. (1975). *Cancer 36*, 1936–1944.
Durant, J. R., Gams, R. A., Bartolucci, A. A., and Dorfman, R. F. (1977). *Cancer Treat. Rep. 61*, 1085–1096.
Durant, J. R., Gams, R. A., Velez-Garcia, E., Bartolucci, A., Wirtshafter, D., and Dorfman, R. (1978). *Cancer 42*, 2101–2109.
Gams, R. A., Durant, J. R., Omura, G. A. (1979). *Blood 54*, 187a.
Gehan, E. A. (1965). *Biometrica 52*, 203–233.
Kaplan, E. L., and Meier, P. (1958). *J. Am. Stat. Assoc. 53*, 457–481.
Lessner, H. E. (1968). *Cancer 22*, 451–456.
Lessner, H. (1970). *Proc. Am. Assoc. Cancer Res. 11*, 48.
Valeriote, F. A., Bruce, W. R., and Meeker, B. E. (1974). *Cancer Chemother. Rep. 58*, 407–411.

Chapter 16

CCNU IN COMBINATION CHEMOTHERAPY FOR SMALL-CELL LUNG CANCER

Robert L. Comis
Sandra J. Ginsberg
Thomas F. Pajak
L. Herbert Maurer

I. INTRODUCTION AND BACKGROUND

Small-cell anaplastic lung cancer is an extremely virulent disease that is highly responsive to combination chemotherapy. A variety of chemotherapeutic agents are capable of inducing objective responses when they are used as single agents (Table I). Included in the list of agents that have demonstrable activity is CCNU, the most extensively studied nitrosourea, which has a 14% objective response rate (Livingston, 1978). The other nitrosoureas, BCNU and methyl CCNU, have been studied in fewer patients, and both have shown response rates of 26% (Broder et al., 1976).

Although there appears to be a low order of activity for CCNU when cumulative data are analyzed, CCNU has been shown to significantly increase the objective response rate when it is combined with cyclophosphamide. Edmonson et al. (1976) have reported the results of a randomized comparison of the combination of CCNU and cyclophosphamide to cyclophosphamide alone. The objective response rate for the combination of agents was 43% versus 22% for the single drug. The complete response rate was also significantly increased when

TABLE I. Single Agents in Small-Cell Carcinoma[a]

Drug	No. patients	Response (%)	CR (%)
Cyclophosphamide	189	52 (28)	4
Adriamycin	36	11 (31)	–
Methotrexate	73	22 (30)	–
Vincristine	43	18 (42)	7
VP-16	167	75 (45)	9
CCNU	76	11 (14)	4
Hexamethylmelamine	69	21 (30)	7
HN$_2$	55	24 (44)	–

[a] Adapted from Livingston (1978).

the combination was compared to the single agent: 12% versus 3%, respectively. Twenty percent of the patients treated were alive at 1 year in the combination therapy group as opposed to 15% of those treated with cyclophosphamide alone. These observations have led to a variety of combination chemotherapy programs that have included CCNU and cyclophosphamide combined with other agents.

Extensive reviews of the combination chemotherapy of small-cell anaplastic lung cancer are available (Broder *et al.*, 1976; Livingston, 1978). The purpose of this report is to review the experience of both the SUNY Upstate Medical Center and the Cancer and Acute Leukemia Group B relative to the use of CCNU in combination chemotherapy programs.

II. MATERIALS AND METHODS

We have recently reported on the long-term results of a series of studies in 72 patients treated at the SUNY Upstate Medical Center (King *et al.*, 1977; Ginsberg *et al.*, 1979). The three treatment programs employed a combination of CCNU, cyclophosphamide, and vincristine (CCV) with irradiation (2400 rads/8 sessions) to primary lesion in all patients (Regimen A); an alternating combination chemotherapy program including hexamethylmelamine, adriamycin, and vincristine alternating with CCV (HAV–CCV, Regimen B), which received a truncated evaluation because of decreased patient compliance secondary to severe nausea, vomiting, and neurological toxicity; and a high-dose CCV–AV program (Regimen C), which alternated the combination of CCV with adriamycin plus vincristine. Radiation therapy to the primary tumor was delivered only in patients with limited disease in the latter two studies. The details of each regimen are presented in Table II.

In 1977, the Cancer and Acute Leukemia Group B (CALGB) initiated two protocols that included CCNU in combination with other agents. The details of the extensive and limited disease studies are presented in Fig. 1. The limited disease study, CALGB 7781, compared MACC plus split-course radiotherapy to

TABLE II. Small-Cell Lung Cancer Studies: SUNY Upstate Medical Center

CCV Regimen (A) 8/74–1/76

	Weeks 0	Weeks 3	Weeks 6	
CCNU 70 mg/m²	X	X	X	
Cyclophosphamide 700 mg/m²	X	X	R	2400 rads/8 sessions
Vincristine 2.0 mg	X	X	T	

HAV–CCV Regimen (B) 1/76–8/76

	0	3	6	9	12	15
Hexamethylmelamine 8 mg/kg/d × 21						
Adriamycin 60 mg/m²		X	X		X	
Vincristine 2 mg		X	X	X	X	
CCNU 70 mg/m²			X			
Cyclophosphamide 700 mg/m²			X	X		

CCV–AV Regimen (C) 8/76–10/77 Limited Disease

	0	3	6	8	11	14	17	20	21
CCNU 100 mg/m²	X			70[a]		X			
Cyclophosphamide 1.0 gm/m²	X	X		700[a]	700[a]	X	X		
Vincristine 2 mg	X	X		X		X	X	X	X
Adriamycin 75 mg/m²								X	X
2250 rads/10 sessions			X		X				

CCV–AV Regimen (C) Extensive Disease

	0	3	6	9	12	21
CCNU 100 mg/m²	X				X	
Cyclophosphamide 1.0 gm/m²	X	X			X	X
Vincristine 2.0 mg	X	X	X	X	X	X
Adriamycin 75 mg/m²			X	X		

[a] mg/m².

SMALL CELL LUNG CANCER PROTOCOL CLB-7781

(Limited Disease)

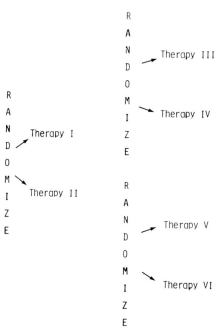

Fig. 1. Small-cell lung cancer protocol CALGB 7781 (limited disease). *Therapy I:* methotrexate (30 mg/m² iv), adriamycin (35 mg/m² iv), CCNU (30 mg/m² po), cyclophosphamide (400 mg/m² iv). The four drugs are given on days 0, 21, 105; adriamycin was not given with the other three drugs on day 84. Radiotherapy (3000 rads over 12 days) was given between days 42 and 84 as a split course with a 2-week and 1-week rest period prior to chemotherapy on day 84. *Therapy II:* Scheduling of chemotherapy and radiotherapy was as in therapy I. Chemotherapy consisted of cyclophosphamide (700 mg/m² iv, on days 0 and 84); CCNU (70 mg/m² po on days 0 and 84); vincristine (1.0 mg/m² iv, on days 0, 21, 84, 105); adriamycin (50 mg/m² iv on day 21 and 105). *Therapy III:* Same drugs and dosages as in therapy I given on day 136 and every 21 days for four doses. *Therapy IV:* Same as therapy III with the addition of MER immunotherapy (200 μg/site intradermally at 5 sites) to each chemotherapy dose. *Therapy V:* Same drugs and dosages as in therapy II given on day 136 (cyclophosphamide, CCNU, vincristine) alternating every 21 days with adriamycin and vincristine for 2 cycles. *Therapy VI:* Same as therapy V with the addition of MER immunotherapy (200 μg/site intradermally at 5 sites) on day 136 and 178. Continue courses of chemotherapy every 5 weeks until maximum dose of adriamycin (365 mg/m² for Therapy I, III, IV and 300 mg/m² for Therapy II, V, VI) is reached. After completing adriamycin continue other drugs every other month for 2 years from onset of protocol.

a variation of our original CCV–AV program plus the same radiotherapy. Elective cranial irradiation was administered to all patients. After induction with combined chemotherapy and radiation therapy, patients were randomly allocated to treatment with the methanol-extracted residue of BCG (MER). The extensive disease study, CALGB 7782 (Fig. 2) compared MACC chemotherapy to MACC plus irradiation to the primary lesion (3000 rads/10 sessions) to CCV–AV therapy. Elective cranial irradiation was not employed.

In all studies, patients were staged prior to treatment with a complete history and physical examination, radionuclide liver, bone, and brain scans, and iliac crest bone marrow aspirates and biopsies. Patients were required to have normal hematologic, hepatic, and renal function prior to entry. If liver abnormalities were found, patients could be treated with appropriate dose adjustments if the abnormalities were secondary to hepatic involvement with tumor. Patients with bone marrow involvement received full doses of therapy.

Limited disease was defined as disease limited to the primary lesion and the ipsilateral or contralateral draining lymph nodes within the chest and supraclavicular areas. Patients with malignant pleural effusions were considered to have extensive disease in CALGB protocols 7781 and 7782. Extensive disease was defined as involvement more estensive than the limited disease criteria.

In CALGB 7781 and 7782, patients were stratified according to performance status prior to randomization. The CALGB studies excluded patients who were bedridden (performance status 4).

Complete response is defined as a complete disappearance of all measurable

SMALL CELL LUNG CANCER PROTOCOL CLB-7782

(Extensive Disease)

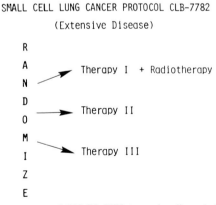

Fig. 2. Small-cell lung cancer protocol CALGB 7782 (extensive disease). *Therapy I:* methotrexate (30 mg/m^2), adriamycin (40 mg/m^2), CCNU (30 mg/m^2), cyclophosphamide (400 mg/m^2) on day 0, 21, 63 (no adriamycin given on this day) and every 21 days thereafter for 5 doses. Radiotherapy (3000 rads over 12 days) was administered to the primary lesion and draining lymph nodes beginning on day 42. Therapy I was then continued until relapse. *Therapy II:* same drugs and dosages as in therapy I given every 21 days for 9 doses and then continued until relapse. *Therapy III:* CCNU (70 mg/m^2 days 0, 64, 168), cyclophosphamide (700 mg/m^2 days 0, 21, 64, 105, 168), vincristine (2 mg iv, day 0 and every 21 days for 9 doses). Therapy III was continued with alternating regimens until relapse.

disease and a resolution of all symptoms related to the disease. *Partial regression* is a $\geqslant 50\%$ decrease in the sum of the products of the perpendicular diameters of all measurable disease for at least 1 month. *Improvement* represents a 25–49% decrease in the sum of the products of the perpendicular diameters of all measurable disease for at least 1 month. *Relapse* and *progression* represent at $\geqslant 25\%$ increase in the size of a previously responding or nonresponding lesion, respectively.

In the CALGB studies, survival was plotted by the life method (Armitage 1973), and differences in survival were evaluated by the Breslow modification of the Kruskall-Wallis test (Breslow, 1980).

III. THERAPEUTIC RESULTS

A. SUNY Upstate Medical Center Studies

The rationale for the sequential development of the three therapeutic programs reflects similar approaches that occurred in the mid 1970s in many institutions. Vincristine was added to the CCNU plus cytoxan regimen because of its activity (Dombernowsky *et al.*, 1976) and lack of significant myelotoxicity. We, as did most others, employed it on an every three week schedule, in spite of the fact that its activity had been demonstrated on a weekly schedule. After completing the CCV program, an attempt was made to devise a unique alternating combination chemotherapy program including hexamethylmelamine, used at a dose and schedule that had demonstrated antitumor activity, and adriamycin. Significant dose reductions of hexamethylmelamine were necessary, and even at 2 mg/kg/d $\times$ 21 days, patient compliance was so poor that the regimen was discontinued. Finally, an attempt was made to increase the dose of CCV, based upon data showing an increased complete response rate with higher doses of cyclophosphamide, and CCNU (Cohen *et al.*, 1977) and to alternate the CCV regimen with adriamycin at 75 mg/m² and vincristine.

Patients were treated with each regimen for a total of 16–18 months, at which time they were extensively restaged, as previously reported (Ginsberg *et al.*, 1979), and therapy was discontinued. The complete response rates for each program and the percentage and number of patients alive and disease free at 16–18 months is presented in Table III. There was no significant difference in complete response among the three regimens.

Of particular interest is the lack of an apparent increase in the complete response rate when the doses of CCNU and cyclophosphamide were increased from 70 and 700 mg/m² to 100 and 1000 mg/m², respectively. We did not conduct a randomized prospective trial of the two dose levels as did the NCI-VA, who compared CCNU and cylophosphamide (combined with methotrexate) at doses of 50 mg/m² and 500 mg/m² to doses of 100 mg and 1000 mg/m², respectively. But it is possible that the asymptote in complete response frequency with cy-

TABLE III. Small-Cell Carcinoma: SUNY Upstate Medical Center
Results of Nitrosourea-Containing Programs

| | Regimen | | | | | |
| | A | | B | | C | |
	Lim.	Ext.	Lim.	Ext.	Lim.	Ext.
Patient no.	20	19	4	6	12	11
% CR	70	21	75	50	64	30
% Alive 16–18 mos. (no.)	20 (4)	5 (1)	50 (2)	0	17 (2)	0

clophosphamide and CCNU may occur at around 700 mg/m^2 and 70 mg/m^2, respectively.

Nine patients were alive and disease free at 16–18 months, representing 13% of the entire group and 22% of the limited disease patients. Only one patient with extensive disease (3%) proven by liver biopsy is a long-term survivor. The fate of these patients is presented in Table IV. One patient died within 4 months of an extensive negative evaluation. A second patient relapsed in the site of her primary lesion and liver at 61 months and died. At autopsy she was found to have an adenocarcinoma. The original diagnosis of an intermediate type small-cell anaplastic carcinoma was corroborated by Dr. Mary Matthews. Two patients are alive with second primary tumors, a metastatic prostate cancer, and a Clark's level III malignant melanoma. Five of the original 72 patients (7%) are alive without cancer from 3.5–5 years, representing 3% and 11% of our extensive disease and limited disease patients, respectively.

These long-term follow-up data corroborate the caution expressed by Brigham et al. (1978) in assuming that patients with disease-free survival times of less than 5 years are cured of their disease. Our data, plus that presented by Bradley et al. (1980), which showed that the majority of long-term survivors post chemotherapy exhibit aneuploidy, indicate that second turmors and leukemia may be a significant problem in this group of long-term surviving patients.

Finally, the problem of local relapse has been of substantial concern, institutionally. Even the most aggressive programs combining radiotherapy with chemotherapy are associated with an extremely high incidence of local relapse within the irradiated intrathoracic tumor volume (Johnson et al., 1978). We have recently reported on the possible role of surgery as the preferred local modality in

TABLE IV. SUNY Upstate Medical Center: Long-Term Results of Small-Cell Lung Cancer Therapy

NED at 16–18 mos.	9	(8 limited, 1 extensive)
Died of small-cell	1	(4 mos.)
Died of cancer	1	(61 mos., adenocarcinoma of lung)
Alive with second cancer	2	(prostate 42 mos., melanoma 48 mos.)
NED	5	(3.5–5 years)

TABLE V. SUNY Upstate Medical Center: Adjuvant Surgery Plus Combination Chemotherapy in Small-Cell Lung Cancer

		Survival (Mos.)		
Stage	Therapy	From initial Tx	NED from final tx	Died with disease
T_1N_0	S $\rightarrow$ CCV	27	18	
T_2N_0	S $\rightarrow$ CCV	18	4	
T_2N_1	S $\rightarrow$ CCV + XRT	62	47	
T_2N_1	S $\rightarrow$ CCV-AV	42	26	
T_2N_1	S $\rightarrow$ CCV-AV	40	25	
T_2N_2	CCV $\rightarrow$ S	19	6	
T_2N_2	CCV $\rightarrow$ XRT $\rightarrow$ S	—	—	8
T_3N_2	CCV $\rightarrow$ XRT $\rightarrow$ S	—	—	10

certain situations (Meyer *et al.*, 1979). Our series has been expanded, and the results of the combination of surgical resection and nitrosourea-containing combination chemotherapy are presented in Table V. All five patients with stage I and II disease are alive and disease free with a median follow-up time of 40 months. Two of the three patients with stage III disease did not respond to chemotherapy preoperatively and died at 8 and 10 months, respectively. The third patient, who developed a partial remission prior to surgical resection, is alive and disease free 19 months after lobectomy, mediastinal node dissection, and 12 months of CCV chemotherapy.

B. CALGB Studies

The schema for the limited disease study, CALGB 7781 (H. Maurer, Chairman) is presented in Fig. 1. The study was initiated in September 1977. The

TABLE VI. CALGB 7781: Comparability of Treatment Groups

	Coded treatment	
	P	Q
Total evaluated	77	82
Percent male	73	72
Mean Age	59	59
% 70+	9	7
Initial performance score		
Percent 0	43	46
Percent 1	30	31
Percent 2	22	18
Percent 3	5	5

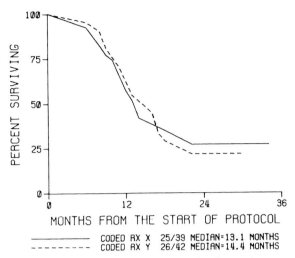

Fig. 3. Treatment programs for limited disease compared; only complete responses shown.

results presented represent an interim analysis with response data as of April 1979 and survival data on the same patients through October 1979. The data are still coded, and the prognostic factors of each group are presented in Table VI.

The complete response frequency is 57% in each group. Nine percent and 11% of patients in each treatment arm have continuing partial remission and are potential complete responders. The median survival for all limited disease patients are 10.6 and 11.7 months, respectively, and approximately 25% of patients developing a complete remission are expected to be alive at 24–36 months (Fig. 3).

Sixty-nine patients were randomized after induction to receive or not receive MER. There was no significant advantage to the addition of MER to the chemotherapeutic regimen. Because of these data, excessive local toxicity, and poor patient and physician compliance, the MER randomization has been discontinued.

At the time of this writing approximately 300 patients have been entered onto this investigation, and the study is nearing completion.

The schema for CALGB protocol 7782 is also presented in Fig. 2. As with the previous study, the treatment arms are still coded. The dates for activation, response, and survival analysis are as stated above. The prognostic factors for each treated group are presented in Table VII.

The overall response frequency is 43, 40, and 36%, respectively. Complete responses have been identified in 10–14% of patients. The median survivals of the treated groups range from 6.6–7.5 months. The effect of response on survival is presented in Fig. 4. It is interesting to note that the quantitative and qualitative characteristics of the actuarial curves of complete responders with extensive disease are quite comparable to those observed in the two regimens for limited disease.

TABLE VII. CALGB 7782: Comparability of Treatment Groups

	Coded Treatment		
	P	Q	R
Total evaluable	55	49	51
Percent male	71	80	76
Mean age	57	59	58
Initial performance score			
Percent 0	26	28	19
Percent 1	40	38	50
Percent 2	21	19	17
Percent 3	13	15	14

IV. TOXICITY

A. SUNY Upstate Medical Center Studies

As noted above, the third study, high dose CCV-AV (Table II), was designed to administer intensive nitrosourea-containing combination chemotherapy with split-course radiation therapy to patients with limited disease and intensive CCV–AV therapy to patients with extensive disease. Figure 5 illustrates the problems encountered in delivering high-dose intermittent nitrosourea therapy. In the first instance delayed and cumulative myelotoxicity lead to continuing dose reductions in limited disease patients treated with radiotherapy. On the other

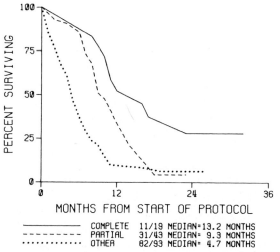

————	COMPLETE	11/19 MEDIAN=13.2 MONTHS
– – – – – –	PARTIAL	31/43 MEDIAN= 9.3 MONTHS
· · · · · · · · · · ·	OTHER	82/93 MEDIAN= 4.7 MONTHS

Fig. 4. Extensive disease; comparison of response types.

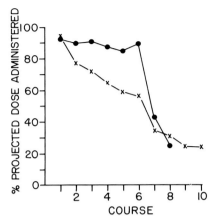

Fig. 5. High-dose CCV-AV therapy; x—x = chemotherapy with split-course radiation therapy; ●—● = chemotherapy without radiation therapy.

hand, patients with extensive disease who received alternating high-dose therapy without radiation received close to full doses for six courses, but ultimately developed dose-limiting cumulative and delayed toxicity.

The cumulative and delayed nature of nitrosourea therapy has been well described (Wasserman *et al.*, 1975), and the results presented above attest to the difficulties in delivering intensive combination chemotherapy that includes the nitrosourea, CCNU, administered on a high-dose intermittent schedule.

B. CALGB Studies

Israel *et al.* (1973) have shown that less myelotoxicity is encountered when the nitrosoureas are administered on a lower dose and more frequent schedule as opposed to a high-dose intermittent schedule. Similar results have been obtained by others. These data led to the development of the MACC chemotherapy program by Chahinian *et al.* (1979) and to the incorporation of this program into both the limited and extensive disease CALGB studies. A recent report (Vogl *et al.*, 1979) has indicated that the MACC regimen is excessively toxic. This has not been the experience in the hundreds of patients with small-cell anaplastic lung cancer treated in CALGB studies.

In the limited disease study, CALGB 7781, although still coded, the incidence of severe and/or life threatening leukopenia and thrombocytopenia was 5% in the MACC-containing regimens with treatment-related mortality in 2%. Since the MACC chemotherapy program is a well-tolerated and effective regimen, it will continue to be used as the control arm in subsequent chemotherapy studies in patients with extensive small-cell anaplastic carcinoma.

V. CONCLUSION

The results of studies completed by the SUNY Upstate Medical Center, Syracuse, and the ongoing studies of the CALGB that employ CCNU in combination chemotherapy programs have been reviewed. The results of these studies indicate that CCNU can be safely and effectively integrated into combination chemotherapy programs. Unfortunately, the results of these programs do not appear to be superior to other non-nitrosourea-containing combination chemotherapy programs. Only the minority of limited-disease patients survive longer than 2 years, and over 90% of extensive-disease patients are dead within 16–18 months. Thus, although a highly treatable disease, small-cell anaplastic lung cancer cannot be considered a routinely curable disease. These findings emphasize the point that the delivery of conventional, noncurative therapies will only provide palliation to the vast majority of patients and that continued intensive clinical investigations into the therapy of this disease should be the goal and the aim of all oncologists. In this way, this highly treatable disease may become a highly curable disease.

REFERENCES

Armitage, P. (1973). *In* "Statistical Methods in Medical Research," pp. 134–135. John Wiley and Sons, New York.

Bradley, E., Matthews, M.J., Cohen, M.H., Fossieck, B., Bunn, P., Ihde, D., and Minna, J. (1980). *Proc. Am. Soc. Clin. Oncol. 21*, 321.

Breslow, N. (1980). *Biometrika 57*, 579–594.

Brigham, P. A., Bunn, P. A., Jr., Minna, J.D., Cohen, M.H., Ihde, D.C., and Shackney, S.E. (1978). *Cancer 42*, 2880–2886.

Broder, L.E., Cohen, M.H., Selawry, O.S., Perlia, C.P., Bennett, J.M., Muggia, F.M., Wampler, G., Brofovsky, H.S., Horton, J., Colsky, J., Mansour, E.G., Creech, R., Stolbach, L., Greenspan, E.M., Levitt, M., Israel, L., Ezdinli, E.Z., and Carbone, P.P. (1976). *Cancer Treat. Rep. 60*, 925–932.

Chahinian, A.P., Mandel, E.M., Holland, J.F., Jaffrey, I.S., and Teirstein, A.S. (1979). *Cancer 43*, 1590–1597.

Cohen, M.II., Creaven, P.J., Fossieck, B.F., Broder, L.E., Selawry, O.S., Johnston, A.V., Williams, C.L., and Minna, J.D. (1977). *Cancer Treat. Rep. 61*, 349–354.

Dombernowsky, P., Hansen, H.H., Sorensen, P.G., and Hainau, B. (1976). *Cancer Treat. Rep. 60*, 239–242.

Edmonson, J.H., Lagokos, S., Selawry, O.S., Perlia, C.P., Bennett, J.M., Muggia, F.M., Wampler, G., Brodovsky, H.S., Horton, J., Colsky, J., Mansour, E.G., Creech, R., Stolbach, L., Greenspan, E.M., Levitt, M., Israel, L., Ezdinli, E.Z. and Carbone, P.P. (1976). *Cancer Treat. Rep. 60*, 925–932.

Ginsberg, S.J., Comis, R.L., Gottlieb, A.J., King, G.B., Goldberg, J., Zamkoff, K., Elbadaivi, A. and Meyer, J.A. (1979). *Cancer Treat. Rep. 63*, 1347–1349.

Israel, L., Chahinian, P., and Accard, J.L. (1973). *Eur. J. Cancer 9*, 799–802.

Johnson, R.B., Brereton, H.D., and Kent, C.H. (1978). *Ann. Thoracic Surg. 25*, 510–515.

King, G., Comis, R., Ginsberg, S., Goldberg, J., Dale, H.T., Brown, J., Dalal, P., Chungtaik, C., and Gottlieb, A.J. (1977). *Radiol. 125*, 529–530.

Livingston, R.B. (1978). *Semin. Oncol. 5*, 299–308.

Meyer, J.A., Comis, R.L., Ginsberg, S.J., Ikins, P.M., Burke, W.A., and Parker, F.B. Jr. (1979). *J. Thoracic Cardiovasc. Surg. 77*, 243–248.

Vogl, S.E., Mehta, C.R., and Cohen, M.H. (1979). *Cancer 44*, 864–868.

Wasserman, T.H., Slavik, M., and Carter, S.K. (1975). *Cancer 36*, 1258–1268.

Chapter 17

POCC VERSUS POCC/VAM THERAPY FOR SMALL-CELL ANAPLASTIC (OAT CELL) LUNG CANCER

Branimir Ivan Sikic
John R. Daniels
Linda Chak
Michael Alexander
Marsha Kohler
Stephen K. Carter

I. INTRODUCTION

The POCC regimen (procarbazine, vincristine, cyclophosphamide, CCNU) was devised in 1972 by collaborative discussions among investigators at Stanford University. The first randomized study of small-cell lung cancer at Stanford compared POCC to COM (cyclophosphamide, vincristine, methotrexate) combination chemotherapy (Alexander et al., 1977). All patients in this study received radiation therapy to all sites of involvement documented at initial staging, but not prophylactic cranial radiation. Median survival of the POCC group was 14 months compared to 11 months for COM ($p = 0.055$). Despite the presence of two drugs, CCNU and procarbazine, which enter the central nervous system (CNS), CNS relapse occurred in 3 of the first 10 patients on POCC, as well as 3 of the first 9 patients treated with COM. The CNS was the initial site of relapse in two of the patients on POCC. Thereafter, so-called prophylactic cranial irradiation became a standard feature of small-cell lung cancer protocols at Stanford and the Northern California Oncology Group (NCOG). Thrombocytopenia was the major dose-limiting toxicity of POCC, resulting in treatment delays in 5 of 11 patients.

In this study, 5 of 8 initial relapses in the chest occurred in previously irradiated sites. Therefore, the second small-cell lung cancer trial at Stanford tested the role of involved field irradiation in extensive disease (Williams *et al.*, 1977a; 1977b). Twenty-seven patients with extensive stage disease (defined as T=3, or N=3, or M=1) were all treated with POCC for three cycles, then randomized to either continued chemotherapy or interruption for involved-field radiotherapy as in the previous study. There was no significant difference in response rates or survival between the 2 groups, with a median survival of 11 months for POCC alone and 9 months for POCC plus involved-field irradiation. Furthermore, involved-field radiotherapy did not alter the pattern of relapse, with 20 of 22 patients initially relapsing in a previously involved site.

In 1977, NCOG initiated protocol #2061 for small-cell lung carcinoma, a phase III randomized trial of POCC chemotherapy versus a seven-drug regimen consisting of induction with 3 cycles of VP-16, adriamycin, and methotrexate (VAM), then alternating cycles of POCC and VAM. This study was closed in September 1979 for extensive disease and in May 1980 for limited disease. Results of patients randomized from October 1977 to October 1979 have been reported (Daniels *et al.*, 1980). The central question posed in this study was whether it was advantageous to use two combinations of drugs early in treatment (alternating regimen) rather than to begin with one combination and use the second combination after the first was no longer effective (sequential treatment).

II. METHODS

The organization, membership, clinical trials procedures, and data management system of the Northern California Oncology Group have been previously described (monograph available on request from the NCOG Operations Office, 1801 Page Mill Road, Palo Alto, CA 94304).

Patients were eligible for the study if they had a histologic diagnosis of small-cell undifferentiated lung carcinoma, Karnofsky performance status greater than or equal to 20%, and no prior chemotherapy or radiotherapy other than emergency irradiation. Excluded were patients over age 70 and those with serious liver or cardiac impairments as indicated by a bilirubin of greater than 1.5 mg%, SGOT greater than twice normal, or a recent history of myocardial infarction or congestive heart failure. Staging work-up included a chest x ray, blood counts and chemistries, liver, bone, and brain scans, and a bone marrow biopsy. Extensive stage disease was defined as T=3 or N=4 or M=1.

Patients were randomized to receive either POCC or the seven-drug regimen, which included three initial cycles of VAM followed by alternating cycles of POCC and VAM. The drug doses and schedules for POCC are listed in Table I.

TABLE I. The POCC Regimen[a]

Procarbazine, 100 mg/m² po daily, d. 2-15
Vincristine, 2.0 mg total iv, d. 1 and 8
Cyclophosphamide, 600 mg/m² iv, d. 1 and 8
CCNU (lomustine), 60 mg/m² po, d. 1

[a] No treatment, days 16-28. New cycle begins on day 29 if blood counts are adequate.

VAM consisted of VP-16, 150 mg/m², adriamycin, 50 mg/m², and methotrexate, 25 mg/m², all given iv on day 1 of a 21-day cycle. A progressive dose attentuation schema was used if the white blood count was lower than 3500/mm³ or the platelets were less than 150,000/mm³ on the day of scheduled treatment (Table II). Chemotherapy could be delayed for 2 or 3 weeks if blood counts did not permit delivery of at least 75% of the calculated drug dosage.

All patients who completed the initial two cycles of POCC or three cycles of VAM received 3000 rads of "prophylactic" cranial irradiation over 2 weeks. Patients with extensive stage disease received no further radiotherapy, whereas patients with limited disease received 5000-5500 rads irradiation to the primary tumor, as well as to the hila, mediastinum, and supraclavicular areas. Patients with extensive disease generally received cranial irradiation without interruption of their cyclical chemotherapy, whereas in limited disease the planned radiotherapy required a 5-week delay in chemotherapy. Additional delays were frequently necessary because of hematologic and other radiation-induced toxicities. Following radiotherapy, patients either resumed POCC or began alternating cycles of POCC and VAM. Patients who failed treatment with POCC alone were to be treated with VAM, thus providing a comparison of "sequential" treatment with seven drugs versus "alternating" seven-drug chemotherapy earlier in the course of treatment. Maximum duration of therapy for patients maintaining a complete remission was 18 months.

Results of the study are analyzed for objective complete and partial response

TABLE II. Dose Modification Parameters for Hematologic Toxicity

	Platelets			
WBC	≥150,000	100,000-149,000	75,000-99,000	<75,000
>3500	100%	75%	50%[a]	0
3000-3499	75%	50%[a]	0	0
2500-2999	50%[a]	0	0	0
<2500	0	0	0	0

[a] Initiation of a cycle may be delayed up to 3 weeks if adjusted doses are less than 75% of calculated doses.

rates, durations of response, and survival time of patients. A complete response (CR) was defined as complete disappearance of previously evident disease, and a partial response (PR) as a greater than 50% decrease in the product of the two largest perpendicular diameters of a tumor mass. The statistical analysis of survival time was patterned after the Cox–Breslow hazard rate regression model for censored survival data with covariates (Cox, 1972; Breslow, 1974).

III. RESULTS

A total of 161 patients were randomized from October 1977 to October 1979. Of these, 25 patients are not evaluable for therapeutic response because of ineligibility or major protocol violations. Therefore, this report presents the results of treatment for 135 patients, of whom 89 had extensive stage disease and 47 had limited disease. Stratification resulted in a satisfactory balance between the two treatment arms for age, sex, stage of disease, Karnofsky performance status, and sites of involvement.

The seven-drug alternating regimen was superior to POCC in prolonging survival of patients with extensive disease, with a median survival in the POCC/VAM arm of 300 days compared to 209 days for POCC ($p < 0.01$) (Fig. 1). Median survival in limited-stage patients was 488 days and 324 days for the seven-drug and four-drug regimens, but because of the small number of patients involved, this difference is not significant ($p > 0.25$).

Median durations of response were also significantly longer for POCC/VAM (6 months) compared to POCC alone (4 months) in patients with extensive disease and were 11 months in both arms for limited-stage patients.

Overall response rates were similar in both arms in extensive stage disease:

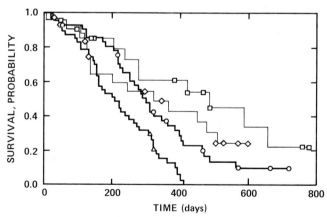

Fig. 1. Kaplan-Meier survival curves for patients treated with POCC versus POCC/VAM. ◇ = limited stage, POCC: □ = limited stage, POCC/VAM: △ = extensive stage, POCC; ○ extensive stage, POCC/VAM.

61% for POCC/VAM and 52% for POCC. However, complete response (CR) rates were higher for the seven-drug alternating regimen: 40% versus 20% ($p < 0.05$).

The design of the study, including three cycles of VAM prior to alternation, permitted a comparison of the relative initial effects of VAM versus POCC. The three-drug and four-drug regimens each resulted in a 20% complete remission rate early on. However, when POCC was added to VAM in the alternating schedule, an additional 20% of patients achieved complete remissions, resulting in the superior 40% CR rate of the seven-drug regimen.

In addition to treatment arm, two other covariates independently correlated with improved survival by Cox analysis in patients with extensive disease: a favorable Karnofsky status ($p = 0.02$) and negative rather than positive bone marrow biopsy ($p < 0.001$). The POCC/VAM regimen was superior to POCC in patients with extensive disease who were analyzed separately on the basis of distant metastases (M=1) as well as in patients with "extensive" intrathoracic disease without distant metastases (M=0) (Table III).

Among patients with limited disease, three patients in each arm are currently off treatment and in a sustained complete remission, for an 18-month estimated disease-free survival rate of 15-20%. Two patients with extensive stage in the POCC/VAM arm are in complete remission 2 years after initiation of treatment.

Seven deaths in this study were attributed to toxicity, including four due to radiation pneumonitis in patients with limited-stage disease. Three patients with extensive-stage disease died with leukopenia and sepsis, two of whom had chemotherapy shortly after emergency radiotherapy. Thus, six of the seven deaths were associated with radiotherapy to involved fields. Hematologic toxicities were comparable for POCC/VAM versus POCC (Table IV). As expected, the increased frequency of vincristine administration resulted in a higher incidence of severe peripheral neuropathy in patients with POCC alone. Involved-field radiotherapy for patients with limited-stage disease resulted in a substantial delay in chemotherapy after the initial three cycles of VAM or two cycles of POCC, with a mean delay of 88 days (range 57-168 days).

Fifteen patients who failed treatment with POCC alone were subsequently treated with VAM. Only one of these patients achieved even a partial response to VAM. Twenty-one other patients who failed POCC refused or were considered inappropriate for further chemotherapy.

TABLE III. Effect of POCC/VAM versus POCC on Median Survival Time in Days

	POCC	POCC/VAM	(p-value)
Limited stage	324	488	($p = 0.28$)
All extensive stage	209	300	($p < 0.001$)
Extensive, M = 0	251	325	($p = 0.02$)
Extensive, M = 1	201	290	($p = 0.02$)

The analysis of sites of relapse in patients who initially responded to therapy is currently ongoing.

Because of the delayed and potentially cumulative hematologic toxicity of the nitrosourea CCNU (lomustine) in the POCC regimen, a detailed analysis was performed of the drug dosage reductions and treatment delays necessitated by low blood counts (Fig. 2 and 3).

Patients treated with POCC alone required progressive drug-dosage reduction over time, so that by the fifth cycle a mean of only 49% of the initial calculated dose was delivered (Fig. 2). By contrast, patients in the alternating POCC/VAM arm received a mean of 67% of their projected doses at the temporally equivalent

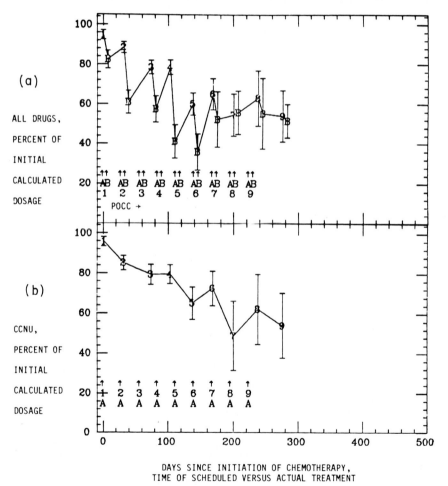

DAYS SINCE INITIATION OF CHEMOTHERAPY,
TIME OF SCHEDULED VERSUS ACTUAL TREATMENT

Fig. 2. A: percentage of total drug dosages actually delivered over time for patients receiving POCC alone. Arrows in the lower part of the figure indicate projected times for treatment cycles in the absence of hematologic toxicity. B: percentage of CCNU dose delivered in patients on the POCC arm.

cycle (POCC #2) in their treatment (Fig. 3). To analyze the possibility that increased hematologic tolerance contributed to the superiority of POCC/VAM rather than the presence of three additional drugs per se, survival was compared in subgroups that differed markedly in their tolerance for chemotherapy. In particular, a subset of 13 patients was identified receiving POCC alone who had significant cumulative myelosuppression and received less than 70% (Mean = 48%) of their projected dosage at the third cycle of treatment. However, the survival of these patients was *not* inferior to the 17 patients on the same treatment arm who were able to receive a mean of 87% of their treatment dose. When analyzed separately, the 7-drug arm of POCC/VAM was superior to POCC alone for both subgroups of patients who received greater than 70% or less than 70% of

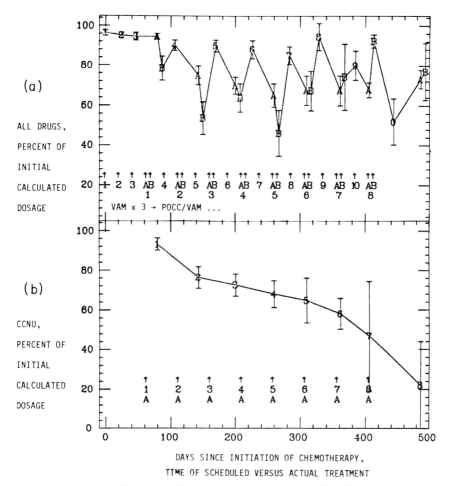

Fig. 3. A: percentage of total drug dosages actually delivered over time for patients receiving POCC/VAM. B: percentage of CCNU (lomustine) dose delivered in patients on the POCC/VAM arm.

TABLE IV. Percent of Patients Developing Major Toxicities
during Treatment

	POCC (%)	POCC/VAM (%)
WBC < 2,000	35	39
Platelets < 50,000	20	27
Grade 2–3[a]	29	13
> 10% weight loss	51	48

[a] Grade of peripheral neuropathy on an 0–4 rating.

their projected doses. Both treatment arms resulted in similar lengths of delay in treatment over time (Figs. 2 and 3).

In the subset of 17 patients on POCC who received 87% of their total drug dose at cycle 3, the mean delay in initiation of cycle 3 due to low blood counts was 10 days. By contrast, the 13 patients who received only 48% of their total drug dose had a mean treatment delay of 18 days. Thus treatment delays do not account for higher dose tolerance in the former group.

Patients with documented bone marrow involvement at presentation did not have reduced hematologic tolerance for chemotherapy as compared to patients with extensive stage and negative bone marrow biopsies.

IV. DISCUSSION

The major conclusion of this study is that a seven-drug alternating regimen is superior to the four-drug regimen of POCC in the treatment of small-cell lung cancer. This was true in patients with extensive-stage disease for complete response rate, duration of response, and overall survival, and was independent from other covariates.

The non–cross-resistance of early alternating treatment with seven drugs is demonstrated by the temporally related increase in CR rate from 20% to 40% when POCC was added to VAM. However, attempted salvage or sequential treatment with VAM after failure of POCC alone did not increase survival, and in fact only one of fifteen patients achieved even a partial remission. Whether this apparent cross-resistance to VAM after several cycles of POCC is due to biologic evolution of tumor cells, clinical deterioration of patients, or a combination of factors is not known.

The survival of patients in the POCC/VAM arm compares favorably with results from other recent studies in small-cell lung cancer (Broder, 1977; Livingston *et al.*, 1978; Cohen *et al.*, 1979). Although the definition of extensive stage in this study includes patients with T_3 or N_4 disease without distant metastases, survival of patients with distant metastases was not significantly inferior to other extensive-stage patients treated with POCC/VAM (290 versus

325 days median survival). By contrast, patients with limited-stage disease in this arm had a median survival of 488 days (Table III). The lack of superiority in survival with POCC/VAM compared to POCC in limited disease (488 versus 324 days, but $p > 0.25$) is partly due to the relatively small number of patients with this stage, but may also be a "dilutional" effect of the long delay in chemotherapy after 5000–5500 rads of involved-field irradiation.

The advantage of early treatment with 7 drugs versus 4 drugs may best be explained by the concept of clonal heterogeneity of individual tumors.

There is abundant evidence in animal tumor models that individual tumor stem cells may undergo spontaneous mutations and differentiation into clones of cells with widely varying biological properties, including susceptibility to cytoxic drugs. Such heterogeneity has recently been directly demonstrated in the drug-sensitivity profiles of clones of cells from several individual human cancers (Chu *et al.*, 1979; Dexter *et al.*, 1979; Shapiro and Yung, 1980). Phenotypic heterogeneity has been shown in cell cultures in human small-cell lung cancers (Radice and Dermody, 1980).

Thus, treatment with a larger number of agents might be expected to result in a larger fractional killing of cells within a tumor, provided that the drugs used are active as single agents and drug dosage is not severely compromised by overlapping toxicities. Small-cell lung cancer is a reasonable model on which to test this hypothesis since it is sensitive to most of the commonly used anticancer agents. The advantage of combination chemotherapy with four drugs in this disease versus the same drugs used sequentially was shown years ago (Alberto *et al.*, 1976). Two-drug therapy with cyclophosphamide and CCNU was superior to cyclophosphamide alone (Edmonson *et al.*, 1976). Improved treatment results were also shown for three- versus two-drug combinations (Hansen *et al.*, 1976).

The non–cross-resistant alternating six-drug combinations of CMC/VAP also appears to be superior to the three-drug CMC regimen (Cohen *et al.*, 1979). Whether earlier exposure to a larger number of active agents will actually result in increased disease-free survival has not yet been determined.

Patients on alternating POCC/VAM treatment were able to receive a higher percentage of their initial calculated drug doses throughout treatment than patients on POCC alone (Figs. 2 and 3). The less frequent administration of the nitrosourea CCNU (lomustine), every 7 weeks for POCC/VAM versus every 4 weeks in POCC, appears to have resulted in less cumulative myelosuppression for the seven-drug alternating arm (Figs. 2B and 3B).

The fact that survival of patients with good hematologic tolerance for POCC alone was not superior to that of patients with poor tolerance indicates that the biological sensitivity of tumors to individual drugs may be a more important determinant of patient survival than dose-response effects. This also supports the hypothesis that the use of a greater number of active drugs early in the course of treatment accounts for the superiority of the POCC/VAM arm.

The current NCOG study in small-cell lung cancer, protocol #2091, was activated in May 1980. For patients with extensive-stage disease, the best arm of protocol #2061 (VAM × 3 → POCC/VAM . . .) will be compared to two other arms: immediate alternation of POCC and VAM (POCC/VAM . . .) and a nine-drug regimen with a third alternating cycle consisting of cisplatin and hexamethylmelamine (POCC/VAM/HP . . .). The activity of hexamethyl-melamine in small-cell lung carcinoma is well recognized (Broder *et al.*, 1977), and recent phase II studies of cisplatin at Stanford indicate appreciable activity of this agent as well (Rosenfelt *et al.*, 1980).

In limited-stage disease, all patients receive 9 months of intensive chemotherapy with alternating POCC/VAM . . . (5 cycles of each). Patients with a complete response will then be randomized to no further treatment or to 5000 rads of radiotherapy to the originally involved field. This study is, therefore, testing the possible role of irradiation as a modality for consolidative treatment after maximum chemotherapy has been delivered, in patients with minimal tumor burdens. Two randomized studies have shown no advantage in limited-stage patients for early radiotherapy plus chemotherapy versus chemotherapy alone, when the irradiation is delivered in the first few months of treatment (Stevens *et al.*, 1979; Hansen *et al.*, 1979). The single study that attempted to deliver intensive chemotherapy and radiotherapy together at the outset of treatment achieved a substantial long-term complete remission rate, but at the expense of a high rate of lethal toxicities which would be unacceptable in a cooperative group study (Johnson *et al.*, 1976). Nevertheless, most patients who achieve clinical complete remissions with small-cell lung cancer eventually die from their disease, and almost all patients treated with chemotherapy alone relapse initially at the sites of original involvement (Alexander *et al.*, 1977). The success of "prophylactic" cranial irradiation in eradicating microscopic or minimal metastases provides further rationale for testing the effect of late consolidative radiotherapy in patients who have achieved a sustained complete remission with chemotherapy (Broder *et al.*, 1977; Moore *et al.*, 1978).

REFERENCES

Alberto, P., Brunner, K. W., Martz, G., Obrecht, J-P., and Sonntag, R. W. (1976) *Cancer 38:* 2208–2216.

Alexander, M., Glatstein, E. J., Gordon, D. S., and Daniels, J. R. (1977) *Cancer Treat. Rep. 61:* 1–6.

Breslow, N. (1974) *Biometrics 30:* 89.

Broder, L. E., Cohen, M. H., and Selawry, O. S. (1977) *Cancer Treat. Rev. 4:* 219–260.

Chu, M. Y., Takeuchi, T., Yeskey, K. S., Bogaars, H., and Calabresi, P. (1979) *AACR/ASCO 29:* 151.

Cohen, M. H., Ihde, D. C., Bunn, Jr., P. A., Fossieck, Jr., B. E., Matthews, M. J., Shackney, S. E., Johnston-Early, A., Makuch, R., and Minna, J. D. (1979) *Cancer Treat.Rep. 63:* 163–170.

Cox, D. R. (1972) *J. Royal Statistical Soc. B34.*

Daniels, J. R., Chak, L., Alexander, M., Kohler, M., Friedman, M., Torti, F., and Carter, S. K. (1980) *AACR/ASCO 346.*

Dexter, D. L., Fligiel, Z., Vogel, R., and Calabresi, P. (1979) *AACR/ASCO 20:* 1979.

Edmonson, J. H., Lagakos, S. W., Selawry, O. S., Perlia, C. P., Bennett, J. M., Muggia, F. M., Wampler, G., Brodovsky, H. S., Horton, J., Colsky, J., Mansour, E. G., Creech, R., Stolbach, L., Greenspan, E. M., Levitt, M., Israel, L., Ezdinli, E. Z., and Carbone, P. P. (1976) *Cancer Treat. Rep. 60:* 925–932.

Hansen, H. H., Selawry, O. S., Simon, R., Carr, D. T., Van Wyk, C. E., Tucker, R. D., and Sealy, R. (1976) *Cancer 38:* 2201–2207.

Hansen, H. H., Dombernowsky, P., Hansen, H. S., and Rorth, M. (1979) *AACR/ASCO 20:* 277.

Johnson, R. E., Brereton, H. D., and Kent, C. H. (1976) *Lancet 2:* 289–291.

Livingston, R. B., Moore, T. N., Heilbrun, L., Bottomley, R., Lehane, D., Rivkin, S. E., and Thigpen, T. (1978) *Ann. Intern. Med. 88:* 194–199.

Moore, T. N., Livingston, R., Heilbrun, L., Eltringham, J., Skinner, O., White, J., and Tesh, D. (1978) *Cancer 41:* 2149–2153.

Radice, P. A., and Dermody, W. C. (1980) *AACR/ASCO 41.*

Rosenfelt, F. P., Sikic, B. I., Daniels, J. R., and Rosenbloom, B. E. (1980) *AACR/ASCO 21:* 449.

Shapiro, W. R., and Yung, W-K. (1980) *AACR/ASCO 21:* 147.

Stevens, E., Einhorn, L., and Rohn, R. (1979) *AACR/ASCO 20:* 435.

Williams, C., Alexander, M., Glatstein, E. J., and Daniels, J. R. (1977a) *Cancer Treat. Rep. 61:* 1427–1431.

Williams, C. J., Alexander, M., Glatstein, E., and Daniels, J. R. (1977b) *In* "Adjuvant Therapy of Cancer," (S. E. Salmon and S. E. Jones, eds.), pp. 237–244. Elsevier/North-Holland Biomedical Press, Amsterdam.

Chapter 18

NITROSOUREA-CONTAINING COMBINATIONS IN SMALL-CELL LUNG CANCER

Paul A. Bunn, Jr.

I. INTRODUCTION

There are 20,000–25,000 new cases of small-cell lung cancer (SCCL) in the United States each year, and the incidence continues to rise at a rapid rate. The biological characteristics, natural history, and response of SCCL to various therapeutic modalities make it distinct from the other lung cancer cell types (Bunn et al., 1977; Cohen and Matthews, 1978). The poor survival following curative surgical resection (1% at 5 years), the autopsy demonstration of distant metastases in 70% of SCCL patients dying within 30 days of curative surgical resection, and the Medical Research Council trial demonstrating that radiotherapy was superior to surgery have led to an abandonment of surgery as the initial treatment (Mountain, 1978; Matthews et al., 1973; Fox and Scadding, 1973). Radiation therapy alone produces high response rates in SCCL, and there was a great deal of enthusiasm for radiotherapy in early trials showing good local control in a majority of patients. However, few if any patients were cured by radiation therapy, and as survival was prolonged with the introduction of effective combination chemotherapy, it became apparent that radiation therapy controlled local (chest) disease in a minority of patients. Thus, chemotherapy is the mainstay of treatment for SCCL, and the role of radiotherapy is controversial and under investigation (Bunn and Ihde, 1981; Bunn et al., 1981). This chapter will consider the place of the nitrosoureas in the chemotherapy of SCCL. Studies conducted at the National Cancer Institute and Finsen Institute will be emphasized.

II. RESULTS

A. Single-Agent Chemotherapy

To understand the role of nitrosoureas in SCCL the single-agent results with these drugs as well as with other agents must be considered first. In 1969, the Veterans Administration Lung Group reported that three courses of intravenous cyclophosphamide (CTX) significantly prolonged survival in extensive stage SCCL from 1.5 to 4 months when compared to a placebo (Green et al., 1969). This trial provided a major impetus to the study of chemotherapeutic agents in SCCL. Since that time, a large number of agents have been studied, and these trials have been reviewed by Broder et al. (1977). Table I shows the agents with established activity in SCCL based on an overall response rate in excess of 15% in 40 or more patients in at least two studies. It should be emphasized that complete responses are a prerequisite for prolonged tumor-free survival in SCCL and that even the best single agents infrequently produce complete responses (Table I).

The single-agent response rates for CCNU, BCNU, and methyl CCNU are shown in Table II. The response rates to BCNU and methyl CCNU are similar to those reported for other active single agents, but few patients have been studied. CCNU has been evaluated in more patients, but the overall response rate is somewhat less than for the other two agents. Nevertheless, some activity for CCNU has been demonstrated in several trials, and CCNU was the only one of these three nitrosoureas that significantly improved survival when compared to

TABLE I. Active Single Agents,[a] 1980[b]

Drug	No. patients	% response	(%CR)[c]
1. Cyclophosphamide	363	38	
2. Mechlorethamine	55	44	
3. Doxorubicin	53	30	(4)
4. Methotrexate	73	30	
5. Vincristine	43	42	(7)
6. VP-16	201	44	(7)
7. Hexamethylmelamine	69	30	(7)
8. CCNU	76	14	(4)

[a] Overall response ≥ 15% in 40 or more patients in at least two studies.
[b] Adapted from Broder et al. (1977).
[c] CR = Complete response.

TABLE II. Single-Agent Nitrosoureas[a]

Nitrosourea	No. patients	% response	% complete response
1. BCNU	15	26	7
2. CCNU[b]	76	14	4
3. Methyl CCNU	19	26	—

[a] Adapted from Broder et al. (1977).
[b] Significantly improved survival in extensive stage compared with placebo.

placebo in extensive stage SCCL patients studied by the Veterans Administration Lung Cancer Study Group (Wolf, 1976). The results of this study and the combination chemotherapy results reported later in this chapter suggest that CCNU is the nitrosourea of choice in SCCL at present.

More recently, some of the newer nitrosoureas have been evaluated in SCCL with disappointing results (Table III). Streptozotocin (STZ) was an especially attractive nitrosourea since it lacks the myelosuppressive properties of other nitrosoureas and is active against many other tumors with APUD properties. Unfortunately, no activity was seen with streptozotocin in two trials from two institutions (Bunn et al., 1978; Kane et al., 1978). Chlorozotocin has similarly been shown to lack activity in SCCL (Killen, 1981).

B. Combination Chemotherapy

1. Optimal Drug Number

A series of randomized trials have combined active single agents to determine the optimal drug number, and these trials are summarized in Table IV (Wolf, 1976; Edmonson et al., 1976; Hansen et al., 1976; Hansen et al., 1978; Alexander et al., 1977). Edmonson et al. (1976) studied the value of adding CCNU to cyclophosphamide. The two-drug (CCNU and CTX) combination had a significantly improved overall (complete plus partial) and complete response rate and an improvement in survival of borderline significance ($p = 0.07$).

Hansen et al. (1976) conducted a randomized two- versus three-drug trial to determine the value of adding CCNU to CTX + MTX (methotrexate). The three-drug nitrosourea-containing combination produced higher response rates,

TABLE III. Results with Newer Nitrosoureas

Reference(s)	Nitrosourea	No. patients	% response	% complete response
Bunn et al. (1978) Kane et al. (1978)	Streptozotocin	26	0	0
Killen (1981)	Chlorozotocin	29	0	0

TABLE IV. Randomized Trials of Optimal Drug Number with Nitrosoureas[a]

Reference	Drug(s)	No. patients	% CR + PR	% CR	2 yr DFS	MS
Wolf (1976)	Placebo	NR	NR	NR	NR	NR[b]
	CCNU	NR	NR	NR	NR	NR
Edmonson et al.	CTX	112	22	3	NR	4.0[c]
(1976)	CCNU + CTX	106	43	12	NR	4.5
Hansen et al.	CTX + MTX	29	38	NR	0	5.5[d]
(1976)	CTX + MTX + CCNU	33	56	NR	0	7.5
Hansen et al.	CTX + MTX + CCNU	52	69	NR	0	6[e]
(1978)	CTX + MTX + VCR + CCNU	53	74	NR	1	7.5
Alexander et al.	CTX + VCR + MTX	11	70	20	NR	11.0[f]
(1977)	CTX + VCR + PCZ + CCNU	12	67	25	NR	14.0

[a] CR = complete response; PR = partial response; 2 yr DFS = 2-year disease-free survival; MS = median survival in months.

[b] Survival not reported but significantly improved with CCNU, extensive disease only.

[c] CR and CR + PR rates significantly improved with CTX + CCNU, survival borderline improved ($p = 0.07$) with CTX + CCNU.

[d] No significant difference in response rate, survival borderline improved ($p = 0.17$) with three drugs, extensive disease only.

[e] No significant difference in response rate, survival significantly improved with four drugs, extensive disease only.

[f] No significant difference in CR or CR + PR rates, survival borderline improved ($p = 0.055$) with four drugs, patients also given chest irradiation.

although this was not statistically significant. Survival in the three-drug combination was also superior, again with borderline significance ($p - 0.17$).

There are two randomized trials with a four-drug nitrosourea-containing combination versus a three-drug combination (Hansen *et al.*, 1978; Alexander *et al.*, 1977). Hansen *et al.* (1978) found that the addition of vincristine (VCR) to CTX + MTX + CCNU significantly improved median survival from 6 to 7.5 months but did not affect response rates. Alexander *et al.* (1977) found a four-drug combination containing CCNU (CTX + VCR + PCZ [procarbazine] + CCNU) to be superior to a 3 drug combination of CTX + MTX + VCR in survival ($p = 0.055$), but there were no differences with respect to response rates (Alexander *et al.*, 1977).

C. Optimal Drug Dose

At the same time that the trials with varying drug numbers were being conducted, Cohen *et al.* (1977a) designed a randomized trial to determine whether the intensity of therapy using a three-drug combination of CTX + MTX + CCNU was important. This trial was based on the premise that long-term survivors were exclusively patients who had a complete response and that the complete response rate was low when standard-dose three-drug regimens were

TABLE V. Randomized Trial of Optimal Induction Drug Dose with Nitrosoureas[a]

Dose/m^2							
CTX[b]	MTX[c]	CCNU[d]	No. patients	% CR + PR[e]	% CR	MS	% 2 yr DFS
500	10	50	9	44	0	5.0	0
1000	15	100	23	96	30	10.5	9

[a] Adapted from Cohen et al. (1977a).
[b] iv d 1,22.
[c] iv biw.
[d] po d 1.
[e] CR = complete response; PR = partial response; MS = median survival (months); 2 yr DFS = 2 year disease free survival.

used. The results are summarized in Table V. Patients were randomized to receive a standard 6-week induction cycle consisting of 500 mg/m^2 of CTX, 10 mg/m^2 of MTX, and 50 mg/m^2 of CCNU or high-dose therapy consisting of 1000 mg/m^2 of CTX, 15 mg/m^2 of MTX and 100 mg/m^2 of CCNU. Maintenance therapy doses consisted of the lower dose schedule in both groups. Patients receiving the high-dose therapy had higher response rates and longer median survivals than patients receiving the lower doses. Complete responses and long term (2 year) disease-free survival were seen only in patients receiving the high-dose therapy. While there was significantly more myelosuppression in the group receiving the higher dose therapy, there were no infectious deaths in either group.

More recent studies have escalated the induction dosage of cyclophosphamide to higher levels. Unfortunately this dose escalation appears to have increased toxicity to unacceptable levels without dramatic improvement in survival (Bunn et al., 1981).

The randomized trials for four drugs versus three drugs utilized lower drug doses than in the three-drug trial of Cohen et al. (1977a). It would appear, then, that three- or four-drug combinations employing relatively high drug doses during the first 6–8 weeks provide the best results at present.

D. Alternating Drug Combinations

In their high-dose three-drug combination of CTX + MTX + CCNU, Cohen et al. (1977a) noted that maximal tumor response was observed within the first 6 weeks of therapy in all cases. The same group also demonstrated that an alternative three-drug combination of VCR + ADR (adriamycin) + PCZ produced high response rates in patients failing primary therapy with CTX + MTX + CCNU (Cohn et al., 1977b). They then asked whether the addition of a second and/or third non–cross-resistant regimen would increase complete response rate and prolong remission duration and survival by introduction of the second regimen

TABLE VI. Trials of Alternating Regimens with Nitrosoureas[a]

Reference	Drug combinations	No. patients	% CR + PR	% CR	Median survival	% 2 yr DFS
Cohen et al. (1977a)	CTX + MTX + CCNU	23	96	30	10.5	9
Cohen et al. (1979)	CTX + MTX + CCNU/VCR + ADR + PCZ/ ± VPIF ± THY	61	93	48	10.7	7
Aisner et al. (1980)[b,c]	CTX + ADR + VP16	42	76	38	10+	NR
	CTX + ADR + VP16/CCNU + MTX + PCZ + VCR	43	84	44	10+	NR
Daniels et al. (1980)[b,d]	CTX + VCR + PCZ + CCNU	134	70/53[e]	35/20	8/7	NR
	VP16 + ADR + MTX/CTX + VCR + PCZ + CCNU		39/55	33/35	16/10	NR
Dombernowsky et al. (1979)[b,f]	CTX + MTX + VCR + CCNU	73	81	NR	9	NR
	CTX + MTX + VCR + CCNU/ADR + VP16	73	90	NR	9	NR

[a] See prior tables for abbreviations.
[b] Randomized trial.
[c] No statistical difference in response rates or survival.
[d] No statistical difference in response rates. Survival significantly prolonged in extensive disease.
[e] Limited disease/extensive disease.
[f] No statistical difference in response rates or median survival. Significantly prolonged median duration of response in alternating regimen.

238

after initial tumor response and prior to disease progression. The results of this trial and other alternating regimens employing nitrosoureas are shown in Table VI. Cohen *et al.* (1979) found that the addition of the second non–cross-resistant regimen during weeks 6 to 12 increased the complete response rate from 30% at 6 weeks to 48% at 12 weeks. This advantage was not seen in survival, however, as both median survival and 2-year disease-free survival were similar in both groups. This was explained by the shorter duration of remission for those achieving complete response at 12 weeks as compared to 6 weeks. The addition of a third alternative non–cross-resistant regimen at week 12 had no impact on response rate or survival.

The results of 3 randomized trials employing alternating regimens are also shown in Table VI. Aisner *et al.* (1980) found that a second non–cross-resistant regimen of CCNU + MTX + PCZ + VCR did not influence response rate, remission duration, or survival when compared with continuous administration of a single three-drug regimen (CTX + ADR + VP-16[1]). Dombernowsky *et al.* (1979) reported that alternating a combination of ADR + VP-16 with CTX + MTX + VCR + CCNU lengthened response duration but had no impact on response rate or median survival when compared with continuous CTX + MTX + VCR + CCNU. Daniels *et al.* (1980) have found the most benefit from alternating regimens. They found that alternating CTX + VCR + PCZ + CCNU with VP-16 + ADR + MTX produced significantly longer survival in extensive stage disease patients, compared to a continuous regimen of CTX + VCR + PCZ + CCNU. In this trial limited-stage patients all received chest radiotherapy. Although there was some survival advantage for limited-stage patients receiving alternating combinations, it was not statistically significant. In summary, alternating regimens in some instances have increased response rates and lengthened remission duration or even survival, but do not appear to have made a major impact, particularly on 2-year disease-free survival.

E. Drug Combinations with and without Nitrosoureas

Thus far, all the trials discussed in this chapter have employed nitrosoureas. Before concluding that nitrosoureas are useful in these combinations, we should compare the results of nitrosourea-containing combinations with combinations containing only nonnitrosoureas. Table VII is a compilation of results in the literature of drug combinations with and without nitrosoureas. The details of the studies in this compilation can be found in the review of Bunn and Ihde (1981). The table is organized by trials containing 2, 3, 4, or more drugs; by the stage of disease; and by whether or not radiation therapy was employed. In each group there are no significant differences in response rate, complete response rate, and median or long-term survival between those treated with or without a

---

[1] Demethyl-ethylidene-glucopyranosyl-epipodophyllotoxin.

TABLE VII. Comparison of Tabulated Results from the Literature in Trials with and without Nitrosoureas[a]

Drug no.	Stage of disease	Nitrosourea	Radiation therapy	No. pts.	% CR + PR	% CR	MS	2 yr DFS
2	ED	yes	no	122	46	16	4.5-7.5+	TE
2	ED	no	no	71	56	38	5.5-14.5	0/52
3	ED	yes	no	117	71	27	5.0-10.5	2/1.7
3	ED	no	no	247	59	29	5.0+-9.5	0/27
3	ED	yes	yes	18	94	78	13.5	3/18
3	ED	no	yes	227	80	59	9.0-17	21/119[b]
4+	LD & ED	yes	no	429	85	40	6.5-14	10/184[b]
4+	LD & ED	no	no	162	75	34	7.0-10.0+	0/82
4+	LD & ED	yes	yes	103	81	40	5.5-14	6/77
4+	LD & ED	no	yes	125	66	21	6.0+ 10.0	0/11
Grand totals								
Nitrosourea				789	76%	35	4.5-14	21/396
No nitrosourea				832	69	39	5.0+-17	21/291

[a] CR = complete response; PR = partial response; MS = median survival; 2 yr DFS = 2-year disease-free survival; ED = extensive stage disease; LD = limited stage disease; TE = too early.
[b] More patients, too early.

nitrosourea-containing combination. Similarly, if one totals the results, as shown at the bottom of the table, no difference is seen. Obviously, one must be careful in interpreting summaries of literature reviews, but results appear to be equivalent when any combination employing at least three active agents is used.

F. Central Nervous System Relapse

As early as 1973, Hansen and co-workers at the National Cancer Institute recognized that central nervous system (CNS) metastases were more frequent in SCCL than in the other histologic types of lung cancer and that the CNS was a frequent site of relapse after remission (Hansen *et al.*, 1978). Nitrosoureas are lipid soluble and therefore may penetrate the blood-brain barrier to a greater extent than other chemotherapeutic agents. For these reasons it was hoped that CCNU would reduce the frequency of CNS metastases. Unfortunately, this does not appear to be the case (Table VIII). The frequency of CNS relapse appears to be as high in trials employing a nitrosourea as in trials not employing a nitrosourea. In addition, intracerebral, leptomeningeal, and spinal cord relapses are all common, even in studies employing CCNU (Nugent *et al.*, 1979).

G. Combined Modality Therapy

In studies employing radiation therapy alone it was obvious that even in limited-stage disease more than 95% of patients died of SCCL within 5 years,

TABLE VIII. Frequency of CNS Relapse in Trials with and without Nitrosoureas[a]

Reference	Number of patients	Number of CNS relapses	(%)
	Nitrosoureas		
Abeloff et al. (1976)	34	6	(18)
King et al. (1977)	37	3	(8)
Alexander et al. (1977)	23	6	(26)
Trowbridge et al. (1978)	15	4	(27)
Nugent et al. (1979)	146	47	(32)
Broder et al. (1978)	42	2	(51)
Total	297	68	(23)
	No nitrosoureas		
Eagan et al. (1973)	45	9	(20)
Holoye and Samuels (1975)	39	12	(31)
Nixon et al. (1975)	13	2	(15)
Choi and Carey (1976)	58	17	(29)
Einhorn et al. (1976)	29	5	(17)
Donavan et al. (1976)	35	2	(6)
Holoye et al. (1977)	45	8	(18)
Tenczynski et al. (1978)	22	6	(27)
Total	286	61	(21)

[a] None of these trials employed prophylactic cranial irradiation.

usually as a result of distant metastases. In contrast, in trials employing chemotherapy alone, the chest is the predominant site of initial relapse. Thus, it was logical to combine these therapeutic modalities. Unfortunately, results of combined therapy have not thus far established a superiority for the combined modality approach. Several issues are clear from these studies while many issues are unresolved. It is apparent that toxicity, including acute and chronic pulmonary toxicity, esophagitis, myelosuppression, and possibly skin, CNS, and cardiac toxicity, are more severe in the combined modality trials (Bunn and Ihde, 1981; Bunn et al., 1981). Retrospective reviews of the literature comparing nonrandomized trials show similar response rates and median survivals in those treated with chemotherapy alone and those treated with combined therapy (Bunn and Ihde, 1981). There is also a suggestion that the dose and timing of the radiotherapy may be critical and that large doses given concurrently with chemotherapy over 3 weeks yield the best results (Catane et al., 1981).

Randomized trials employing nitrosourea combinations with or without chest irradiation are summarized in Table IX. Hansen et al. (1979) at the Finsen Institute conducted a trial of CTX + MTX + VCR + CCNU with or without chest irradiation in patients with limited-stage disease. The total dose of radiation was 4000 rads, which was delivered as a split course between chemotherapy cycles. There were no differences in response rate or survival in this study.

TABLE IX. Randomized Combined Modality Studies Employing Chemotherapy and Radiation Therapy

Reference	Chemotherapy	Radiation therapy	% CR + PR	% CR	Median survival
Hansen et al. (1979)[a]	CTX + VCR + MTX + CCNU	4000 rads split course[b]	88	NR	11[c]
		None	91	NR	14
Cohen et al. (1980)[a]	CTX + MTX + CCNU/VCR + ADR + PCZ	4000 rads continuous[d]	NR	79	NR[e]
		None	NR	43	NR
Williams et al. (1977)[f]	CTX + VCR + PCZ + CCNU	3000 rads continuous[g]	69	38	9[h]
		None	92	42	11

[a] Limited stage only.
[b] Between chemotherapy cycles.
[c] No significant difference in response or survival.
[d] Concurrent beginning day 1.
[e] Trend in survival favoring combined modality; study still in progress.
[f] Extensive stage.
[g] Radiation given between chemotherapy cycles and also given to sites of known metastatic disease.
[h] No significant difference in response or survival.

Williams et al. (1977) at Stanford have reported a study of CTX + VCR + PCZ + CCNU with or without irradiation in patients with extensive-stage disease. A continuous course of 3000 rads of radiation was delivered to the chest and also to known sites of metastatic disease between chemotherapy cycles. There was no difference in response or survival in the two groups. In addition, the sites of relapse were similar with chest being the most frequent site of relapse in both groups.

Cohen and co-workers (1980) from the National Cancer Institute are conducting an ongoing trial in patients with limited disease. Patients are randomized to receive CTX + MTX + CCNU alternating with VCR + ADR + PCZ alone or with radiation therapy. The radiation therapy is a continuous course of 4000 rads/15 fractions/3 weeks given concurrently with the chemotherapy beginning on day 1. This radiation therapy schedule gave the best results in an earlier trial (Catane et al., 1981). The results of this trial are too preliminary to draw firm conclusions, although the early results favored the combined modality treatment.

III. CONCLUSIONS

At least one nitrosourea, CCNU, has established activity in SCCL, although the single-agent response rate may be lower than with some other active agents. BCNU and methyl CCNU may also be active, but they have been studied in relatively small numbers of patients. Streptozotocin and chlorozotocin do not appear to be active in SCCL.

Drug combinations containing CCNU give better results than single-agent results with any active agent. Three- or four-drug combinations would appear to give the best results, and 3- or 4-drug combinations containing a nitrosourea give results equivalent to those not containing a nitrosourea. High-dose induction therapy (CCNU = 100mg/m^2) producing moderate to severe leukopenia (WBC ≤ 1000/μl) during the first 6-8 weeks gives better results than standard- or low-dose therapy (CCNU = 50mg/m^2).

Alternating non-cross-resistant drug combinations may improve the complete remission rate and duration of remission slightly but do not make a major impact on long-term disease-free survival.

Nitrosoureas are not effective in reducing the frequency of CNS relapse. At present, CCNU-containing drug combinations produce results that are equivalent to those achieved by combined modality therapy with the same drugs plus irradiation therapy, although further trials on this issue are needed.

REFERENCES

Abeloff, M. D., Ettinger, D. S., Baylin, S. B., and Hazra, T. (1976). *Cancer 38*, 1394-1401.

Aisner, J., Whitacre, M., Van Echo, D. A., Esterhay, R. J., Jr., and Wiernik, P. H. (1980). *Proc. AACR and ASCO 21*, 453.

Alexander, M., Glatstein, E. J., Gordon, D. S., and Daniels, J. R. (1977). *Cancer Treat Rep. 61*, 1-6.

Broder, L. E., Cohen, M. H., and Selawry, O. S. (1977). *Cancer Treat. Rev. 4*, 219-260.

Broder, L. E., Selawry, O. S., Bagwell, S. P., Silverman, M. A., and Charyulu, K. N. (1978). *Proc. Am. Assoc. Cancer Res. 17*, 71.

Bunn, P. A., and Ihde, D. C. (1981). *In* "Lung Cancer: Advances in Research and Treatment" (W. L. McGuire, ed.). Martinus Nijhoff, Boston. In press.

Bunn, P. A., Cohen, M. H., Ihde, D. C., Fossieck, B. E., Matthews, M. J., and Minna, J. D. (1977). *Cancer Treat. Rep. 61*, 333-342.

Bunn, P. A., Ihde, D. C., Cohen, M. H., Shackney, S. E., and Minna, J. D. (1978). *Cancer Treat. Rep. 62*, 479-481.

Bunn, P. A., Lichter, A. S., Glatstein, E., and Minna, J. D. (In press). *In* "Small Cell Lung Cancer" (A. Greco, R. Oldham, and P. Bunn, eds.) Grune and Stratton, New York.

Catane, R., Lichter, A., Lee, Y. S., Brereton, H. D., Schwade, J. G., Kent, C. H., Johnson, R. E., and Glatstein, E. (1981). *Cancer*. In press.

Choi, C. H., and Carey, R. W. (1976). *Cancer 37*, 2651-2657.

Cohen, M. H., and Matthews, M. J. (1978). *Semin. Oncol. 5*, 234-243.

Cohen, M. H., Creaven, P. J., Fossieck, B. E., Broder, L. E., Selawry, O. S., Johnston, A. V., Williams, C. L., and Minna, J. D. (1977a). *Cancer Treat. Rep. 61*, 349-354.

Cohen, M. H., Broder, L. E., Fossieck, B. E., Bull, M., Ihde, D. C., and Minna, J. D. (1977b). *Cancer Treat. Rep. 61*, 485-487.

Cohen, M. H., Ihde, D. C., Bunn, P. A., Fossieck, B. E., Matthews, M. J., Chackney, S. E., Johnson-Early, A., Makuch, R., and Minna, J. D. (1979). *Cancer Treat. Rep. 63*, 163-170.

Cohen, M. H., Lichter, A. S., Bunn, P. A., Glatstein, E. J., Ihde, D. C., Fossieck, B. E., Jr., Matthews, M. J., and Minna, J. D. (1980). *Proc. Am. Assoc. of Cancer Res. 21*, 448.

Daniels, J. R., Chak, L., Alexander, M., Kohler, M., Friedman, M., Tori, F., and Carter, S. (1980). *Proc. AACR and ASCO 21*, 346.

Dombernowsky, P., Hansen, H. H., Sorenson, S., Rørth, M. (1979). *Prox. AACR and ASCO 20*, 277.

Donavan, M., Baxter, D., Sponzo, R., Cunningham, T., and Horton, J. (1976). *Proc. AACR and ASCO 17*, 100.

Eagan, R. T., Maurer, H., Forcier, R. J., and Tulloh, M. (1973). *Cancer 32*, 371–379.

Edmonson, J. H., Lagakos, S. W., Selawry, O. S., Perlia, C. P., Bennett, J. M., Muggia, F. M., Wampler, G., Brodovsky, H. S., Horton, J., Colsky, J., Mansour, E. G., Creech, R., Stolbach, L., Greenspan, E. M., Levitt, M., Israel, L., Ezdinli, E. Z., and Carbone, P. P. (1976). *Cancer Treat. Rep. 60*, 925–932.

Einhorn, L. H., Fee, W. H., Farber, M. O., Livingston, R. B., and Gottlieb, J. A. (1976). *JAMA 235*, 1225–1229.

Fox, W., and Scadding, J. G. (1973). *Lancet 2*, 63–65.

Green, R. A., Humphrey, E., Close, H., and Patno, M. E. (1969). *Am. J. Med. 46*, 516–525.

Hansen, H. H., Selawry, O. S., and Simon, R. (1976). *Cancer 38*, 2201–2207.

Hansen, H. H., Dombernowsky, P., Hansen, M., and Hirsch, F. (1978). *Ann Intern. Med. 89*, 177–181.

Hansen, H. H., Dombernowsky, P., Hansen, H. S., Sorensen, S., and Østerlind, K. (1979). *Proc. AACR and ASCO 20*, 277.

Holoye, P. Y., Samuels, M. L., Lanzotti, V. J., Smith, T., and Barkley, H. T. (1977). *JAMA 237*, 1221–1224.

Holoye, P. Y., and Samuels, M. L. (1975). *Chest 67*, 675–679.

Kane, R. C., Bernath, A. M., and Cashdollar, M. R. (1978). *Cancer Treat. Rep. 62*, 477–478.

Killen, J. (1981). *In* "Small-Cell Lung Cancer" (A. Greco and R. Oldham, eds.). Grune and Stratton, New York.

King, G. A., Comis, R. L., Ginsberg, S., Goldberg, J., Dale, H. T., Brown, J., Dalal, P., Chung, C., and Gottlieb, A. (1977). *Radiology 125*, 529–530.

Matthews, M. J., Kanhouwa, S., Pickren, J., and Robbinette, D. (1973). *Cancer Chemo. Rep. (Part 3) 4*, 63–67.

Mountain, C. F. (1978). *Semin. Oncol. 5*, 272–279.

Nixon, D. W., Carey, R. W., Suit, H. D., and Aisenberg, A. C. (1975). *Cancer 367*, 867–872.

Nugent, J. L., Bunn, P. A., Matthews, M. J., Ihde, D. C., Cohen, M. H., Gazdar, A., and Minna, J. D. (1979). *Cancer 44*, 1885–1893.

Tenczynski, T. F., Valdivieso, M., Hersh, E., Khalil, K. G., Mountain, C. F., and Bodey, G. P. (1978). *Proc. AACR and ASCO 19*, 376.

Trowbridge, R. C., Kennedy, B. J., and Vosika, G. J. (1978). *Cancer 41*, 1704–1709.

Williams, C., Alexander, M., Glatstein, E. J., and Daniels, J. R. (1977). *Cancer Treat. Rep. 61*, 1427–1431.

Wolf, J. (1976). *Cancer Treat. Rep. 60*, 753–756.

Chapter 19

NITROSOUREA-CONTAINING CHEMOTHERAPY REGIMENS IN THE TREATMENT OF NON-SMALL-CELL LUNG CANCER

Martin H. Cohen

I. INTRODUCTION

The introduction of BCNU, CCNU, and MeCCNU into cancer clinical trials generated considerable interest and excitement within the oncologic community. These drugs as single agents were highly active, both systemically and intracerebrally in a wide variety of rodent tumor systems. In addition, therapeutic synergism between nitrosoureas and several previously identified anticancer drugs, especially cyclophosphamide, was evident. For these reasons, phase I trials for each nitrosourea were quickly completed, and phase II trials were initiated in non-small-cell lung cancer as well as in many other human neoplasms. This report is concerned with phase II single-agent nitrosourea trials and with nitrosourea-containing combination chemotherapy studies in non-small-cell bronchogenic carcinoma.

II. MATERIALS AND METHODS

Clinical trials reported in the English language literature were reviewed. The data abstracted from these trials concerned only patients with non-small-cell lung cancer, that is, squamous cell, adenocarcinoma, and large-cell anaplastic lung cancer. For a patient to be considered as a responder to chemotherapy there had to be at least a 50% reduction in size of one or more indicator lesions. Responses were classified as either partial or complete, using standard definitions. Data on median survival was not considered in this report because of large differences in study populations from one trial to another, especially as regards prognostic factors such as performance status and extent of prior therapy. Where possible, however, it was determined whether responding patients survived significantly longer than nonresponding patients.

III. RESULTS

A. Preclinical Studies and Rationale for Human Nitrosourea Clinical Trials

Nitrosourea trials in non–small-cell lung cancer patients were considerably influenced by results of therapeutic experiments performed in murine Lewis lung carcinoma. In that system MeCCNU was the most active nitrosourea followed by CCNU and then by BCNU (Schabel, 1976). Cyclophosphamide was less active against the primary tumor mass than was MeCCNU but was approximately as effective as MeCCNU against systemic metastatic disease. Combinations of MeCCNU and cyclophosphamide were therapeutically synergistic, and if debulking surgery was also performed, even moderately advanced tumors could be cured (Mayo *et al.*, 1972; Schabel, 1976).

Dose ratios of MeCCNU and cyclophosphamide (based on the LD_{10} dose for each drug) yielding optimal control of Lewis lung tumor masses of 400 mg are indicated in Table I. From these results two tentative conclusions can be drawn. First, large total drug doses may be required for high cure rates to be achieved (compare lines 1 and 3). Second, when combining MeCCNU and cyclophosphamide, dose ratios favoring MeCCNU, the more active drug in the combina-

TABLE I. Lewis Lung Carcinoma:
Cyclophosphamide Plus Methyl CCNU Treatment

% of LD_{10} doses[a] MeCCNU/CTX	Cure rate of 400 mg tumor (%)[b]
67/50	80
50/50	55
67/33	30
50/33	5

[a] LD_{10} doses: MeCCNU 36 mg/kg; cyclophosphamide 312 mg/kg.
[b] Adapted from Mayo *et al.* (1972).

tion, appeared to give best therapeutic results (compare lines 1 and 2) (Mayo *et al.*, 1972; Schabel, 1976). To translate this data for human clinical trials, it is first necessary to determine a moderately toxic dose for each single agent and to determine the drug's therapeutic activity (phase I and phase II trials). Subsequently, when drugs are combined, the dose of the most active drug in the combination should be as close as possible to the optimal dose of that drug when used alone. Doses of less active drugs in the combination should be decreased proportionately so that, in general, the combination is not significantly more toxic than optimal dose single-agent treatment.

B. Phase II Single-Agent Nitrosourea Trials

Table II lists phase II treatment results for BCNU, CCNU, MeCCNU, and for chlorozotocin. Streptozotocin, a nitrosourea for which considerably less data exist, is not included in this review. Unfortunately, there is little information from randomized trials comparing the nitrosoureas in non–small-cell lung cancer. In one trial, in which patients were randomized to CCNU or to MeCCNU, response rates of 2 and 4% were obtained (Eagan *et al.*, 1974). Neither of the two BCNU trials listed in Table II had response rates above 10%. In contrast, two of four CCNU trials and one of four MeCCNU trials had response rates of 20% or more. Initial results with chlorozotocin do not suggest that this drug is more active in non–small-cell lung cancer than are previously tested nitrosoureas. Since no combination data have been published as yet for chlorozotocin, this drug will not be further considered here. In comparison to either CCNU or MeCCNU, cyclophosphamide probably has less activity in non–small-cell lung cancer, with response rates of 10% or less in several trials (Cohen and Perevodchikova, 1979). Adriamycin yields comparable, or some-

TABLE II. Phase II Nitrosourea Studies

Drug	Dose & schedule mg/m²	No. of patients	Response rate (%)	Reference
BCNU	75–100/day × 3 every 6 weeks	21	10	Ahmann *et al.* (1972)
	185 every 6 weeks	18	0	Olshin *et al.* (1972)
CCNU	130 every 7 weeks	55	2	Eagan (1979)
	130 every 6 weeks	78	5	Eagan *et al.* (1974)
	130 every 6 weeks	42	21	Stolinsky *et al.* (1975)
	75 every 3 weeks	10	20	Israel *et al.* (1979)
MeCCNU	150–200 every 7 weeks	53	4	Eagan *et al.* (1974)
	150 every 6 weeks	18	11	Richards *et al.* (1973)
	150 every 6 weeks	9	0	Richards *et al.* (1976)
	200 every 6 weeks	15	27	Bodey *et al.* (1973)
Chlorozotocin	30–40/day × 5 every 6 weeks	30	7	Casper and Gralla (1979)
	125 every 5 weeks	31	3	Creagan *et al.* (1979)

what higher, response rates as compared to the nitrosoureas (Cohen and Perevod-chikova, 1979).

In designing a CCNU or MeCCNU plus cyclophosphamide combination based on the aforementioned data, there should be a higher percentage of the optimal single dose of the nitrosourea with a corresponding decrease in optimal cyclophosphamide dosage. A CCNU–cyclophosphamide combination with 100 mg/m^2 of CCNU every 6 weeks (approximately 75% of optimal CCNU dosage) with 500-600 mg/m^2 of cyclophosphamide every 3 weeks (50-60% of optimal single dose) seems reasonable. For a CCNU plus adriamycin combination 90 mg/m^2 of the former drug every 6 weeks and 40 mg/m^2 of the latter every 3 weeks (approximately 67% of optimal single doses for each agent) seems appropriate.

C. Nitrosourea Combinations with One Additional Drug

Table III lists studies in which a nitrosourea was combined with one additional drug. Usual dose levels for every 6 week administration of a nitrosourea alone are BCNU 200-240 mg/m^2, either as a single dose or given in divided doses for 2 to 3 days; CCNU 100-130 mg/m^2 as a single dose; and MeCCNU 200-225

TABLE III. Nitrosourea-Containing Two-Drug Combinations

Drug doses & schedules		No. of patients	Histologies studied	Response rate (%)	Survival[a] benefit	Reference
CCNU	70 every 6 weeks+	173	S,A,L[b]	3	—	Wolf et al. (1979)
CTX	700 every 3 weeks					
CNNU	70 every 6 weeks+	83	A	12 (3)[c]	yes	Edmonson et al. (1976a)
CTX	700 every 3 weeks					
CCNU	70 every 6 weeks+	114	S,L	6	yes	Edmonson et al. (1976b)
HN$_2$	10 every 3 weeks					
BCNU	100 every 3 weeks+	37	S,A,L	16 (2)	yes	Curtis (1974)
CTX	1000 every 3 weeks					
MeCCNU	100 every 6 weeks+	10	—	30	—	Bodey et al. (1973)
CTX	1000 every 6 weeks					
CCNU	100 every 6 weeks+	32	S,A,L	25 (2)	yes	Trowbridge et al. (1978)
Adria	60 every 6 weeks					
CCNU	70 every 6 weeks+	174	S,A,L	5	yes	Wolf et al. (1979)
Adria	40 every 3 weeks					
MeNU[d]	300 mg twice weekly+	19	S	5	—	Perevodchikova (1976)
CTX	400 mg three times weekly.					
	Every 3 weeks					
MeCCNU	150 every 6 weeks	26	S,A,L	0	—	Richards et al. (1976)
VCR	1 day 1,8,15					
	Repeat every 6 weeks					
CCNU	100-130 every 6 weeks	24	S,A	13	—	Hoogstraten et al. (1975)
Bleo	10 twice weekly × 10 weeks					

[a] Responders to therapy survived significantly longer than did nonresponders.
[b] S = squamous cell carcinoma, A = adenocarcinoma, L = large-cell anaplastic carcinoma.
[c] ( ) = number of complete responders.
[d] MeNU = methyl nitrosourea.

mg/m^2 as a single dose (Wasserman *et al.*, 1975). An alternate drug schedule for CCNU administration would be to administer CCNU 75 mg/m^2 every 3 weeks (Israel *et al.*, 1973) or 130 mg/m^2 every 4 weeks, for 3 or more courses (Broder and Hansen, 1973). A standard cyclophosphamide dose and schedule is 1000–1100 mg/m^2 every 3 weeks; and for intermittent adriamycin administration, 60–75 mg/m^2 either as a single dose or divided over 3 days is often given.

Table III lists two CCNU plus cyclophosphamide studies and one study in which mechlorethamine (nitrogen mustard, HN_2) replaced cyclophosphamide (based on Veterans Administration Lung Group data suggesting a superiority for mechlorethamine in patients with squamous cell carcinoma of the lung (Green *et al.*, 1969). In each of these trials approximately 54% of the optimal dose of CCNU was employed along with 70% of the optimal single dose of cyclophosphamide or mechlorethamine. The highest objective response rate in any of these three trials was 12% (Edmonson *et al.*, 1976a; Edmonson *et al.*, 1976b; Wolf *et al.*, 1979). One trial each studied BCNU plus cyclophosphamide and MeCCNU plus cyclophosphamide. The former trial utilized high doses of both drugs and obtained a 16% response rate, with 2 complete responses among 37 patients (Curtis, 1974). The latter trial, a pilot study including only 10 patients with unspecified histologies, used smaller quantities of drugs while achieving a 30% response rate (Bodey *et al.*, 1973).

Two trials evaluated CCNU plus adriamycin combinations. The study of Trowbridge *et al.* (1978) employed higher doses of CCNU and lower doses of adriamycin than did the study of Wolf *et al.* (1979). The former trial obtained a 25% response rate with two complete responses among 32 patients as compared to a 5% response rate in the latter trial. Factors other than drug dosage that might have influenced the outcome of these two studies include differences in the patient populations under study, differences in the amount of therapy that patients received prior to entry into study, and the fact that the latter trial was a multiinstitution protocol. Other combinations of a nitrosourea with a second drug (listed in Table III) do not appear especially promising.

D. Nitrosourea Combinations with Two Additional Drugs

Table IV lists chemotherapy regimens that combine a nitrosourea with two additional drugs. It is obvious that as one combines 3 myelosuppressive drugs the percentage of the optimal single dose of any one of the drugs in the combination necessarily decreases. Despite this, however, response rates to these three-drug combinations appear to be generally higher than response rates to the various component drugs as single agents or to two-drug combination chemotherapy. Several studies in this table deserve comment. In the initial study listed, patients with adenocarcinoma were randomly allocated to methotrexate plus cyclophosphamide alone or with CCNU (Hansen *et al.*, 1976). Addition of CCNU resulted in a significant increase in both the objective response rate and in median sur-

TABLE IV. Nitrosourea-Containing Three-Drug Combinations

Drug doses & schedule (mg/m^2)		No. of patients	Histologies[a] studied	Response rate (%)	Survival[b] benefit	Reference
CCNU	50 every 6 weeks	28	A	38	yes	Hansen *et al.* (1976)
MTX	10 twice weekly					
CTX	500 every 3 weeks					
	Randomized trial					
CCNU	50 every 6 weeks	46	S,L	8	yes	Hansen *et al.* (1976)
MTX	10 twice weekly					
HN$_2$	10 every 3 weeks					
	Randomized trial					
CCNU	50 every 4 weeks	28	A	25	yes	Richards *et al.* (1979)
MTX	10 twice weekly					
CTX	500 every 2 weeks					
	Randomized trial					
CCNU	50 every 6 weeks	88	A,L	16 (2)[c]	yes	Vincent *et al.* (1980)
MTX	10 twice weekly					
CTX	500 every 3 weeks					
	Randomized trial					
CCNU	80 every 5 weeks	23	S,A,L	22	yes	Alberto *et al.* (1975)
Adria	20/day × 2					
HU	1000/day × 5					
	Repeat every 5 weeks					
CCNU	65 every 8 weeks	115	S	16	yes	Livingston *et al.* (1977)
Adria	40 every 4 weeks					
HN$_2$	8 every 4 weeks					
	Randomized trial					
CCNU	50 every 4 weeks	53	S	19	yes	Richards *et al.* (1979)
Adria	30 every 2 weeks					
MTX	10 twice weekly					
	Randomized trial					
CCNU	70–100 every 4–7 weeks	8	S,A,L	75	yes	Wilson *et al.* (1976)
Adria	15 weekly					
HMM	225 daily					
CCNU	40/day × 2	23	A	22 (1)	yes	Bedikian *et al.* (1979)
CTX	200/day × 4					
5-FU	cont. infusion or oral.					
	Randomized trial					
CCNU	90 every 6 weeks	15	S,A,L	13	no	DiBella *et al.* (1975)
MTX	8 twice weekly × 3 weeks					
HMM	150 daily × 3 weeks except on MTX days					
CCNU	50–100 day 1	13	S,A,L	0	—	Mintz *et al.* (1978)
BLEO	15u weekly × 3					
HMM	300 daily × 3 weeks					
	Repeat every 6 weeks					
MeCCNU	100 every 6 weeks	30	S,A,L	0	—	Richards *et al.* (1976)
MTX	6 twice weekly × 3w					
VCR	1 days 1,8,15					
	Randomized trial					
MeCCNU	150 every 6 weeks	19	A,L	0	—	Bernath *et al.* (1976)
5-FU	500 weekly					
VCR	1 every 6 weeks					

[a] S = squamous cell, A = adenocarcinoma, L = large-cell anaplastic carcinoma.
[b] Responders to therapy survived significantly longer than did nonresponders.
[c] ( ) = number of complete responders.

vival. In contrast, the addition of CCNU to methotrexate plus mechlorethamine failed to improve response rate or median survival in patients with squamous cell or large-cell anaplastic lung cancer. In two other randomized trials, including adenocarcinoma patients alone or with large-cell anaplastic carcinoma patients, the CCNU, methotrexate, cyclophosphamide combination was also superior to alternative, non–nitrosourea-containing drug combinations (Richards *et al.*, 1979; Vincent *et al.*, 1980). Combinations containing CCNU plus adriamycin with a variable third drug appear to give comparable response rates to the CCNU, methotrexate, cyclophosphamide combination (Alberto *et al.*, 1975; Livingston *et al.*, 1977; Richards *et al.*, 1979; Wilson *et al.*, 1976). It would be of interest if another group confirmed the high response rate seen with CCNU, adriamycin, and hexamethylmelamine treatment (Wilson *et al.*, 1976).

E. Nitrosourea Combinations with Three Additional Drugs

The two four-drug combination chemotherapy regimens most studied in non–small-cell lung cancer patients are the methotrexate, adriamycin, CCNU, cyclophosphamide (MACC) regimen originally reported, and recently updated, by Chahinian and colleagues (Chahinian *et al.*, 1977; Chahinian *et al.*, 1979) and the cyclophosphamide, vincristine, MeCCNU, bleomycin (COMB) regimen initially reported by Livingston *et al.* (1975a). As is evident in Table V, none of the confirmatory studies of MACC have reproduced the initial high response rates. Closest to the original results is the combination of MACC plus vincristine, with a 28% response rate (Lyman *et al.*, 1980). All MACC studies, with one exception, however (Vogl *et al.*, 1979), do demonstrate significant survival benefit for responders to MACC chemotherapy.

The COMB regimen was probably never therapeutically useful in that the initial study demonstrated no significant survival benefit for responders to therapy compared to nonresponders (Livingston *et al.*, 1975a). The response rate of 5% to COMB treatment reported by Bodey *et al.* (1977) probably eliminated any remaining interest in this combination.

F. Nitrosourea Combinations with Four or More Additional Drugs

Nitrosourea-containing regimens of 5 or 6 drugs are included in Table VI. The BACON regimen (bleomycin, adriamycin, CCNU, vincristine, and mechlorethamine) and the PACCO regimen (cisplatin, adriamycin, cyclophosphamide, CCNU and vincristine), the former for patients with squamous cell lung cancer, the latter for all non–small-cell lung cancer histologies, both had high response rates in pilot studies. In a confirmatory Southwest Oncology Group study of BACON the objective response rate decreased to 21% (Livingston *et al.*, 1977). Similarly, early results of a confirmatory PACCO trial yielded a 17% response rate.

TABLE V. Nitrosourea-Containing Four-Drug Combinations

Drug dose & schedule (mg/m^2)	No. of patients	Histologies[a] studied	Response rate (%)	Survival & benefit	Reference
MTX 30–40 every 3 weeks Adria 30–40 CTX 400 CCNU 30	68	S,A,L	S 36 A 58 (1)[c] L 35	yes	Chahinian et al. (1979)
MTX 40 every 3 weeks Adria 40 CTX 400 CCNU 30	43	S,A	S 15 A 6	no	Vogl et al. (1979)
MTX 40 every 3 weeks Adria 40 CTX 400 CCNU 30	49	S,A,L	18	yes	Halpern et al. (1980)
MTX 30–35 every 3 weeks Adria 30–35 CTX 300–350 CCNU 30–35	72	S,A,L,S	<22 includes small-cell	yes	Cohen and Perevodchikova (1979)

TABLE V. Nitrosourea-Containing Four-Drug Combinations (*Continued*)

Drug dose & schedule (mg/m²)		No. of patients	Histologies[a] studied	Response rate (%)	Survival & benefit	Reference
MTX	20	38	S,A,L	18	yes	Eagan (1979)
Adria	20					
CTX	200					
CCNU	30					
MTX	days 1 & 8 every 4 weeks	25	S,A,L	28 (2)	yes	Lyman *et al.* (1980)
Adria	Dose & schedule not stated					
CTX						
CCNU						
VCR						
CTX	800–1000 every 4–6 weeks	61	S,A	S 33 A 11	no	Livingston *et al.* (1975a)
VCR	0.75–1.0 mg twice weekly × 12–48 doses					
MeCCNU	100 every 4–6 weeks					
Bleo	7.5–30 mg with VCR					
CTX	1000	20	S	5	no	Bodey *et al.* (1977)
VCR	0.5 twice weekly × 24					
MeCCNU	100					
Bleo	8 twice weekly × 24					
Randomized trial						

[a] S = squamous cell, A = adenocarcinoma, L = large-cell carcinoma.
[b] Responders to therapy survived significantly longer than did nonresponders.
[c] ( ) = number of complete responders.

TABLE VI. Nitrosourea-Containing Combinations with Five or More Drugs

Drug dose & schedule (mg/m²)		No. of patients	Histologies[a] studied	Response rate (%)	Survival[b] benefit	Reference
Bleo	30 mg weekly × 6	50	S	45	yes	Livingston et al. (1975b)
Adria	40 every 4 weeks					
CCNU	65 every 4–8 weeks					
VCR	.75–1.0 mg weekly × 6					
HN₂	8 every 4 weeks					
Bleo	30u weekly × 6	116	S	21	yes	Livingston et al. (1977)
Adria	40 every 4 weeks					
CCNU	65 every 8 weeks					
VCR	.75–1.0 mg weekly × 6					
HN₂	8 every 4 weeks					
Cis-PT	50 every 4–6 weeks	35	S,A,L	S 87 (2)[c]	yes	Takita and Brugarolas (1973)
Adria	50			A 64 (1)		
CTX	300			L 33		
CCNU	50					
VCR	1.4					
Cis-PT	50 every 4–6 weeks	18	S,A,L	17 (1)	—	Whitehead et al. (1980)
Adria	50					
CTX	300					
CCNU	50					
VCR	1.4					
MeCCNU	150 every 6 weeks	28	S,A,L	7	—	McMahon et al. (1975)
CTX	250/day × 4					
VCR	2 mg day 7					
MTX	1 mg days 21,28,35					
Bleo	15 mg days 21,28,35					
CCNU	65 day 1 every 8 weeks	33	S,A,L	18 (2)	yes	Miller et al. (1979)
CTX	500 day 1					
Adria	30 day 2					
Procarb	100 days 8–18					
HMM	250 days 8–12					
MTX	20 days 29,33,36,40					

[a] S = squamous cell, A = adenocarcinoma, L = large-cell carcinoma.
[b] Responders to therapy survived significantly longer than did nonresponders
[c] ( ) = number of complete responders.

G. Surgical Adjuvant Studies Employing a Nitrosourea Alone or in Combination

Nitrosoureas alone or in combination have been utilized in surgical adjuvant trials. Table VII. contains a listing of these studies. Of the four trials listed, the initial two studies are closed to patient entry whereas the latter studies continue to accrue patients. Preliminary analyses of the first two studies fail to show survival benefit for chemotherapy treated patients (Mountain et al., 1979; Higgins and Shields, 1979). These results are not surprising in view of the low response rates associated with CCNU alone or with CCNU plus hydroxyurea combinations in patients with more advanced disease.

TABLE VII. Nitrosoureas in Non–Small-Cell Lung Cancer: Randomized Surgical Adjuvant Trials

Chemotherapy arm drug dose and schedule (mg/m²)		No. evaluable patients	Results to date	Reference
CCNU	130 every 6 weeks for 2 years	72	10 recurrences in the CCNU arm; 8 recurrences in the no-treatment arm	Mountain *et al.* (1979)
CCNU	70 every 6 weeks + Hydroxyurea 1000 twice weekly for 1 year	471	3 year survival in 54.3% in chemotherapy arm and 56.6% in the no-treatment arm	Higgins and Shields (1979)
CCNU CTX MTX	70 every 4 weeks + 1000 every 4 weeks + 40 every 4 weeks ± BCG	392	Too early	Israel *et al.* (1979)
CCNU Adria CTX MTX Repeat × 12	30 every 4 weeks 35 400 30	—	Too early Cancer and Leukemia Group B study	—

H. Current Chemotherapy Trials in Non–Small-Cell Lung Cancer

One method for determining the current status of nitrosourea-containing combinations in the treatment of non–small-cell lung cancer is to review chemotherapy abstracts submitted to the 1980 meeting of the American Society of Clinical Oncology (ASCO). Data from this source is presented in Table VIII. Of the 20 studies employing combination chemotherapy, five studied combinations containing a nitrosourea. The MACC regimen (methotrexate, adriamycin, cyclophosphamide, CCNU) was utilized in two studies, MACC plus vincristine in a third study, MAC without cyclophosphamide in a fourth trial, and PACCO in the final study. The median response rate for these five studies was 19%. For the 15 combination chemotherapy studies that did not include a nitrosourea the

TABLE VIII. Non–Small-Cell Lung Cancer: 1980 ASCO Abstracts

Total number of studies	25
Phase II single agents	5
Combination regimens	20
Nitrosourea-containing combinations	5
MACC (2)	
MACC + VCR	
MAC	
PACCO	
Median response rate	(range)
Nitrosourea regimens	19 (17–28)
Nonnitrosourea regimens	30 (15–73)

median response rate was 30%. In contrast to the nitrosourea-containing regimen studies, which were confirmatory trials of regimens that had previously been reported to be active, the remaining 15 studies represented the initial trial of a new drug combination. Studies by other groups of these latter regimens will undoubtedly be reported within the next year or two.

IV. CONCLUSIONS

This overview of nitrosourea-containing chemotherapy regimens in non-small-cell lung cancer indicates that all of these regimens have, at best, only modest therapeutic activity. Very few complete responders to treatment result from any of these regimens. The synergism of nitrosoureas with other drugs that was evident in preclinical tumor systems has not yet been recorded in human non-small-cell lung cancer trials. At best, additive effects of the various drugs in combination have been seen; at worst, the various nitrosourea combinations tested to-date may only be slightly better than single-agent treatment. This is not to say that nitrosoureas may not be important in future lung cancer combinations as new active drugs are identified. While it is possible to criticize the design of many of the nitrosourea-containing drug combinations with regard to dose ratios of the various drugs in the combination, and while there is evidence to suggest that alternate nitrosourea schedules should have been more vigorously tested, it appears at present that further testing along these lines does not have a high priority. Current efforts in non-small-cell lung cancer should still be devoted to finding more active drugs for use in future combination chemotherapy studies.

REFERENCES

Ahmann, D. L., Carr, D. T., and Hahn, R. G. (1972). *Cancer Chemother. Rep. 56*, 401-403.

Alberto, P., Barrelet, L., Chaputs, B., and Garcia, B. (1975). *Eur. J. Cancer 11*, 795-799.

Bedikian, A. Y., Staab, B., Livingston, R., Valdivieso, M., Burgess, M. A., and Bodey, G. P. (1979). *Cancer 44*, 858-863.

Bernath, A. M., Cohen, M. H., Ihde, D. C., Fossieck, B. E., Matthews, M. J., Minna, J. D. (1976). *Cancer Treat. Rep. 60*, 1393-1394.

Bodey, G. P., Gottlieb, J. A., Livingston, R., and Frei, E. III. (1973). *Cancer Chemother. Rep. 4*, 227-229.

Bodey, G. P., Lagakos, S. W., Gutierrez, A. C., Wilson, W. E., Selawry, O. S. (1977). *Cancer 39*, 1026-1031.

Broder, L. E., and Hansen, H. A. (1973). *Eur. J. Cancer 9*, 147-152.

Casper, E. S., and Gralla, R. J. (1979). *Cancer Treat. Rep. 63*, 549-550.

Chahinian, A. P., Arnold, D. J., Cohen, J. M., Purpora, D. P., Jaffrey, I. S., Teirstein, A. S., Kirschner, P. A., and Holland, J. F. (1977). *JAMA 237*, 2392-2396.

Chahinian, A. P., Mandel, E. M., Holland, J. F., Jaffrey, I. S., and Teirstein, A. S. (1979). *Cancer 43*, 1590-1597.

Cohen, M. H., and Perevodchikova, N. I. (1979). *In* "Lung Cancer: Progress in Therapeutic Research" (F. Muggia and M. Rozencweig, eds.), pp. 343-374. Raven Press, New York.

Cohen, J. M., Weisberg, S. R., and Kanner, S. P. (1980). *Proc. Am. Soc. Clin. Oncol. 21*, 450.

Corkey, J., Wilkinson, J., Zipoli, T., Greene, R., and Lokich, J. (1980). *Proc. Am. Soc. Clin. Oncol. 21*, 449.

Creagan, E. T., Eagan, R. T., Flemming, T. R., Frytak, S., Kvols, L. K., and Ingle, J. M. (1979). *Cancer Treat. Rep. 63*, 2105-2106.

Curtis, J. E. (1974). *Cancer Chemother. Rep. 58*, 883-888.

DiBella, N. J., Nelson, R. A., and Norgard, M. J. (1975). *Oncol. 32*, 82-85.

Eagan, R. T. (1979). *In* "Lung Cancer: Progress in Therapeutic Research" (F. Muggia and M. Rozencweig, eds.), pp. 383-391. Raven Press, New York.

Eagan, R. T., Carr, D. T., Coles, D. T., Cines, D. E., and Ritts, R. E., Jr. (1974). *Cancer Chemother. Rep. 58*, 913-918.

Edmonson, J. H., Lagakos, S., Stolbach, L., Perlia, C. P., Bennett, J. M., Mansour, E. G., Horton, J., Regelson, W., Cummings, F. J., Israel, L., Brodsky, I., Shnider, B. I., Creech, R., and Carbon, P. P. (1976a). *Cancer Treat. Rep. 60*, 625-627.

Edmonson, J. H., Lagakos, S. W., Selawry, O. S., Perlia, C. P., Bennett, J. M., Muggia, F. M., Wampler, G., Brodovsky, H. S., Horton, J., Colsky, J., Mansour, E. G., Creech, R., Stolbach, L., Greenspan, E. M., Levitt, M., Israel, L., Ezdinli, E. Z., and Carbone, P. P. (1976b). *Cancer Treat. Rep. 60*, 925-932.

Green, R. A., Humphrey, E., Close, H., and Patno, M. E. (1969). *Am. J. Med. 46*, 516-525.

Halpern, J., Catane, R., Sulkes, A., Brufman, G., Rizel, S., Gez, E., Weshler, Z., Biran, S., and Fuks, Z. (1980). *Proc. Second World Conf. on Lung Cancer 2*, 131.

Hansen, H. H., Selawry, O. S., Simon, R., Carr, D. T., Van Wyk, C. E., Tucker, R. D., and Seally, R. (1976). *Cancer 38*, 2201-2207.

Higgins, G. A., and Shields, T. W. (1979). *In* "Lung Cancer: Progress in Therapeutic Research" (F. Muggia and M. Rozencweig, eds.), pp. 433-442. Raven Press, New York.

Hoogstraten, B., Haas, C. D., Haut, A., Talley, R. W., Rivkin, S., and Isaacs, B. L. (1975). *Med. Ped. Oncol. 1*, 95-106.

Ihde, D. C., Cohen, M. H., Bunn, P. A., Bernath, A. M., Broder, L. E., and Minna, J. D. (1978). *Cancer Treat. Rep. 62*, 155-157.

Israel, L., Chahinian, P., Accard, J. L., Choffel, C., Combes, P. F., Canrigal, A., Germouty, J., Migueres, J., Schaerer, R., and Sotto, J. J. (1973). *Eur. J. Cancer 9*, 789-797.

Isracl, L., Bonadonna, G., and Sylvester, R. (1979). *In* "Lung Cancer: Progress in Therapeutic Research" (F. Muggia and M. Rozencweig, eds.), pp. 443-452. Raven Press, New York.

Livingston, R. B., Einhorn, L. H., Bodey, G. P., Burgess, M. A., Freireich, E. J., Gottlieb, J. A. (1975a) *Cancer 36*, 327-332.

Livingston, R. B., Einhorn, L. H., Burgess, M. A., Freireich, E. J., Gottiieb, J. A. (1975b) *Cancer Chemother. Rep. 6*, 361-367.

Livingston, R. B., Heilbrun, L., Lehane, D., Costanzi, J. J., Bottomley, R., Palmer, R. L., Stuckey, W. J., Hoogstraten, B. (1977) *Cancer Treat. Rep. 61*, 1623-1629.

Lyman, G. H., Colledge, P., Johnson, D., Williams, C. C. (1980) *Proc. Am. Soc. Clin. Oncol. 21*, 460.

Mayo, J. G., Laster, W. R., Jr., Andrews, C. M., Shabel, F. M., Jr. (1972) *Cancer Chemother Rep. 56*, 183-195.

McMahon, L. J., Jones, S. E., Durie, B. G. M., Salmon, S. E. (1975) *Cancer Letters 1*, 97.

Miller, C. F., Weltz, M. D., Heim, W. J., Blom, J. (1979) *Cancer Treat. Rep. 63*, 1351-1352.

Mintz, U., Bitran, J. D., Cooksey, J. A., Desser, R. K., DeMesster, T. R., Golomb, H. M. (1978) *Cancer Treat. Rep. 62*, 567-569.

Mountain, C. F., Vincent, R. G., Sealy, R., Khalil, K. G. (1979) in "Lung Cancer: Progress in Therapeutic Research" (F. Muggia and M. Rozencweig, eds.) Raven Press, New York, pp. 421-431.

Olshin, S., Siddiqui, S., and Firat, D. (1972). *Cancer Chemother. Rep. 56*, 259-261.

Perevodchikova, N. I. (1976). *Cancer Treat. Rep. 60*, 1479-1481.

Richards, F. II., Pajak, T. F., Cooper, M. R., and Spurr, C. L. (1973). *Cancer Chemother. Rep. 57*, 419-422.

Richards, F. II, Cooper, M. R., Muss, H. B., White, D. R., and Spurr, C. L. (1976). *Cancer 38*, 1077-1082.

Richards, F. II, White, D. R., Muss, H. B., Jackson, D. V., Stuart, J. J., Cooper, M. R., Rhyne, L., Spurr, C. L. (1979). *Cancer 44*, 1576-1581.

Samson, M. K., Baker, L. H., Fraile, R. J., Izbicki, R. M., and Vaitkevicius, V. K. (1977). *Cancer Treat. Rep. 61*, 59-64.

Schabel, F. M., Jr. (1976). *Cancer Treat. Rep. 60*, 665-698.

Stolinsky, D. C., Bull, F. E., Pajak, T. F., and Bateman, J. R. (1975). *Oncol. 31*, 288-292.

Takita, H., and Brugarolas, A. (1973). *J. Nat. Cancer Inst. 50*, 49-53.

Takita, H., Marabella, P. C., Edgerton, F., and Rizzo, D. (1979). *Cancer Treat. Rep. 63*, 29-33.

Trowbridge, R. C., Kennedy, B. J., Vosika, G. J. (1978). *Cancer 41*, 1704-1709.

Vincent, R. G., Mehta, C. R., Tucker, R. D., Mountain, C. F., Cohen, M. H., Wilson, H. E., and Vogel, C. (1980). *Cancer 46*, 256-260.

Vogl, S. E., Mehta, C. R., and Cohen, N. M. (1979). *Cancer 44*, 864-868.

Wasserman, T. H., Slavik, M., and Carter, S. K. (1975). *Cancer 36*, 1258-1268.

Whitehead, R., Crowley, J., and Carbone, P. P. (1980). *Proc. Am. Soc. Clin. Oncol. 21*, 458.

Whittington, R. M., Fairly, J. L., Majima, H., Patno, M. E., and Prentice, R. (1972). *Cancer Chemother. Rep. 56*, 739-743.

Wilson, W. L., Andrew, W. C., Frekick, R. W., Nealon, T. F., Bick, R. L., and Adams, T. (1976). *Cancer Treat. Rep. 60*, 269-271.

Wolf, J., Hyde, L., Phillips, R. W., and Mietlowski, W. (1979). *In* "Lung Cancer: Progress in Therapeutic Research" (F. Muggia and M. Rozencweig, eds.), pp. 375-382. Raven Press, New York.

Chapter 20

CHEMOTHERAPY OF RECURRENT BRAIN TUMORS[1]

Victor A. Levin

I. INTRODUCTION

Despite the prospects of a decade ago, chemotherapy has proved to be disappointingly ineffective as a treatment modality for brain tumors. Many reviews summarize the results of single- and multiple-agent chemotherapy regimens for recurrent brain tumors (Levin and Wilson, 1975; Wilson *et al.*, 1976; Edwards *et al.*, 1980); while recent studies show that response and/or stabilization of disease can be achieved in many patients who harbor recurrent malignant gliomas (Levin *et al.*, 1978; Levin *et al.*, 1980) or medulloblastoma (Crafts, et al., 1978), the median time to tumor progression (MTP) is only approximately 6 and 10 months, respectively. Chemotherapy for primary and secondary (metastatic) tumors of the central nervous system is still a palliative measure only because the chemotherapeutic armamentarium for brain tumors is limited; despite all efforts, it has changed little in the past 6 years. The nitrosoureas, particularly BCNU and CCNU, and procarbazine are the mainstays of treatment.

This review of nitrosourea therapy for recurrent brain tumors will be limited to a brief outline of methods for the diagnosis of recurrence and a brief discussion of various treatment plans and their results. Glioblastoma multiforme (GM), the most malignant of the glial tumors, will be considered separately. For ease of

[1]Supported in part by National Cancer Institute Project Program Grant CA-13525 and American Cancer Society Faculty Research Award FRA-155.

presentation, we established a category of nonglioblastoma multiforme malignant gliomas (NGM) that includes predominantly anaplastic astrocytoma and mixed malignant glioma. This latter category is somewhat arbitrary from a histologic standpoint because of the heterogeneity of tumor types within the NGM group. Tumors in the NGM group have many common biological features and respond similarly to radiation therapy and chemotherapy. Tumors such as medulloblastoma, ependymoma, and malignant oligodendroglioma will be discussed briefly, which reflects the limited literature concerning nitrosourea therapy for these tumors.

II. METHODS

A. Diagnosis of Tumor Recurrence

The diagnosis of tumor recurrence is easiest for supratentorial tumors and somewhat more difficult for infratentorial tumors, particularly those that metastasize along the neural axis. Sometimes neurological deterioration is caused by radiation necrosis, hydrocephalus, seizures, cerebrovascular accidents, and other factors, and not by tumor growth. To avoid the pitfall of assuming tumor regrowth when there is none, patients with presumed recurrence undergo a full range of diagnostic studies to establish an unequivocal diagnosis; infrequently, operative determination of the pathologic process may be necessary.

As a rule, tumor recurrence can be diagnosed with confidence if the following criteria are met:

1. Progressive deterioration of neurological signs and symptoms that are not attributable to other causes.
2. Radiographic evidence of increasing tumor mass as shown by at least two of the following procedures: angiography, pneumoencephalography, computerized tomography (CT) brain scan, or radionuclide (RN) brain scan.
3. An interval of at least 3 months after the completion of radiation therapy to allow for delayed benefit and for the disappearance of radiation-induced dysfunction. The delay is advisable because 28% of patients who worsen during this period do so for reasons other than tumor recurrence (Hoffman et al., 1979).

Certain recurrent tumors require an exception to these general criteria, but they are encountered rarely. The most common occur in patients in whom a well-differentiated astrocytoma transforms to an anaplastic astrocytoma; in this instance, the RN scan may show an increase in the permeability defect before a change is seen on the CT scan, and there may or may not be clinical deterioration. In another instance, the CT scan shows decalcification within the tumor that coincides with dedifferentaition of oligodendroglia, ependymoma, or astrocytoma tumors; changes can be seen on the CT scan before they are evident on RN scans and from clinical findings. In both instances, a documented decline in only one diagnostic study warrants the institution of, or change in, therapy. Although it is uncommon, radiation necrosis can present as an intracranial mass. Necrosis can sometimes be differentiated from recurrent tumor on arteriograms and CT scans; occasionally, surgical exploration is required to make this diagnosis unequivocal.

B. Measurements of Response to Therapy

Reduction in tumor mass is the major criterion of drug effect for recurrent tumors. If the effects of concurrently administered glucocorticoids are taken into account, the tumor's response to chemotherapy can be qualitatively determined by comparing serial neurological examinations, CT brain scans, and RN brain scans (Levin *et al.*, 1977). Our patients are evaluated with these modalities before each course of therapy (approximately every 6–8 weeks). The result of each test is compared with each of the others to determine response or disease progression. To designate relative changes, the neurological examination and both scans are rated on a scale of 0 to either -3 or $+3$, in which 0 is defined as no change in any category, -1 is possibly worse, -2 is definitely worse, and -3 is markedly worse; the $+$ values indicate comparable levels of improvement.

Response is defined as a positive change in two of the three tests ranked $+2$ or $+3$ in the same evaluation period, or when two of the three tests are ranked $+2$ or $+3$ in consecutive evaluation periods, provided that the patient is receiving a stable or decreasing dose of glucocorticoids.

Progression is defined as a decline in two of the three tests ranked -2 or -3 within the same evaluation period, or when two of the three tests are ranked -2 or -3 in consecutive evaluation periods, provided that the patient's dosage of glucocorticoids has not been decreased since the last evaluation.

Stable disease (synonymous with "probable responder" from previous studies) is defined as either no change or possible (-1 or $+1$) changes in all three tests, or when only one test shows a definite change; in either instance, patients should be receiving stable or decreasing doses of glucocorticoids.

C. Chemotherapy

The various treatment schedules in use for nitrosoureas given as single agents and in combination with other agents is shown in Table I. These schedules are current and represent typical schedules that can be used; they do not constitute an exhaustive review of published treatment schedules.

TABLE I. Nitrosourea Schedules Used to Treat Recurrent Brain Tumors

Drug	Dose (mg/m²)	Schedule (day(s) of course)	Course (weeks)
	Single Agent		
BCNU (iv)	80	1-3	6
BCNU (iv)	220	1	6
CCNU (oral)	130	1	6
MeCCNU (oral)	170-200	1	6
PCNU (iv)	100	1	6
	Drug Combinations		
BCNU	200	1	6
+ 5-Fluorouracil (iv)	1000 (continuous)	14-16	
BCNU	100	1	6
+ Procarbazine (oral)	100	2-15	
BCNU	30	3	6
+ Vincristine (iv)	2	weekly	
BCNU	180	1	6
+ 5-Fluorouracil (iv)	1000 (continuous)	15-17	
+ Hydroxyurea (oral)	400 q 6 hr	24-26	
+ 6-Mercaptopurine (oral)	100 q 6 hr	33-35	
CCNU (PCV 3)	110	1	6
+ Procarbazine (oral)	60	8-21	
+ Vincristine (iv)	1.4	8, 29	
CCNU (PCV 2)	75	1	6
+ Procarbazine (oral)	100	8-21	
+ Vincristine (iv)	1.4	8, 29	
CCNU (PCV 1)	75	1	6
+ Procarbazine (oral)	100	1-14	
+ Vincristine (iv)	1.4	1	

III. RESULTS OF TREATMENT

A. Glioblastoma Multiforme and Anaplastic Astrocytomas

BCNU, CCNU, and procarbazine have produced comparable response rates when used as single agents against recurrent cerebral glioblastoma multiforme and anaplastic astrocytomas (Rosenblum *et al.*, 1973; Hilderbrand *et al.*, 1975; Levin and Wilson, 1975; EORTC Brain Tumor Group, 1978); however, BCNU has a somewhat better MTP. Table II summarizes studies with nitrosoureas used in single-agent and combination regimens.

MTPs in the BCNU, CCNU, BCNU–vincristine (VCR), and procarbazine–CCNU–vincristine (PCV 1) studies were based on a decline in both the RN brain scan *and* the clinical neurological status; MeCCNU response was measured by improvements in either the RN scan or clinical status that were sustained after discontinuation of glucocorticoids. The current criteria evaluate patients dif-

TABLE II. Responses to Chemotherapy for Supratentorial Malignant Gliomas

Agent	All malignant gliomas	GM[a]	NGM[a]	MTP[b] (weeks)	Reference
BCNU	20/40 (50%)[c,d]			38/–[d]	Wilson et al. (1970); Levin and Wilson et al. (1975)
CCNU	10/22 (45%)[c,d]			24/–[d]	Fewer et al. (1972); Levin and Wilson (1975)
MeCCNU	8/29 (28%)[c]			44/–[c]	Tranum et al. (1975)
BCNU–VCR	6/12 (50%)[c,d]			17/–	Fewer et al. (1972)
BCNU–PCB	21/52 (40%)[c]	3/3/18	10/5/34	34/20	Levin et al. (1976)
BCNU–5-FU	24/29 (83%)	1/2/4	8/13/25	34/26	Levin et al. (1978)
BCNU–5-FU–HU–6MP	10/15 (67%)	1/1/4	4/4/11	38/22	Unpublished observations
CCNU–PCB–VCR (PCV 1)	18/29 (62%)	1/2/8	11/4/21	30/–	Gutin et al. (1975)
CCNU–PCB–VCR (PCV 3)	16/19 (84%)	0/1/1	8/7/18	31/25	Levin et al. (1980)
CCNU–PCB–VCR (PCV 3)	12/27 (44%)	2/3/11	2/5/16		

[a] Responders/stable disease/total group.

[b] Median time to tumor progression from beginning of treatment for responders/stable disease patients.

[c] Combined GM and NGM groups; responders and stable disease patients combined. No previous chemotherapy given to patients in these studies; exception [ ] = previous chemotherapy failures. Patients failing to receive a second course of chemotherapy were excluded.

[d] These values are overestimated because previous response criteria required clinical deterioration together with RN scan evidence of tumor enlargement.

[e] Criteria of response not comparable to other studies; improved in either RN scan or clinical status required for designation of response.

ferently and in many ways are more sensitive. In the earlier studies, the time to tumor progression would have been documented earlier in nearly 25% of patients; thus the MTPs for BCNU, CCNU, BCNU–VCR, and PCV 1 are overestimated; and because the MeCCNU results require discontinuation of glucocorticoids, they may be underestimated. Because the studies are ongoing, BCNU–5-fluorouracil (5-FU)–hydroxyurea (HU)–6-mercaptopurine (6-MP) data are approximate.

In attempts to improve the efficacy of BCNU, protocols were developed that combined BCNU with VCR, procarbazine, 5-FU, and the combination of 5-FU, HU, and 6-MP. For reasons that are unknown, BCNU combined with VCR produced results that were inferior to BCNU alone; patients treated with BCNU–procarbazine had a disappointing response rate (defined as a response and stabilization rate). We had better success with the BCNU–5-FU combination, a protocol developed from animal data and human glioma cell kinetic studies. 5-FU was administered 2 weeks after BCNU at a time when tumor cell repopulation was calculated to be maximal. Disease progression was temporarily arrested in 83% of patients (producing either improvement or stability) (Levin et al., 1978). A dose–response relationship was noted because the patients designated as "responders" tolerated higher dosages of BCNU during their first five courses of chemotherapy than did patients who were categorized as having "stabilized disease." The MTP for the entire group was 26 weeks, and that for the responders was 34 weeks. Unfortunately, only 17% of patients were alive at 1 year.

In an effort to improve this therapy, we added HU and 6-MP to the BCNU–5-FU combination. Preliminary data from 34 patients show no significant difference between BCNU–5-FU and the new combination of BCNU–5-FU–HU–6-MP (unpublished observation). CCNU combinations with procarbazine and vincristine (PCV) are well tolerated and have been used against a large spectrum of tumors (Gutin *et al.*, 1975; Shapiro and Young, 1976). Our schedules for PCV for patients with recurrent gliomas has changed over the years (PCV 1 to PCV 3). Using the results of animal studies, we have been able to reduce systemic toxicity by instituting treatment with procarbazine a week after beginning therapy with CCNU, and by reducing the dosage of procarbazine (PCV 3, Table I). PCV 3 was evaluated in a mixed patient population, among whom 59% had received prior chemotherapy (Levin *et al.*, 1980). Of the patients who had not received prior chemotherapy, 42% responded and 42% had disease stability; patients who had received prior chemotherapy showed only a 15% response rate and a 30% rate of disease stabilization. Interestingly, among the patients who showed either response or disease stabilization, the MTP differed little between those who had received prior chemotherapy and those who had not; the MTP was 31 weeks for responders and 25 weeks for stable disease patients. Thirty percent of patients were alive at 1 year.

In both the PCV 3 and BCNU–5-FU studies, patients harboring GM invariably did considerably worse (MTP of 12 to 14 weeks) than did patients who harbored NGM (MTP of 26 to 27 weeks). The most important common feature

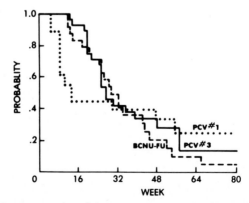

Fig. 1. Kaplan–Meier representation of time to tumor progression for patients who responded to therapy or who had disease stabilization. PCV 3 (——), BCNU–5-FU (FU) (- - -), and PCV 1 ($\cdots$) are compared. PCV 1 was worse than the other two treatment regimens ($p < 0.06$ by the Gehan modification of the Wilcoxon rank sum analysis). Reproduced from Levin *et al.* (1980).

between the BCNU–5-FU and PCV 3 combinations is an ability to halt disease progression; unfortunately, long-term remission was not achieved by the new protocols when the results were compared, respectively, to those with either BCNU alone (Wilson *et al.*, 1970) or PCV 1 (Gutin *et al.*, 1975) (Fig. 1).

Thus, over the past decade the improvement gained using nitrosoureas in combination with other agents has been minimal. We may be able to stabilize disease more consistently than in the past, but we cannot appreciably alter the ultimate prognosis for either GM or NGM patients if the tumor recurs.

B. Oligodendroglioma

Experience with recurrent oligodendroglioma is limited and mostly anecdotal. Our impression has been that PCV 3 is active against these tumors and that, in general, no therapeutic distinctions can currently be drawn between recurrent oligodendrogliomas and recurrent anaplastic astrocytomas.

C. Brain Stem Tumors

These tumors are similar to glioblastoma multiforme and anaplastic astrocytomas and should respond to the same oncolytic agents. Unfortunately, the chemotherapeutic treatment of recurrent brain stem tumors has been disappointing with respect to both the frequency and duration of response. We recently reviewed our experience with 22 patients who received radiation therapy alone after diagnosis, but at the time of subsequent tumor progression received either CCNU, BCNU, or procarbazine alone, or were given a combination of CCNU, 5-FU, HU, and 6-MP, or procarbazine and either BCNU or CCNU (Fulton *et al.*, 1980). The MTP was 13 weeks after the initiation of chemotherapy, and the median survival was 13–28 weeks (Table III). In terms of time to tumor progression, it did not seem to make much difference if the drug schedules were nitrosourea-prominent or procarbazine-prominent single agent or drug combinations: Uniformly these patients did not respond, and all patients with malignant astrocytomas died.

D. Ependymoma

From our experience with 15 patients harboring recurrent ependymomas who were treated with BCNU, we believe that palliation is possible with either 3-day or single-dose BCNU therapy administered every 6 to 8 weeks. For either schedule, BCNU slowed tumor progression and improved neurological signs and symptoms in patients with spinal or fourth ventricular ependymomas. Precise evaluation of the chemotherapeutic efficacy against recurrent ependymomas is difficult because of tumor location in the spinal cord (which makes the need for reoperation necessary) and the slow growth of these tumors if they go untreated.

TABLE III. Results of Chemotherapy for Recurrent Brain Stem Gliomas

Treatment	N	MTP (weeks)	Median survival (weeks)
BCNU or CCNU and one or more cell-cycle-specific drugs	9	14	28
BCNU or CCNU and procarbazine	12	11	13
		$p = 0.04^a$	$p = 0.02^a$

[a]Gehan modification of Wilcoxon Rank Sum analysis. Adapted from Fulton *et al.* (1980).

E. Medulloblastoma

A number of chemotherapeutic agents can be used to palliate patients harboring medulloblastoma at the time of recurrence. To our knowledge there are no published reports of single agent nitrosourea therapy programs for recurrent medulloblastoma. Most published studies have used CCNU plus vincristine (Mealey and Hall, 1977; Pompili *et al.*, 1978; Ward, 1978) in addition to our use of PCV (Crafts *et al.*, 1978). One program included vincristine, BCNU, and methotrexate (Cohen *et al.*, 1978).

Common to all the combination studies is an ability to induce a high rate of remission for variable periods, sometimes in excess of 40 months. Chemotherapy given at recurrence may be palliative only because prior cranial–spinal irradiation damages the bone marrow stem cells, which compromises efforts to treat recurrent medulloblastoma tumors with myelotic drugs such as the nitrosoureas.

Our experience with PCV 1 and PCV 2 therapies is indicative of the general experience. Table IV summarizes our experience with 22 patients treated on these two programs; 21/22 of patients treated showed stabilization or improvement, and 25% of patients were still responding to therapy at 33 months.

Because few patients tolerated chemotherapy without marked myelotoxicity that required severe dose reduction or delays in retreatment, we are advocating a reduction in spinal radiation doses to 2500 rads in patients without demonstrable spinal metastases at the time of initial irradiation. It is hoped that with this change in therapy, adjuvant chemotherapy can be given at higher doses; and if there is a need to treat medulloblastoma at recurrence, higher drug doses can be administered.

TABLE IV. CCNU, Procarbazine, and Vincristine Chemotherapy for Recurrent Medulloblastoma

	Response	Stable	MTP (weeks)	25th quartile (weeks)
PCV 1	10/16	5/16	45	142
PCV 2	2/6	4/6	63	145

IV. CONCLUSION

The nitrosoureas, particularly BCNU and CCNU, remain the mainstays of brain tumor chemotherapy, despite the fact that they are myelotoxic and occasionally are toxic to lung and kidney. At the time of tumor recurrence, BCNU and CCNU alone or in combination with other agents offer patients harboring primary glial, ependymal, or medulloblastoma tumors the best potential for palliation without undue drug-induced discomfort or risk.

ACKNOWLEDGMENT

The author is deeply indebted to physicians in private practice and at the Kaiser-Permanente Medical Centers of the Greater Bay Area for referring patients to us. The author thanks Beverly J. Hunter for manuscript preparation and Neil Buckley for editorial assistance.

REFERENCES

Cohen, M. E., Duffner, P. K., and Freeman, A. (1978). *Ann. Neurol.* 2, 257 (abstract).

Crafts, D. C., Levin, V. A., Edwards, M. S., Pischer, T. L., and Wilson, C. B. (1978). *J. Neurosurg.* 49, 589-592.

Edwards, M. S., Levin, V. A., and Wilson, C. B. (1980). *Cancer Treat. Rep.* In press.

EORTC Brain Tumor Group. (1978). *Eur. J. Cancer* 14, 851-856.

Fewer, D., Wilson, C. B., Boldrey, E. B., and Enot, K. J (1972). *Cancer Chemother. Rep.* 56, 421-427.

Fulton, D. S., Levin, V. A., and Wilson, C. B. (1980). *J. Neurosurg.* In press.

Gutin, P. H., Wilson, C. B., Kumar, A. R. V., Boldrey, E. B., Levin, V. A., Powell, M., and Enot, K. J. (1975). *Cancer* 35, 1398-1404.

Hildebrand, J., Brihaye, J., Wagenmechl, L., Michael, J., and Kenis, Y. (1975). *Eur. J. Cancer* 11, 585-587.

Hoffman, W. F., Levin, V. A., and Wilson, C. B. (1979). *J. Neurosurg.* 50, 624-628.

Levin, V. A., and Wilson, C. B. (1975). *Semin. Oncol.* 2, 63-67.

Levin, V. A., Crafts, D. C., Wilson, C. B., Schultz, M. J., Boldrey, E. B., Enot, K. J., Pischer, T. L., Seager, M., and Elashoff, R. M. (1976). *Cancer Treat. Rep.* 60, 243-249.

Levin, V. A., Crafts, D., Norman, D., Wilson, C. B., Hoffer, P., and Spire, J. P. (1977). *J. Neurosurg,* 47, 329-335.

Levin, V. A., Hoffman, W. F., Pischer, T. L., Seager, M. L., Boldrey, E. B., and Wilson, C. B. (1978). *Cancer Treat. Rep.* 62, 2071-2076.

Levin, V. A., Edwards, M. S., Wright, D., Seager, M. L., Pischer, T. L., and Wilson, C. B. (1980). *Cancer Treat. Rep.* 64, 237-242.

Mealy, J. and Hall, P. V. (1977). *J. Neurosurg.* 46, 56-64.

Pompili, A., Cianfriglia, F., and Occhipinte, E. (1978). *Minerva Pediatr.* 30, 1721-1724.

Rosenblum, M. L., Reynolds, A. F., Smith, K. A., Rumack, B. H., and Walker, M. D. (1973). *J. Neurosurg.* 39, 306-314.

Shapiro, W. R., and Young, D. F. (1976). *Trans. Am. Neurol. Assoc.* 101, 217-220.

Tranum, B. L., Haut, A., Rivkin, S., Weber, E., Quagliana, J. M., Shaw, M., Tucker, W. G.,
    Smith, F. E., Samson, M., and Gattlieb, J. (1975). *Cancer 35*, 1148–1153.
Ward, H. W. C. (1978). *Med. Pediatr. Oncol. 4*, 315–320.
Wilson, C. B., Boldrey, E. B., and Enot, K. J. (1970). *Cancer Chemother. Rep. 54*, 273–281.
Wilson, C. B., Gutin, P., Kumar, A. R. V., Boldrey, E. B., Levin, V. A., Enot, K. J., Fewer, D.,
    Renaudin, J., Calogero, J., and Crafts, D. C. (1976). *Arch. Neurol. 33*, 739–744.

Chapter 21

LIPID-SOLUBLE NITROSOUREAS IN THE MANAGEMENT OF CHILDREN WITH PRIMARY AND RECURRENT BRAIN TUMORS

Jeffrey G. Rosenstock

I. INTRODUCTION

A discussion of the role of the lipid-soluble nitrosoureas in childhood brain tumors requires first a definition of the problem of childhood brain tumors, which are the second most common malignancy of childhood (Miller, 1969). Koo's review of the experience in Vienna highlights the distribution of pediatric brain tumors (Koos and Miller, 1971). In that series, 50% of the 700 tumors were infratentorial, with 120 tumors being medulloblastomas and 115 being cystic cerebellar astrocytomas. These two tumors are the most common types of pediatric brain tumors, but represent only 33% of all pediatric brain tumors. Patients with cystic astrocytomas have an excellent long-term prognosis following surgical intervention alone (Hirsch *et al.*, 1979). To focus simply on the problems of children with medulloblastoma would exclude a very large number of children with less frequent types of brain tumors. In a series from Columbia University, 54% of the children had supratentorial tumors with only 18% living 5 years past diagnosis (Low *et al.*, 1965). It is very easy to exclude these other children because of the low incidence of the individual types, but when confronted clinically with the management of such a child, data as to the efficacy and toxicity of each modality of therapy available is critical. A large percentage of the supratentorial tumors are gliomas of varying grades of malignancy. Gliomblastoma, the most common brain tumor of the adult, represents only 5% of pediatric brain tumors, but other lower grade astrocytomas excluding cerebellar astrocytomas represent 10% of all pediatric brain tumors (Koos and Miller, 1971; Low *et al.*, 1965).

A review of our experience at the Children's Hospital of Philadelphia confirms the complexity and challenge of the large number of types of childhood brain tumors (Table I). Gliomas represented 57% of all the brain tumors. The single most common site was the brain stem. As reported by Littman *et al.* (1980), our results show only approximately a 20% long-term survival rate because of local progression despite irradiation. In contrast to the highly malignant brain stem gliomas are the optic gliomas, which occur only slightly less commonly in our experience than medulloblastoma. Some of these tumors involve either both nerves and/or the chiasm and are surgically inaccessible. Radiation is able to control the vast majority, but especially in the very youngest patients, long-term effects on vision, intellect, and hormonal regulation are not good. Local control in this group of children with less morbidity is the goal.

Approximately one-third of the gliomas are in sites other than the common sites of optic tracts, cerebellum, and brain stem, and over half of these tumors are high grade. Few are classified as glioblastoma multiforme, but they are still quite aggressive and commonly are not cured by local radiation therapy. Surprisingly, we also have a significant number of low-grade gliomas in other sites. These are predominantly located deep in the midbrain. They are not surgically accessible and recurrences with death are common.

TABLE I. Central Nervous System Tumors: Children's Hospital of Philadelphia, 1975–1979

Tumor type	No. of patients	Total no. patients
Glioma		91
brain stem	26	
optic	13	
cerebellar		
low grade	14	
high grade	2	
cord	5	
other sites		
low grade	13	
high grade	18	
Ependymoma		18
posterior fossa	11	
supratentorial	6	
spinal cord	1	
Medulloblastoma		17
Malignant Neuroectodermal tumor		6
Other		28
		160

The ependymomas occur with almost the same frequency as the medulloblastomas and are treated similarly. The ependymomas are supposedly radiosensitive. Unfortunately, our recent experience has been very poor. Of 18 children seen here with ependymomas from 1975 to 1979, 10 have already expired, and 5 of the remaining 8 children are alive postrecurrence. Only 3 children remain without recurrence, and 1 of these 3 had a spinal primary, which has a good prognosis. Many of these children have received adjuvant chemotherapy or chemotherapy at time of recurrence.

Medulloblastoma is the classic pediatric brain tumor, and a very large literature exists about its natural history and also its responsiveness to therapy. Only 17 of the 160 tumors seen in this 5 year period were medulloblastomas. Traditionally, these tumors have been very radiosensitive, but recurrences are common with long-term survival in the range of 40% being reported in the past few years (Bloom, 1979). Our own recent experience has shown continued improvement in the median survival and disease-free survival of these children. The quality of survival of these children has become a significant issue (Hirsch et al., 1979). A balancing of therapeutic modalities may shortly become the prime goal of future clinical research programs.

Against this background of a large variety of pediatric brain tumors are the problems related to local recurrence of high-grade lesions and long-term toxicity from radiation of childrens' brains. Therapy in the present era will need to approach these problems creatively. Cure of the tumor with good long-term functional results is the realistic goal. Until recently, most people approached the child with a brain tumor despondently. There was little expectancy that the child would survive; if he or she survived, the child was not expected to be competitive. Physicians, parents, patients, and committee approached the patient with this attitude. A close inspection of current results shows that this concept is only partially true. At this time, we need to define the goal of therapy for each type of brain tumor in each site and to consider the particular risks of each host. The tolerance to radiation appears to be age dependent, with the greatest neurotoxicity occurring in the youngest child.

These refinements of therapy have been approached for most childhood malignancies outside the central nervous system. Until recently, the study of brain tumors in childhood lagged behind. All of the phase II data below were obtained by individual institutions. No phase II cooperative group trial data for the nitrosoureas are available. Even the largest individual institutions can obtain only gross ideas of response in these very uncommon tumors. Fortunately, very sophisticated phase III trials have recently been begun, showing that the time has come to answer the specific questions about the more common pediatric brain tumors.

Evaluation of response of brain tumors is extremely difficult, as has been stressed. Many patients with clinical signs of progressive neurologic deterioration may not have recurrent tumor but problems related to prior surgery or radiation. Response to therapy often means stopping progression of damage, but recovery of neurologic deficit may not occur. A drug may very well eliminate a tumor, but a patient may have so much fixed neurologic damage that death rapidly ensues, and clinical response of the tumor to the drug is not detected (Crafts and Wilson, 1955; Seiler, 1979). Fortunately, in the past few years good, noninvasive neurodiagnostic evaluation of tumor response to therapy has become widely available with the advent of computerized tomography. Most data studying the efficacy of the nitrosoureas in children with recurrent brain tumors predate this objective assessment so that the quality of extant data is suboptimal.

II. CLINICAL RESULTS

A. Brain Stem Glioma

Thirteen children with clinical recurrence are reported to have been given varying dose schedules of BCNU with or without other agents. There were five responses, two of which lasted 7+ and 8+ months (Walker and Hurtz, 1970; Wilson et al., 1970; Wilson et al., 1976). Only five children with clinical recurrence are reported as having received CCNU at the standard 130 mgm/m² every 6 weeks. Three responded, including responses of 8+ and 13+ months (Fewer et al., 1972a; Rosenblum et al., 1973; Ward, 1978). One small adjuvant study has reported using CCNU with vincristine and radiation for histologically proven malignant brain stem glioma (Reigel et al., 1979). All four patients died within 7.5 months of diagnosis. This is very discouraging in the light of reported efficacy in recurrent disease. An in-progress large cooperative group trial under the aegis of the Children's Cancer Study Group (CCSG) will help to better define the use of CCNU with vincristine as adjuvant therapy.

B. Ependymomas

Even fewer patients with recurrent ependymomas have been reported. Six children at the University of California in San Francisco received BCNU on a 6 weekly cycle, and three of the six showed good responses with an 11 month average duration of response (Fewer et al., 1972a; Fewer et al., 1972b; Wilson et al., 1970; Wilson et al., 1976). Five children have been reported to have received CCNU, with only one response (Shapiro, 1973; Wilson et al., 1976).

A prospective randomized trial using vincristine weekly during radiation and CCNU and vincristine every 6 weeks following radiation for 1 year is in progress. At present there are too few patients available for an evaluation of efficacy.

C. Medulloblastoma

The response of recurrent medulloblastoma has been studied more extensively than that of other tumors (Table II). Only three children treated for recurrence

TABLE II. Recurrent Medulloblastoma: Response to Nitrosourea Chemotherapy

Drug	No. patients	Other agents	No. response
CCNU	15	0	11
CCNU	12	procarbazine & vincristine	7
BCNU	3	0	1
MeCCNU	2	0	1

with BCNU are reported with only one response (Walker *et al.*, 1970; Wilson *et al.*, 1970). Fifteen children have been reported to have received CCNU alone for recurrence with 11 responses (Fewer *et al.*, 1972; Garret *et al.*, 1974; Rosenblum *et al.*, 1973; Shapiro, 1973; Ward, 1974; Ward, 1978) and 12 children received CCNU, procarbazine, and vincristine and 7 responded (Levin and Wilson, 1975; Levin and Wilson, 1976). Two children received MeCCNU and one had a short response (Crist *et al.*, 1976; Nathanson and Kovacs, 1978). In general, the responses have been less than 1 year, and cure has not been achieved. The use of repeat surgery, radiation, and multiagent chemotherapy for local recurrence has produced almost universal response, but again death 1 to 2 years later appears inevitable (Mealey and Hall, 1977; Thomas *et al.*, 1980) (Table III).

It is clear from these studies that medulloblastoma is quite sensitive to the lipid-soluble nitrosoureas. No real studies have been done to evaluate in a randomized fashion the relative efficacy of the three agents BCNU, CCNU, and MeCCNU in this tumor. Also few dose schedules have been studied.

With the knowledge that radiotherapy is not universally curative in medulloblastoma and that first recurrences are most commonly in the primary site, Phase III studies have been undertaken both by individual institutions and, most importantly, by cooperative groups. These trials of adjuvant chemotherapy for medulloblastoma following surgery and radiation have been undertaken with both BCNU and CCNU in combination with other agents including in some studies intrathecal methotrexate (IT MTX). Several of the trials are institutional with only historical controls, but the median survival rates have been impressive (Bloom, 1979; Hirsch *et al.*, 1979; Seiler *et al.*, 1978; Thomas *et al.*, 1980). Presently, large randomized cooperative group trials are being conducted by the Childrens Cancer Study Group (CCSG) together with the Radiation Therapy

TABLE III. Recurrent Medulloblastoma: Response to Multimodal Therapy

No. patients	Therapy/surgery	Radiation	Chemotherapy	Response
8	2	4	BCNU, VCR Steroids, MTX IT MTX IV	6 complete 2 partial
2	0	1	BCNU, VCR	2
2	1	1	MeCCNU, VCR	2

Oncology Group and by the International Society for Pediatric Oncology (SIOP) (Bloom, 1979; Evans *et al.*, 1979). Both trials are very similar in their use of vincristine weekly during radiation followed by a 4 week rest. Then maintenance chemotherapy is begun with CCNU 100mg po every 6 weeks for eight cycles and vincristine weekly for the first 3 weeks of each of the eight cycles. Prednisone is given daily for the first 14 days of each cycle of maintenance therapy in the CCSG protocol. Over 200 children have been entered on the CCSG study and 350 on the SIOP study. These studies are in preliminary analysis, but several very important observations are available. In both studies the median disease survival of the children randomized to receive radiotherapy alone is better than projected. If these trials had only historical controls, the response to chemotherapy would have been exaggerated. However, in both studies the over-all survival and disease-free survival in both children who had complete resections and in those who had less extensive surgical extiration of their tumors appears to be increased by the addition of the chemotherapy. The data show either a positive trend or actual significance in favor of the adjuvant chemotherapy. Further and final evaluation of these excellently conducted studies is necessary for scientific conclusions, and such conclusions should be available within the next 2 years.

D. Other Tumors

Possibly the most interesting group of children reported have been those treated with nitrosoureas for recurrent low-grade supratentorial gliomas (Garret *et al.*, 1974; Levin *et al.*, 1976; Levin *et al.*, 1978; Sumer *et al.*, 1978; Ward, 1978; Wilson *et al.*, 1976) (Table IV). The investigators often used multiagent chemotherapy and possibly steroids, though these were not always listed. The overall response rate is approximately 50%, with several long-term responses. These reports suggest that the chemotherapy, including the nitrosoureas, may well have a role in the treatment of the significant number of children who have either primary or recurrent low-grade gliomas.

TABLE IV. Recurrent Astrocytoma, Low Grade, Supratentorial: Response to Nitrosourea Chemotherapy

Nitrosourea	No. patients	Other drugs (No. patients)	No. responses
BCNU	29	5-FU (14)	6
		Procarbazine (6)	6
		Vincristine (2)	1
		Vincristine + IT MTX (4)	2
		(3)	1
			16
CCNU	7		3

III. TOXICITIES

The toxicities of nitrosoureas in childhood do not seem to differ greatly from those in adults. The dominant problem has been bone marrow suppression, and this can be an especially serious problem in children who receive craniospinal irradiation, as in the treatment of medulloblastomas and some ependymomas (Shapiro, 1975). The nitrosoureas cannot be used concomitantly with radiation in children undergoing extended fields. The bone marrow toxicity can be cumulative as in adults. The children who have received the lipid-soluble nitrosoureas for recurrence of this primary tumor are often on the chemotherapy for longer than 1 year, and to date no apparent problems seem to be frequent except the cumulative marrow toxicity. In very young children there does not appear to be any special problems using either CCNU or BCNU.

IV. CONCLUSION

The problems of children with brain tumors are somewhat different from those of adults. High-grade gliomas represent only a small proportion of the tumors that occur in people under age 15. In brain stem gliomas, which are frequently of high grade, approximately 80% of the children die despite maximal radiation therapy. Some data on responses in children with recurrences suggest that the nitrosoureas are active, but the limited adjuvant data show no improvement in the survival rates.

In contrast, frequent and preliminary data on the response of medulloblastomas to nitrosoureas from multiagent adjuvant trials suggest that disease-free and overall survival can be improved by the adjuvant therapy.

Possibly most interesting are the reports of responses to nitrosoureas in patients with recurrent lower grade gliomas. These less common tumors frequently present a therapeutic dilemma, but chemotherapy can at least palliate the child with recurrence.

The nitrosoureas, predominantly CCNU, have a demonstrated role in several pediatric brain tumors. They produce controllable acceptable levels of toxicity, and increase survival and palliate recurrence in medulloblastomas as well as low-grade gliomas. Future studies will better define the relative role of these agents in the therapy of children with both high-grade and low-grade brain tumors.

ACKNOWLEDGMENT

I would like to acknowledge the help of Patricia Jarrett and the Delaware Valley Tumor Registry in the preparation of data on the Children's Hospital of Philadelphia experience, and of Dana Mansor in the preparation of this manuscript.

REFERENCES

Bloom, H.J.G. (1979). *Recent Results Cancer Res. 68,* 412–422.

Crafts, D., and Wilson, C. (1975). *Semin. Oncol. 2,* 15–17.

Crist, W.M., Ragag, A.H., Vietti, T.J., Ducos, R., and Chu, J-Y. (1976). *Am. J. Dis. Child 130,* 639–642.

Evans, A.E., Anderson, J., Chang, C., Jenkin, R.D.T., Kramer, S., Schoenfeld, D., and Wilson, C. (1979). *In* "Adjuvant Chemotherapy for Medulloblastoma and Epenydmoma in Multidisciplinary Aspects of Brain Tumor Therapy," pp. 219–222. Elsevier/North-Holland Biomedical Press, New York.

Fewer, D., Wilson, C.B., Boldrey, E.B., and Enot, J.K. (1972a). *Cancer Chemother. Rep. 56,* 421–427.

Fewer, D., Wilson, C.B., Boldrey, E.B., Enot, K.J., and Powell, M.R. (1972b). *JAMA 222,* 549–552.

Garret, M.J., Hughes, H.J., and Ryall, R.D.H. (1974). *Clin. Radiol. 25,* 183–184.

Hirsch, J.F., Renier, D., Czernichow, L., Beneveniste, L., and Pierre-Kahn, A. (1979). *ACTA Neurochir. 48,* 1–15.

Koos, W.T., and Miller, M.H. (1971). *In* "Intercranial Tumors of Infants and Children." C.V. Mosby, St. Louis.

Levin, V.A., and Wilson, C.B. (1975). *Semin. Oncol.2,* 63–67.

Levin, V.A., and Wilson, C.B. (1976). *Cancer Treat. Rep. 60,* 719–724.

Levin, V.A., Crafts, D.C., and Wilson, C.B. (1976). *Cancer Treat. Rep. 60,* 243–249.

Levin, V.A., Hoffman, W.F., Pischer, T.L., Seager, M.L., Boldrey, E.B., and Wilson, C.B. (1978). *Cancer Treat. Rep. 62,* 2071–2076.

Littman, P., Jarrett, P., Bilanuik, L.T., Rorke, L.B., Zimmerman, R.A., Bruce, D.A., Carabell, S.C., and Schut, L. (1980). *Cancer 45,* 2787–2792.

Low, N.S., Correll, J.W., and Hamill, J.F. (1965). *Arch. Neurol. 13,* 547–554.

Mealey, J. Jr., and Hall, P.U. (1977). *J. Neurosurg. 46,* 56–64.

Miller, R.W. (1969). *J. Ped. 75,* 685–689.

Nathanson, L., and Kovacs, S.G. (1978). *Med. Ped. Oncol. 4,* 105–110.

Reigel, D.H., Scarff, T.B., and Woodford, J.E. (1979). *Childs Brain 5,* 329–340.

Rosenblum, M.L., Reynolds, A.F., Smith, K.A., Rumack, B.H., and Walker, M.D. (1973). *J. Neurosurg. 39,* 306–314.

Seiler, R.W. (1979). *Surg. Neurol. 11,* 97–100.

Seiler, R.W., Imbach, P., Vassella, F., Wagner, H.P. (1978). *Helv. Paediatr. ACTA 33,* 235–239.

Shapiro, W.R. (1973). *Clin. Bulletin 3,* 58–62.

Shapiro, W.R. (1975). *Cancer 35,* 965–972.

Sumer, T., Freeman, A.I., Cohen, M., Bremer, A.M., Thomas, P.R.M., and Sinks, L. (1978). *Surg. Oncol. 10,* 45–54.

Thomas, P.R.M., Duffner, P.K., Cohen, M.E., Sinks, L.F., Tebbi, C., and Freeman, A. (1980).

Walker, M.D., and Hurwitz, B.S. (1970). *Cancer Chemother. Rep. 54,* 283–271.

Ward, H.W.C. (1974). *Br. Med. J. 1,* 642.

Ward, H.W.C. (1978). *Med. Ped. Oncol. 4,* 315–320.

Wilson, C.B., Boldrey, E.B., Enot, K.J. (1970). *Cancer Chemother. Rep. 54,* 273–281.

Wilson, C.B., Gutin, P., Boldrey, E.B., Crafts, D., Levin, V.A. and Enot, K.J. (1976) *Arch. Neurol. 33,* 739–744.

Chapter 22

ADJUVANT THERAPY OF BRAIN TUMORS WITH NITROSOUREAS

Michael E. Walker

I. INTRODUCTION

Brain tumor is a generic term indicating an intracranial growth. There are many types, each of which have different biologic characteristics, cell kinetics, responsiveness to therapy, and metabolic activity as well as histopathologic appearance (Walker, 1981). The most frequently encountered brain tumors are the malignant gliomas which include glioblastoma multiforme (astrocytoma, grade IV), malignant astrocytomas (grade III), and to a lesser extent the ependymomas, oligodendrogliomas, and medulloblastomas. The data reported herein will relate primarily to the malignant gliomas, including glioblastoma multiforme and malignant astrocytomas.

Malignant gliomas are a completely lethal disease. They are refractory to treatment and have a short survival (median, approximately 6 months). If a true "cure" were to be found, it would take but a few patients in order for it to be demonstrated. However, since it is extremely unlikely that a cure will be found in the near future, control of the disease as measured by increasing survival is the focus of multiple studies. Such studies require large numbers of patients in order to demonstrate efficacy (Burdette and Gehan, 1970). In the past, there has been little controversy concerning the necessity for surgical decompression as patients would surely die without it. Surgery is capable of debulking the tumor and bringing about a remarkable reduction of intracranial pressure. It provides tissue for histopathologic diagnosis, but it never can be considered as curative for the treatment of malignant glioma. Its great value is the additional time and biologic information that it provides.

Radiotherapy is a more controversial issue, and after experience of some four decades it was finally subjected to a controlled prospective randomized clinical trial. This study established, without doubt, the efficacy of radiotherapy in the treatment of malignant glioma (Walker *et al.*, 1978). Radiotherapy used in maximal doses (6000 rads) provides another log or two in tumor cell kill but is limited to a maximum cumulative total dose by delayed radionecrosis appearing in brain parenchyma secondary to vascular damage. However, there is also a limit to the increase in median survival that the maximally tolerated dose of some 6000 rads can achieve of approximately 150%.

In discussing the chemotherapeutic approach in general and the nitrosoureas in particular, one must consider them as adjuvant therapy by virtue of the fact that they are used either concomitantly with, or sequentially after, these other modes of treatment. Adjuvant may imply a therapy of lesser importance. However, nothing could be further from the truth when considering the treatment of malignant glioma. Improvement in surgical management of this disease has probably reached its maximum as the intrinsic nature of malignant glioma does not permit its total removal. Similarly, radiotherapy cannot be increased in total dosage, although some variation in use of the type of radiation, as well as time and schedule, could be explored. Thus, chemotherapy is the only remaining form of treatment, and therefore it needs to be carefully explored.

In the design of clinical trials for the treatment of malignant glioma, not only must the type of surgery and radiotherapy received be considered, but a large series of important predictive factors that have significant influence over the survival of patients with this disease must be examined (Gehan and Walker, 1977). These "prognostic factors" include important determinants such as age, performance status, sex, symptoms, and histopathology. At the current state of the art, these factors can have a more significant influence over survivorship than any of the therapeutic modalities that can be applied. Their importance, therefore, must not be underrated.

II. CLINICAL TRIALS

A. Efficacy

The nitrosoureas are ideally suited chemotherapeutic agents as they possess what are considered appropriate pharmacologic characteristics for entrance of drugs through the blood–brain barrier (Rall and Zubrod, 1962). They are all relatively lipid soluble with varying degrees of aqueous solubility. The clinically utilized nitrosoureas have a molecular weight of under 200 daltons and are not ionized. Thus, they are able to penetrate the blood–brain barrier with extraordinary speed and find rapid access from the extracellular space to cell membranes and eventually to the intracellular compartment. The same lipid solubility that permits them to move readily into the brain also allows them to disseminate readily throughout other tissues within the body, and thus rapid distribution takes place and plasma concentrations fall off quickly (Levine *et al.*, 1975). In addition, they undergo rapid metabolic breakdown within the blood stream as a result of both chemical and biologic interactions and are only partially protein bound. The prime toxicity of delayed thrombocytopenia as well as less-pronounced leucopenia and multiple-system delayed toxicity present a sequence that does not make their use any easier. Nevertheless, they remain the single most interesting class of chemotherapeutic agents developed to date that has been evaluated in a wide variety of studies.

Wilson (1970) and Walker (1972) were the first to report Phase II clinical trials utilizing BCNU for the treatment of malignant glioma. These studies, carried out in patients who were symptomatically recurrent, demonstrated a subjective remission rate of approximately 50% with a median duration of response of some 6–9 months. A controlled clinical trial comparing radiotherapy and CCNU alone and together was carried out in 41 patients by Weir *et al.* (1976). In this study the treatment was CCNU (130 mg/m^2/day every 6–8 weeks) or radiotherapy (4000–4500 rads). The dosage of CCNU is clearly within the normal range, whereas the treatment with radiotherapy is somewhat below that commonly employed. A cross-over design was utilized at the time when patients were felt to have demonstrated progressive disease. The median interval to progression for CCNU, radiotherapy, and the combination was 14, 23, and 31 weeks respectively without statistical significance being demonstrated. Median survival was 37, 27, 36 weeks respectively. Of interest is the fact that only 47% of the patients receiving radiotherapy actually crossed over to receive CCNU and did so at a comparatively late time (23 weeks). However, 77% of patients receiving CCNU crossed over to receive radiotherapy and did so considerably earlier (14 weeks). Therefore, the survival figure for CCNU alone is more nearly representative of CCNU followed by radiotherapy, and its similarity in survivorship to the CCNU and radiotherapy arm reflects this as well.

Reagan *et al.* (1976) carried out a similar study utilizing 63 patients in whom the radiation dose was increased to 5500 rads delivered over 5 weeks. His median interval to recurrence for CCNU, radiotherapy, and the combination was 17, 30, and 30 weeks respectively with a median survival time of 28, 50, and 52 weeks. CCNU was felt to be the inferior treatment ($p = 0.02$), and the major efficacy was ascribed to radiotherapy.

In contradistinction to those studies is the evaluation carried out by the EORTC Brain Tumor Group (1978) into which their 104 patients were first stratified into two subsets. The 81 patients who, following surgery and radiotherapy were felt to be neurologically normal and not requiring corticosteroids were randomized to receive CCNU (130 mg/m^2 every 6–8 weeks) either immediately following surgery *or* only upon evidence of recurrence. The symptom-free interval was 34 weeks for both groups, thus indicating no apparent value for CCNU in the early treatment of this disease. However, those patients who were treated with CCNU following recurrence demonstrated a 25% clinical response rate and a median survival of 77 weeks, which was statistically superior to the early CCNU group of 50 weeks ($p = > 0.05$). The second portion of their study evaluated those patients who were symptomatic and required corticosteroids for adequate control. The 23 patients in this subset were randomized to receive CCNU or no CCNU and demonstrated a superior suvivorship for CCNU (31 weeks) as compared to those that received no CCNU (21.5 weeks; $p = 0.01$).

A direct end-to-end comparison of radiotherapy with and without BCNU or CCNU was carried out by Solero et al. (1979). Following surgery, 105 patients received radiotherapy (5000 rads in 5–6 weeks) and were randomized to receive BCNU (80 mg/m^2/day $\times$ 3 days every 6–8 weeks), CCNU (130 mg/m^2 once every 6–8 weeks), or neither BCNU or CCNU. This study contained 102 evaluable patients, and thus it has a statistically better chance of demonstrating the therapeutic value of these treatments. Radiotherapy alone, BCNU + radiotherapy, and CCNU + radiotherapy demonstrated a disease-free interval of 39, 45, and 52 weeks respectively and a median survival of 45, 52, and 69 weeks. Significance was demonstrated between CCNU + radiotherapy versus radiotherapy alone but not between BCNU + radiotherapy versus radiotherapy alone.

In 1972 the Brain Tumor Study Group (BTSG)[1] carried out the first controlled prospective randomized multiinstitutional clinical trial designed to evaluate BCNU and/or radiotherapy as compared to best conventional care without either chemotherapy or radiotherapy (Walker et al., 1978). In this study 303 patients were entered, of which 222 were considered evaluable according to neuropathologic and selection criteria. The patients entered into the four therapeutic arms were considered comparable in analysis of neuropathology, age, initial performance status, interval from symptom to operation, interval from operation to randomization, and sex. The median survival of control patients was 14.0 weeks compared to 18.5 weeks for BCNU, 36.0 weeks for radiotherapy, and 34.5 weeks for BCNU + radiotherapy. Although BCNU failed to demonstrate an additive effect at the median survival point, it was of interest to note that at 18 months virtually all the patients receiving monotherapy had succumbed to their disease, while some 20–25% of those receiving BCNU and radiotherapy were still alive. The toxicity of this form of therapy was quite acceptable, with moderate reversible thrombocytopenia being the major observation.

As a result of this evaluation, an additional study was carried out that utilized radiotherapy (6000 rads in 6–7 weeks) as the control arm and studied methyl CCNU (220 mg/m^2 orally once every 6–8 weeks) and radiotherapy + methyl CCNU as compared to the best treatment arm from the prior study of radiation

[1]List of participants precedes References.

therapy + BCNU (Walker *et al.*, 1980). There were 467 patients randomized into this study, of which 358 met the criteria for evaluation. The four treatment populations were considered as comparable. Methyl CCNU alone had a median survival of 24 weeks as compared to radiotherapy, 36 weeks; methyl CCNU + radiotherapy, 42 weeks; and BCNU + radiotherapy, 52 weeks. While methyl CCNU alone was clearly inferior to the other modes of treatment ($p < 0.005$), the additional value of methyl CCNU or BCNU to radiotherapy was only modest. However, the observation made in the previous trial that monotherapy at 18 months was inferior to combination therapy was again substantiated. Approximately 13% of patients receiving only methyl CCNU or radiotherapy were alive at 18 months, whereas approximately 25% of patients receiving radiotherapy and either one of the nitrosoureas were alive at 18 months. The subsequent trial adopted BCNU and radiotherapy as its best therapeutic arm from prior studies and evaluated high-dose methylprednisolone (400 mg/m^2/day $\times$ 7 days every 4 weeks) versus procarbazine (150 mg/m^2/day $\times$ 28 days every 8 weeks) and the combination of BCNU and methylprednisolone. The study has accrued some 601 patients of which 527 are now evaluable. All patients were operated upon and, in addition, received 6000 rads of radiotherapy during the time they were starting chemotherapy. The single mode of therapy that appears to be inferior is methyl-prednisolone, whereas BCNU, procarbazine, and methylprednisolone + BCNU are all approximately equivalent in their efficacy. Thus in this still ongoing trial, a modest potential value for BCNU and radiotherapy in comparison with radiotherapy alone again appears to be present, but no oncolytic effect for cor-ticosteroids has been demonstrated.

Pediatric brain tumors are the most frequently encountered solid tumor of childhood (Wilson, 1970). An important clinical trial is being carried out by the Children's Cancer Study Group (CCSG) and the International Society for Pediat-ric Oncology (SIOP). Although these are independent trials, they are, in effect, both evaluating the efficacy of radiotherapy with and without CCNU, vincristine, and prednisone in medulloblastomas and ependymomas (Evans *et al.*, 1980) (prednisone is not utilized in the SIOP trial). Over 200 patients have been entered into each of these studies from a wide variety of institutions. Because of the natural history of patients with medulloblastoma and the fact that a median survival of 5 or more years is not unusual, these studies will be slow in coming to a definitive conclusion. However, trends are beginning to develop that indicate adjuvant therapy with CCNU, vincristine, and (prednisone) may indeed be add-ing significantly to the disease-free interval as well as to the survival of these patients.

Two additional studies are being carried out by the Brain Tumor Study Group in which the nitrosoureas play an important role. In the past, the neurooncologist was required to consider cytoreductive surgery for his patient almost immediately in order to bring about a decompression of the brain and a reduction in symp-toms. With the current use of corticosteroids for the control of cerebral edema, it is possible to consider the appropriate sequencing of surgery, radiotherapy, and chemotherapy for these patients. The failure of radiotherapy to completely sterilize the tumor is often blamed upon a small hypoxic residue of cells that are comparatively radioresistant. The neuropathologist points out that following radiotherapy and chemotherapy there is a large necrotic mass of tumor cells that the brain disposes of poorly. Finally, the surgeon has noted that following radiotherapy and chemotherapy there appears to be a more closely demarcated border between tumor and normal brain. Therefore, a protocol has been devised in which patients will be randomized to receive either the usual sequence of care (surgery followed by radiotherapy and chemotherapy with BCNU) or radiotherapy and two courses of BCNU followed by cytoreductive surgery and continued BCNU. All patients will have a provisional diagnosis established that must be later confirmed by histopathology. This study has some 70 patients entered into it thus far; and the only conclusion that can be reached is that the approach is quite feasible and that serious adverse side effects are not obvious.

A further study is being carried out in which radiotherapy plus BCNU is again used as the best treatment arm from prior studies. This will be compared to superfractioned radiotherapy plus BCNU, the use of misonidozole as a radiosensitizor followed by BCNU, or the addition of the new nitrosourea, streptozoticin. This study has also accrued several hundred patients; however, it is too early for analysis. All four therapeutic arms appear to be well tolerated.

B. Toxicities

In reviewing the treatment of malignant tumor with the nitrosoureas, it becomes apparent that there are a number of severe limitations to their employment. First, there is the delayed bone marrow toxicity that appears to involve every element of the bone marrow, although it is more manifest through thrombocytopenia. It is too long to wait 6–8 weeks to carry out additional therapy for brain tumor, yet that is how long we must wait for recovery from the nitrosoureas. There is an additional problem of the cumulative bone marrow toxicity that is seen after multiple doses of the nitrosoureas. This toxicity not only prevents additional full-dose treatment of nitrosoureas but is severely dose limiting in the employment of other marrow toxic agents. It is difficult, if not impossible, to combine full doses of the nitrosoureas with other marrow-suppressing agents, and thus the transition from monodrug to multidrug therapy is extraordinarily difficult to bridge.

All of these factors severely limit the effectiveness of the nitrosoureas, yet they remain the single most potentially effective agent we have. Finally, the long-term chronic complications of prolonged nitrosourea therapy are beginning to become apparent in the form of pulmonary fibrosis, nephrotoxicity, and hepatic degeneration. It must be pointed out that these complications occur in very few patients who have received multiple doses of treatment and have therefore out-survived their expected survival should they not have had treatment. Careful monitoring of patients and observation of pulmonary, renal, and hepatic function can help reduce the occurrence of these untoward side effects.

III. CONCLUSION

In summary, the nitrosoureas have had an impressive trial in the treatment of malignant brain tumor. They appear to be modestly effective agents, but without a clear-cut and obvious choice among them. However, the conflicting reports concerning CCNU leave its efficacy somewhat in doubt, and BCNU has only modest efficacy. Other nitrosoureas (PCNU, streptozotocin, and ACNU) need to be evaluated in controlled clinical trials in order to determine their efficacy and whether they have any distinct superiority from the more conventionally employed nitrosoureas. Until that time, the recommended treatment for patients with malignant glioma following surgery would be radiotherapy and BCNU after the patient has been fully informed of its potential value and complications.

PARTICIPANTS

The following institutions are participants in the Brain Tumor Study Group: Bowman Gray School of Medicine, Duke University Medical Center, Indiana University School of Medicine, Mayo Clinic, Memorial Sloan-Kettering Hospital, National Cancer Institute, New York University Medical Center, Ohio State University Research Foundation, St. Louis University School of Medicine, University of California, Los Angeles, University of California Medical Center, San Francisco, University of Connecticut Health Center, University of Kentucky Medical Center, University of Tennessee Center for Health Sciences.

REFERENCES

Burdette, W.J., and Gehan, E.A. (1970). *In* "Planning and Analysis of Clinical Studies." Charles C. Thomas, Springfield.

EORTC Brain Tumor Group (1978). *Eur. J. Cancer 14*, 851–856.

Evans, A.E., Anderson, J., Chang, C., Jenkin, R.D.T., Kramer, S., Schoenfeld, D., and Wilson, C. (1980). *Aspects of Brain Tumor Ther. 1*, 219–222.

Gehan, E.A., and Walker, M.D. (1977). *In* "Modern Concepts in Brain Tumor Therapy: Laboratory and Clinical Investigation," pp. 189–196. NCI Monograph 46, DHEW.

Levin, V.A., Freeman, M.A., and Landahl, H.D. (1975). *Arch. Neurol. 32*, 785–791.

Rall, D.P., and Zubrod, C.G. (1962). *Ann. Rev. Pharmacol. 2*, 109–128.

Reagan, T.J., Bisel, H.F., Childs, D.S., Jr., Layton, D.D., Rhoton, A.L., Jr., and Taylor, W.F. (1976). *J. Neurosurg. 44:* 186–190.

Solero, C.L., Monfardini, S., Brambilla, C., Vaghi, A., Valagussa, P., Morello, G., and Bonadonna, G. (1979). *Cancer Clin. Trials 2*, 43–48.

Walker, M.D. (1981). *In* "Cancer Medicine" (J.F. Holland and E. Frei III, eds.) Lea & Febiger, Philadelphia, pp. 4079–4164.

Walker, M.D., and Hurwitz, B.S. (1970). *Cancer Chemother. Rep. 54*, 263–271.

Walker, M.D., Alexander, E. Jr., Hunt, W.E., MacCarty, C.S., Mahaley, M.S. Jr., Mealey, J. Jr., Norrell, H.A., Owens, G., Ransohoff, J., Wilson, C.B., Gehan, E.A., and Strike, T.A. (1978). *J. Neurosurg. 49*, 333–343.

Walker, M.D., Green, S.B., Byar, D.P., Alexander, E. Jr., Batzdorf, U., Brooks, W.H., Hunt, W.E., MacCarty, C.S., Mahaley, M.S. Jr., Mealey, J. Jr., Owens, G., Ransohoff, J. II.,

Robertson, J.T., Shapiro, W.R., Smith, K.R., Jr., Wilson, C.B., and Strike, T.A. (1980). *N. Eng. J. Mèd. 303,* 1323-1329.

Weir, B., Band, P., Urtasun, R., Blain, G., McLean, D., Wilson, F., Mielkle, B., and Grace, M. (1976). *J. Neurosurg. 45,* 129-134.

Wilson, C.B., Boldrey, E.B., and Enot, K.J. (1970). *Cancer Chemother. Rep. 54,* 273-281.

Chapter 23

CLINICAL USE OF NITROSOUREAS IN GASTROINTESTINAL CANCER

Paul V. Woolley III

I. INTRODUCTION

The first drugs shown to be of use in the treatment of adenocarcinoma of the gastrointestinal tract were fluorinated pyrimidines such as 5-fluorouracil (5-FU). Clinically active nitrosoureas were developed in the 1960s, and soon their role in treatment of these tumors was recognized. Subsequently several nitrosoureas have been used, singly and in combination with 5-FU and other agents, for treatment of advanced disease of the stomach, pancreas, and colon, as well as in adjuvant studies for cases of surgically resected gastric and colon tumors. This paper reviews the status of the nitrosoureas BCNU, CCNU, and methyl CCNU in the clinical management of gastrointestinal cancer, both as single agents and in combination with other drugs and with radiation.

In approaching this analysis, it is useful to consider separately carcinomas arising in the stomach, the pancreas, and the large bowel. Also the extent of tumor progression affects the therapeutic approach to these diseases. In the case of advanced, widespread or metastatic disease, rediation serves only to treat local complications, and the primary treatment is systemic chemotherapy. A clinically distinct situation is that of locally advanced disease, which operationally means that although the primary tumor cannot be surgically removed, it has not metastasized and can be encompassed within a radiation port of reasonable size.

Finally, in the surgical adjuvant case, the primary has been entirely resected but the patient remains at risk of relapse. These situations are approached separately clinically and will be treated separately in this discussion.

II. RESULTS

A. Advanced Disease

1. Single-Agent Studies

The activites of BCNU, CCNU, and methyl CCNU in gastrointestinal cancer have been investigated both in uncontrolled phase II studies and by randomized comparisons with other agents. In adenocarcinoma of the colon, studies at the Mayo Clinic produced overall response rates of 12.5% for BCNU, 10% for CCNU, and 17.5% for methyl CCNU (Moertel, 1975; Moertel et al., 1976b). Higher response rates were seen in previously untreated cases than in those who had received prior chemotherapy. One randomized comparison between 5-FU and methyl CCNU showed that the nitrosourea was about equal to 5-FU in efficacy when used as initial treatment, although once again its effectiveness fell when it was used in 5-FU failures (Moertel et al., 1976b). The suggestion that methyl CCNU was slightly superior to the other agents led to subsequent studies of the drug in combination, both for advanced disease and in adjuvant trials. However the greater activity of methyl CCNU was not always confirmed (Douglass et al. 1976).

In pancreatic adenocarcinoma the single-agent activity of BCNU, CCNU, and methyl CCNU is minimal (Moertel, 1975). Streptozotocin, a glucose-containing methyl nitrosourea, has produced some objective responses in this disease and in combination with 5-FU (Moertel, 1975). A tolerable dose of this drug is 1.0–1.5 g/m^2 iv weekly, although this will be associated with nausea, vomiting, and in some cases renal tubular damage. The last is reversible with discontinuation of the drug. Streptozotocin also has high activity in the less common islet cell tumors, carcinoid tumors, and other neoplasms of the "apudoma" category (Moertel, 1975).

The single-agent activity of the nitrosoureas is greater in gastric cancer than in pancreatic cancer. In initial trials (Moertel, 1975; Moertel et al., 1976b) a response rate of 18% was seen for BCNU, but CCNU showed no activity in 11 patients. Subsequently a controlled trial of BCNU against 5-FU and the combination of the two drugs showed that BCNU alone did not affect survival of previously untreated gastric cancer patients, whereas both 5-FU alone and the combination of the two drugs produced some survival benefits as compared to clinically matched untreated controls. Further studies by the Eastern Cooperative Oncology Group established single-agent activity of 11% for methyl CCNU (Moertel et al., 1976a).

In conclusion, several nitrosoureas have been shown to have definite, albeit limited, activity in gastrointestinal cancer. Pancreatic cancer is the least responsive site. This then poses the question of the value of combining these drugs with other agents.

2. Combination Drug Studies

Virtually all of the data for combination therapy of gastrointestinal neoplasms involve use of 5-FU. Considering first the colon, several early studies indicated that 5-FU plus a nitrosourea would give response rates that exceeded those of 5-FU alone. Falkson, et al. (1974) reported that the combination of 5-FU, BCNU, DTIC, and vincristine produced a 43% response rate in advanced colorectal carcinoma, as opposed to 25% for 5-FU alone. A similar response rate was found by Moertel et al. (1975) for the combination 5-FU, methyl CCNU, and vincristine. Macdonald et al. (1976) also found a 43% response rate for 5-FU, methyl CCNU, and vincristine, using 5-FU on a weekly intermittent schedule rather than a 5 day loading schedule as had been employed by Moertel. Another randomized trial, by the Southwest Oncology Group, produced a 29% response rate for 5-FU and methyl CCNU compared to 10.6% for 5-FU alone (Baker et al., 1976). Despite these studies, however, subsequent experience has called into question the role of vincristine in this combination, and rates of response to 5-FU and methyl CCNU as low as 10–15% have been reported. Nonetheless, the nitrosourea seems to add 5–10% to the response rate for 5-FU alone, and at present this combination is the closest thing we have to a "standard" therapy for advanced colorectal cancer. However, the survival benefits of these combinations are small at best, and much of the effort in colon cancer is aimed at identifying new active agents through phase II trials. Recent studies by Kemeny (1980) at Memorial Sloan Kettering Cancer Center have suggested that the combination of 5-FU, methyl CCNU, vincristine, and streptozotocin is more active in advanced disease than are 5-FU, methyl CCNU, and vincristine alone, and this needs confirmation. Trials of new agents, including nitrosoureas such as PCNU, are presently ongoing.

The combination of BCNU and 5-FU for pancreatic carcinoma has not met with clinical success. Although the response rate of the combination is greater than that of either drug alone, there is no impact upon patient survival from these responses (Kovach et al., 1974). By contrast, the combination of streptozotocin, mitomycin C, and 5-FU has produced a 43% response rate in a group of 23 patients with advanced pancreatic cancer, and the responders have lived a median of 7.5 months, compared with 3 months for nonresponders (Wiggans et al., 1978).

Combinations of nitrosoureas with other agents have been extensively tested in gastric cancer, and the results are generally more favorable than for pancreatic cancer. Indeed of all sites in the gastrointestinal tract, gastric cancer seems the most responsive to cytotoxic drugs. Historically, the first combinations that were tried were 5-FU plus BCNU and 5-FU plus methyl CCNU. Subsequently, an important role for doxorubicin in treatment of gastric cancer has been defined, and both BCNU and methyl CCNU have been used in combination with 5-FU and doxorubicin.

The combination of 5-FU and BCNU is superior to either drug used alone. A randomized trial (Kovach *et al.*, 1974) of the combination against its two components showed a response rate of 41% for the two drugs, 29% for 5-FU alone, and 17% for BCNU alone. Survival in the group treated with combination was 27% at 18 months, compared with 7% for those treated with 5-FU alone. A comparison of 5-FU and methyl CCNU combinations has been conducted by the Eastern Cooperative Oncology Group (Moertel *et al.*, 1976b). In a four-armed trial, patients were treated either with methyl CCNU alone or 5-FU plus methyl-CCNU, either with or without a cyclophosphamide induction course. The response rate (40%) and survival of the 5-FU and methyl CCNU combination was superior to the other three. Cyclophosphamide induction failed to improve on the results of 5-FU and methyl CCNU, and, if anything, it detracted from them. Clearly, however, either BCNU or methyl CCNU combined with 5-FU produce improvements in clinical results in gastric cancer over single-agent therapy.

Since 1975, it has be realized that the addition of doxorubicin (adriamycin) to drug combinations greatly enhanced the treatment of gastric cancer. At that time the combination of 5-FU, adriamycin, and mitomycin C (FAM) was first described. In the ensuing 5 years of experience, this three-drug combination has produced an overall response rate of 42% in the Georgetown experience, with median survivals of 12.5 months for responders and 3 months for nonresponders (Macdonald *et al.*, 1979; Macdonald *et al.*, 1980). These data have been confirmed in cooperative group studies in the Southwest Oncology Group. Of interest is the relative value of FAM, and two nitrosourea-containing regimens. One randomized trial (O'Connell *et al.*, 1980) has compared (a) 5-FU, adriamycin, and mitomycin C; (b) 5-FU, adriamycin, methyl CCNU; (c) 5-FU, methyl CCNU; and (d) 5-FU, ICRF-159, methyl CCNU prospectively. The FAM and FAMe combinations were superior to both 5-FU plus methyl CCNU and 5-FU, ICRF-159, methyl CCNU in terms of response rate and survival. Another preliminary trial (Karlin *et al.*, 1980) indicated that the four-drug combination of FAM plus methyl CCNU was not superior to FAM. A phase II study of 5-FU, adriamycin, and BCNU (Levi *et al.*, 1979) produced a 52% response rate in 35 patients with a 12.5 month median survival in responders. These results are quite similar to those with FAM. The conclusions of these several trials are that nitrosourea combinations produce responses in gastric cancer that are correlated with improved survival, but that doxorubicin is an important addition to therapy. The FAM, FAMe, and FAB combinations are approximately equivalent, but at the present time the widest experience is with the FAM combination. The combination of 5-FU, doxorubicin, and cisplatin is presently under investigation at Georgetown.

B. Locally Unresectable Disease

Nitrosourea therapy has been incorporated into regimens for treatment of locally advanced gastric cancer. A recent randomized trial (Schein and Novak, 1980) compared chemotherapy with 5-FU and methyl CCNU alone to a regimen of sequential radiotherapy and chemotherapy. The latter consisted of a total of 5000 rads of external-beam irradiation given in two courses with 3 days of 5-FU therapy included in each course. This was followed by 5-FU and methyl CCNU maintenance. Initially the survival curves for these two regimens favored the chemotherapy arm, with a median survival of 70 weeks as opposed to 36 weeks for the combined modality treatment. At later times, however, the survival curves cross, and in the interval from 2 to 3 years, there is a plateau at 20% survival for the combined modality but a continued probability of relapse in the chemotherapy arm. The data indicated a potential for increased early toxicity with very intensive regimens. Further studies of combined radiation and drug therapy with different scheduling would be important.

A role for nitrosourea in locally advanced pancreatic cancer has not as yet been established, although the possibility of using streptozotocin in a combined modality setting is interesting. Controlled trials to date indicate that radiation therapy given in three courses of 2000 rads each with 5-FU on the first 3 days of each course is superior to only two courses of radiation with 5-FU or to radiation alone (Woolley et al., 1977).

C. Surgical Adjuvant Studies

The use of adjuvant chemotherapy to delay or prevent relapse in surgically resected gastrointestinal cancer is an important area of clinical research at this time. In considering the trials that have been done in colorectal cancer, it is important to distinguish rectal cancer from that in the rest of the large bowel because of its greater sensitivity to radiation therapy. While there is a tendency of these trials to show a small benefit in favor of treatment, the differences between control and 5-FU-treated patients have been small and tend to provoke lively debate about their statistical significance (Grage et al., 1977; Higgins et al., 1976; Woolley et al., 1977). A logical sequel of such trials, based on the advanced disease data, is to study 5-FU and methyl CCNU in combination.

Several such trials are currently ongoing in cooperative groups such as the Gastrointestinal Tumor Study Group (GITSG), the Eastern Cooperative Oncology Group, and the Veterans Administration Surgical Adjuvant Group. Given the several years of accrual and follow-up necessary to complete these studies, it is too early for definitive statements. However, in the particular case of the colon, it does not appear that the gains will be large. For example, the GITSG currently is conducting a randomized trial comparing observation, 5-FU plus methyl CCNU, immunotherapy with MER, or chemotherapy plus immunotherapy after resection of Dukes B$_2$ and C lesions. While the study arms are still coded, there is no clear indication out to 195 weeks of follow-up that there are substantial differences in relapse rates or survival between any of these treatment options (Schein, personal communication).

It appears, however, that important advances have been recently achieved in the adjuvant therapy of rectal cancer. The GITSG has conducted a four-armed trial using either observation, 5-FU plus methyl CCNU, local radiation, or combined radiation therapy and chemotherapy following resection (Schein, personal communication). In this instance the control arm proved significantly inferior to all treatment arms and was closed to accrual before the study was complete, while the treatment arms remained open. Further analysis revealed that the combination of radiation plus chemotherapy was most effective in reducing local and systemic recurrences, whereas chemotherapy was least effective in preventing local recurrence. Nonetheless, all treatment arms produced measurable decreases in relapse rates compared to control. This important observation indicates that it is possible to improve on the results of surgery in rectal carcinoma using radiation and chemotherapy.

In gastric cancer that has been surgically resected, there are again several cooperative group studies of 5-FU and methyl CCNU in progress. It is also too early for definitive statements for these trials, but small gains may be forthcoming. Completion of trials with 5-FU and methyl CCNU and institution of adjuvant regimens with FAM are necessary to fully evaluate the role of chemotherapy following surgical resection of gastric carcinoma.

In summary, the nitrosoureas are integral parts of the current trials that test combined modality therapy as adjuvant to surgery in colon, rectal, and gastric cancer. While in many cases the long follow-up necessary for clear statements has not been completed, other cases such as that of rectal cancer appear to be providing useful answers of clinical applicability.

III. FUTURE DIRECTIONS

Nitrosourea development will continue to provide new agents for clincal use. The compound PCNU is discussed in detail elsewhere in this volume. It is a lipid-soluble compound that has completed phase I trials. The drug is currently undergoing phase II testing in colon carcinoma at Georgetown, and initial evidence of activity has been seen. Other specialized approaches such as the intraarterial administration of nitrosoureas alone or in combination may also contribute to the control of regional disease, particularly in the liver. In the area of the gastrointestinal malignancies the nitrosoureas continue to be integral parts of the combined modality approaches to control the disease.

REFERENCES

Baker, L. H., Vaitkevicius, V. K., Gehan, E., and the Gastrointestinal Committee of the Southwest Oncology Group (1976). *Cancer Treat. Rep. 60*, 733–737.

Douglass, H. O., Jr., Lavin, P. T., and Moertel, C. G. (1976). *Cancer Treat. Rep. 60*, 769–780.

Falkson, G., Van Eden, E. G., and Falkson, H. C. (1974). *Cancer 33*, 1207–1209.

Grage, T. B., Metter, G. E., and Correll, G. W. (1977). *Amer. J. Surg. 133*, 298–305.

Higgins, G. A., Humphrey, E., Jules, G. L., LeVeen, H. H., McCaughan, J., and Keene, R. J. (1976). *Cancer 38*, 1461–1467.

Karlin, D. A., Mahal, P. S., Herfitz, L. J., Stroeblin, J. R., and Bennetts, R. W. (1980). *Proc. Amer. Soc. Clin. Oncol. 21*, 169.

Keene, R.J. (1976). *Cancer 38*, 1461–1467.

Kemeny, N., Yagoda, A., and Golbey, R. (1980). *Proc. Amer. Soc. Clin. Oncol. 21*, 417.

Kovach, J. S., Moertel, C. G., Schutt, A. J., Hahn, R. G., and Reitemeier, R. J. (1974). *Cancer 33*, 563–567.

Levi, J. A., Dalley, D. N., and Aroney, R. S. (1979). *Brit. Med. J. 2*, 1471–1473.

Macdonald, J. S., Kisner, D. F., Smythe, T., Woolley, P. V., Smith, L., and Schein, P. S. (1976). *Cancer Treat. Rep. 60*, 1597–1600.

Macdonald, J. S., Woolley, P. V. III, Smythe, T., Ueno, W., Hoth, D. F., and Schein, P. S. (1979). *Cancer 44*, 42–47.

Macdonald, J. S., Schein, P. S., Woolley, P. V., Ueno, W. M., Hoth, D. F., and Smith, F. P. (1980). *Ann. Intern. Med. 93:* 533–536.

Moertel, C. G. (1975) *Cancer 36*, 675–682.

Moertel, C. G., Schutt, A. J., Hahn, R. G., and Reitemeier, R. J. (1975). *J. Nat. Cancer Inst. 54*, 69–71.

Moertel, C. G., Mittelman, J. A., Bakemeier, R. F., Engstrom, P., and Hanley, J. (1976a). *Cancer 38*, 678–682.

Moertel, C. G., Schutt, A. J., Reitemeier, R. J., and Hahn, R. G. (1976b) *Cancer Treat. Rep. 60*, 729–732.

O'Connell, M., O'Fallon, J., Lavin, P., Moertel, C., Bruckner, H., Douglass, H., Liostone, E., Lokich, J., and Mitchell, M. (1980). *Proc. Amer. Soc. Clin. Oncol. 21*, 420.

Schein, P. S., and Novak, J. (1980). *Proc. Amer. Soc. Clin. Oncol. 21*, 419.

Wiggans, R. G., Woolley, P. V. III, Macdonald, J. S., Ueno, W., and Schein, P.S. (1978). *Cancer 41*, 387–391.

Woolley, P. V., Macdonald, J. S., and Schein, P. S. (1977). In "Progress in Gastroenterology" (G. Jerzy Glass, ed.) Vol. 3, pp. 671–692. Grune and Stratton, New York.

Chapter 24

A RANDOMIZED COMPARATIVE TRIAL IN
COLORECTAL CANCER

Joan C. D'Aoust
Archie W. Prestayko
Brian F. Issell

I. INTRODUCTION

5-Fluorouracil (5-FU) has demonstrated limited activity in the treatment of advanced colorectal cancer, with response rates ranging from approximately 10 to 20% (Katz and Glick, 1979). The nitrosoureas BCNU, CCNU, and MeCCNU likewise appear to have limited activity when administered as single agents (Moertel, 1975). Early studies combining 5-FU and nitrosoureas reported higher response rates for the combination of 5-FU + MeCCNU over single-agent 5-FU or MeCCNU therapy (Baker *et al.,* 1976a; Posey and Morgan, 1977). However, it is unclear whether these increased response rates reflect a substantial patient advantage in terms of benefit over risk.

In the Southwest Oncology Group study (Baker *et al.,* 1976a) in which the addition of MeCCNU to weekly injections of 5-FU was examined, therapy with 5-FU alone gave a 9.5% response rate, and the combination therapy gave a significantly better 32% response rate. Because the rate was below 50% and the response durations were generally under 6 months, no significant impact on median survival was shown for the combination. However, despite the inability to show an overall advantage in the total colorectal cancer study group, the survival analysis for the subset of patients with liver metastases appears to show a substantial survival advantage for patients receiving MeCCNU + 5-FU. Statistical significance was not reported for these data.

The choice of MeCCNU as the most appropriate nitrosourea for human colorectal cancer is largely serendipitous. No randomized comparison between MeCCNU, CCNU, and BCNU in this disease has been undertaken. In fact, CCNU is the only drug that has shown superior activity over the other nitrosoureas in prospective randomized clinical comparisons (Maurice *et al.*, 1978; Selawry and Hansen, 1972).

We therefore elected to undertake a three-arm prospective randomized study to determine if a nitrosourea + 5-FU combination may benefit specific subsets of patients with colorectal cancer and to determine whether CCNU may offer an advantage over MeCCNU. To overcome concern that the previously suggested advantage for MeCCNU + 5-FU over 5-FU alone may have been due to the administration of 5-FU alone in a suboptimal way, 5-FU was given on an intensive schedule that was recommended by Roche Pharmaceuticals to be the most effective schedule of administration. A multiinstitutional randomized study was activated in late 1978. Data on approximately 50 patients were analyzed in mid-1980 for this preliminary report.

II. PATIENTS AND METHODS

Eligibility criteria included histologically confirmed colorectal adenocarcinoma with measurable metastatic lesions, normal hematologic and renal functions, and a performance status of $> 60\%$ on the Karnofsky scale. Patients were ineligible if they had received prior chemotherapy or radiotherapy. All patients gave written informed consent before entry to the study.

Patients were randomly assigned to treatment groups in a ratio of 2:2:1; that is, only one of every five patients was assigned to single-agent 5-FU therapy. This was to allow a greater number of patients in the nitrosourea arms since the therapeutic differences between CCNU + 5-FU and MeCCNU + 5-FU were anticipated to be small, requiring the accrual of considerable numbers of patients to statistically prove or disprove comparability.

Base-line studies, in addition to those necessary to fulfill eligibility criteria, included a complete history and physical examination, blood liver function tests, plasma carcinoembryonic antigen (CEA), serum electrolytes, and chest x ray. Other tests such as radionuclide scans, sonograms, and computerized tomography were performed as appropriate to measure extent of disease. On-study analyses included weekly complete blood counts and monthly serum creatinines, urinalyses and blood urea nitrogen (BUN) determinations before each course of chemotherapy. Special attention was also paid to determine if there was any evidence of pulmonary compromise before each further dose of nitrosourea. Other studies appropriate to measure extent of disease were performed at regular intervals.

TABLE I. Randomized Drug Regimens

Regimen	Drug	Dosage	Route	Days	
1	5-FU	12 mg/kg	iv	1–5	q 4 wk
	MeCCNU	125–100 mg/m²	po	1	q 8 wk
2	5-FU	12 mg/kg	iv	1–5	q 4 wk
	CCNU	100 mg/m²	po	1	q 8 wk
3	5-FU	12 mg/kg	iv	1–4	
		6 mg/kg	iv	6, 8, 10, 12	
		15 mg/kg	iv	begin 21	weekly

The treatment schedules to which patients were randomized are shown in Table I. Patients continued on study until disease progression. Subsequent doses of 5-FU, CCNU, and MeCCNU were adjusted according to the lowest recorded white blood counts (WBC) and platelet counts determined during the preceding course of therapy. It was found necessary to reduce the dose of MeCCNU from the proposed starting dose of 125 mg/m² to 100 mg/m² to give comparable acceptable tolerance for both nitrosourea-containing regimens.

A partial response was defined as a 50% or greater decrease in the sum of the products of the perpendicular diameters of most measurable lesions with no increase in size of any lesions nor the appearance of new lesions for at least 4 weeks. Stable disease was defined as less than a 50% tumor regression or no change in tumor measurements over a minimum period of 8 weeks. Survival calculations were determined from the time of initiation of treatment by the Kaplan and Meier method (Kaplan and Meier, 1958).

III. RESULTS

A. Patient Characteristics and Randomized Group Comparability

The characteristics of patients entered into the study and intergroup comparability are shown in Table II. The median age of all patients was 64 years, and most patients had a performance status > 70%. Age, sex, base-line performance status, prior surgery, and the presence of liver metastases were evenly distributed across the three therapy groups.

B. Response

Forty-three of the 47 patients analyzed were evaluable for response. The four inevaluable patients had not received an adequate trial of therapy. Five patients in

TABLE II. Patient Characteristics and Group Comparability

	Total	5-FU	CCNU + 5-FU	MeCCNU + 5-FU
Age				
40–59	16	1	10	5
60–79	30	4	12	14
80–89	1	0	1	0
Median	64	66	61	64
Range	48–80	51–77	48–80	48–78
Sex				
Male	30	4	15	11
Female	17	1	8	8
Baseline performance status				
90–100%	13	1	5	7
70–89%	23	3	13	7
60–69%	3	1	2	0
Not reported	8	0	3	5
Site of disease				
Liver mets only	20 (43%)	4 (80%)	7 (30%)	9 (47%)
Liver and other mets	16 (34%)	0	8 (35%)	8 (42%)
Other mets	11 (23%)	1 (20%)	8 (35%)	2 (11%)
Surgical resection	44	4	12	18
No surgical resection	3	1	1	1

total achieved a partial response. Table III summarizes response and response duration data. There was no significant difference in response rate or duration for the two combination therapy groups. Table IV shows the profile of responding patients. Within the small number of responders, no characteristic demonstrated prognostic significance.

Figure 1 shows projected survival of evaluable patients computed by a life table analysis program. Twenty-five of the 43 patients were still alive and there-

TABLE III. Response to Therapy

	5-FU	CCNU/5-FU	MeCCNU/5-FU
Evaluable patients	4	21	18
Not evaluable for response	1	2	1
Partial response	0	3 (14%)	2 (11%)
Liver metastases	—	2 (13%)	1 (6%)
No liver metastases	—	1 (13%)	1 (50%)
Stable disease	1	9 (21%)	11 (26%)
Liver mets	1 (25%)	7 (47%)	11 (65%)
No liver mets	0	2 (25%)	0
Duration of response (months)			
Median	—	5.1	3.5
Range	—	3.7–7.8	3.2–3.7

TABLE IV. Responders' Profile

Patient	P. status (%)	Site	R_x	Duration (mos.)	Survival (mos.)
1	70	Supraclav nodes	MeCCNU+5-FU	3.5	9
2	90	Liver Pelvic mass	MeCCNU+5-FU	3+	4+
3	70	ABD. mass	CCNU+5-FU	5	16
4	80	Liver	CCNU+5-FU	5	9
5	70	Liver	CCNU+5-FU	3.5	5

fore were censored observations. There appeared to be no difference in the projected survival of patients in the two combination therapy groups. Similarly, there was no difference in the actual median survival times of all evaluable patients in the two combination therapy groups. Actual median survival for the 5-FU + CCNU group was 5.5 months versus 6 months for the 5-FU + MeCCNU group. Median survival of responders was 9.2 months whereas that of nonresponders was 4.9 months.

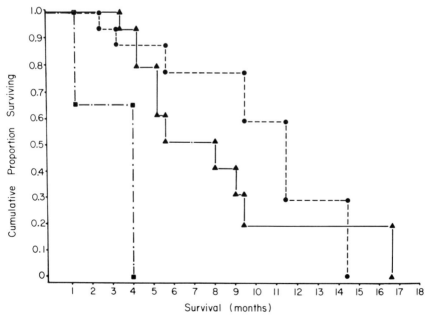

Fig. 1. Predicted survival of the three randomized treatment groups. ■5-FU alone, 2/4 censored observation; ▲ CCNU/5-FU 11/21 censored observation; ● MeCCNU/5-FU 12/18 censored observation. Because of the small numbers of patients there are no statistically significant differences between the groups.

TABLE V. Comparable Tolerance

Toxicity	5-FU	CCNU+5-FU	MeCCNU+5-FU
Leukopenia ($\times\ 10^3$)			
Mild (3.0–4.5)	1	5	6
Moderate (2.0–2.9)	0	3	2
Severe (>1.0–1.9)	0	4	5
Severe (<0.9)	0	3	3
Sepsis	0	5	4
Thrombocytopenia ($\times\ 10^3$)			
Mild (90–130)	0	5	3
Moderate (50–89)	0	4	3
Severe (25–49)	0	3	3
Stomatitis	0	0	4
Mild	0	1	3
Moderate	0	2	2
Severe	0	0	0
Nausea/vomiting	0	2	2
Mild	0	7	4
Moderate	1	2	0
Severe	0	0	0
Ataxia—severe	0	0	1

C. Toxicities

Toxicities experienced by patients in this study were similar to those reported in earlier studies. The incidence and severity of leukopenia and thrombocytopenia were similar for the two combination therapy groups after reduction of the MeCCNU starting dose to 100 mg/m². This followed recognition of an increased myelotoxicity for MeCCNU + 5-FU over CCNU + 5-FU when MeCCNU was initially given at a starting dose of 125 mg/m². Gastrointestinal effects occurred more frequently in the patients treated with 5-FU + CCNU.

Stomatitis occurred more frequently in the patients treated with 5-FU + MeCCNU. Ataxia secondary to 5-FU was experienced by one patient in the 5-FU + MeCCNU group. Table V summarizes the incidence and severities of toxicities. Dose reductions were required by 59% of patients receiving 5-FU + MeCCNU and by 19% of patients receiving 5-FU + CCNU therapy.

IV. DISCUSSION

This report is the preliminary analysis of an ongoing prospectively randomized study. Because of the small number of patients so far evaluable, no conclusions regarding comparable activity can be reached. One observation that can be made from this analysis is that despite an increased potency for CCNU over MeCCNU when compared as single agents (Wasserman, 1976), the addition of 5-FU to each drug appears to result in an equipotency for each nitrosourea in this setting with respect to myelosuppressive toxicity.

ACKNOWLEDGMENT

We would like to acknowledge the following investigators who contributed patients to this study: C. Kardinal, S. Luedke, D. Luedke, N. Oishi, L. R. Morgan, H. Lessner, and J. Mailliard.

REFERENCES

Baker, L. H., Talley, R. W., Matter, R., Lehane, D. E., Ruffner, B. W., Jones, S. E., Morrison, F. S., Stephens, R. L., Gehan, E. A., and Vaitkevicius, V. K. (1976a). *Cancer 38*, 1-7.
Baker, L. H., Vaitkevicius, V. K., Gehan, E. A., and the Gastrointestinal Committee of the Southwest Oncology Group: (1976b) *Cancer Treat. Rep. 60*, 733-737.
Kaplan, E. L., and Meier, P. (1958). *J. Amer. Statist. Assoc. 53*, 457-481.
Katz, M. E., and Glick, J. H. (1979). *Cancer Clin. Trials 2*, 297-316.
Maurice, P., Glidewell, O., Jacquillat, C., Silver, R. T., Carey, R., Ten Pas, A., Cornell, C. J., Burningham, R. A., Nissen, N. I., and Holland, J. F. (1978). *Cancer 41*, 1658-1663.
Moertel, C. G. (1975). *Cancer 36*, 675-682.
Posey, L. E., and Morgan, L. R., (1977). *Cancer Treat. Rep. 61*, 1453-1458.
Selawry, O. S., and Hansen, H. H. (1972). *Proc. AACR 13*, 46.
Wasserman, T. H. (1976). *Cancer Treat. Rep. 60*, 709-711.

Chapter 25

CLINICAL USE OF NITROSOUREAS IN PATIENTS WITH MALIGNANT MELANOMA: A REVIEW

Muhyi Al-Sarraf

I. INTRODUCTION

Malignant melanoma accounts for approximately 1 to 3% of all cancers. Twenty percent of those patients with malignant melanoma initially present with disseminated disease. Once the patient becomes symptomatic from metastases, death usually occurs within a few months; in some instances, however, several years may elapse. Multiple-organ metastases were diagnosed clinically in 78% of all patients and was seen at all autopsies (Amer et al., 1979). Usually these patients with visceral organ involvement, especially liver, had very poor prognosis (Amer et al., 1979).

Chemotherapy for metastatic melanoma has been widely employed, both single agents and, more frequently, a combination of active drugs (Tables I and II) (Luce, 1972; Wasserman et al., 1974, Comis and Carter, 1974; Comis, 1976; Wittes et al., 1978, Amer et al., 1979, Rozencweig et al., 1976).

The most extensively studied single agent in disseminated malignant melanoma has been DTIC, which has produced 208 objective responses in 853 patients treated (24%) in collected published series (Tables I and II). Several currently available commercial agents and investigational drugs have been evaluated, but most of these studies were broad phase II trials not specifically designed for malignant melanoma.

The introduction of the nitrosoureas into clinical trials provided another group

TABLE I. Response Rate to Systemic Chemotherapy: Single Agents

Agents	No. evaluated	No. responders	%
Alkylating agents			
Mechlorethamine	45	3	7
Cyclophosphamide	59	13	22
Melphalan	110	10	9
Thio-TEPA	77	7	9
Chlorambucil	22	2	9
Tic-mustard	108	9	8
Antimetabolites			
5-Fluorouracil	43	1	2
5-FUDR	17	2	12
Methotrexate	25	2	8
6-Mercaptopurine	59	4	7
Arabinosyl cytosine	52	1	2
Vinca alkaloids			
Vinblastine	137	28	20
Vincristine	52	6	12
Antibiotics			
Mithramycin	79	9	11
Actinomycin D	91	15	16
Mitomycin C	68	11	16
Bleomycin	38	4	8
Adriamycin	33	1	3
Other			
DTIC	853	208	24
Hydroxyurea	200	24	12
Pregnenetrione	155	11	7
Procarbazine	50	8	16
Hexamethylmelamine	42	2	5
Dibromodulcitol	18	3	17
TMCA	58	7	12
VP-16	29	1	3
Streptonigrin	13	2	15
Azotomycin	30	3	10
Camptothecin	15	0	0
5-Azacytidine	21	5	24
Cycloleucine	6	1	17

of chemotherapeutic agents possessing biological activity in patients with metastatic malignant melanoma. As a class of compounds, the nitrosoureas have had an adequate evaluation in melanoma patients. There does not appear to be any obvious superiority of one nitrosourea agent over another. The degree of activity, both qualitative and quantitative, appears to be similar to DTIC, although the evaluation of each drug has been less extensive.

TABLE II. Response Rate to Systemic Chemotherapy: Combination Therapy

Agents	No. evaluated	No. responders	%
DTIC combination			
DTIC, VCR	115	22	19
DTIC, VLB	34	6	18
DTIC, actino D	69	14	20
DTIC, CTX	39	7	24
DTIC, procarbazine	94	14	15
DTIC, CTX, VCR	20	5	25
VCR combination			
VCR, actino D	12	6	50
VCR, MTX	9	0	0
VCR, CTX	9	0	0
VCR, MITO C, melphalan	12	1	8
VCR, CTX, MTX, 5-FU	54	15	28
VCR, CTX, MTX, 5-FU, prednisone	23	5	22

II. RESULTS

A. BCNU (NSC-409962)

BCNU was the first of the nitrosoureas available for clinical trial in patients with advanced malignant melanoma. In view of the high incidence of brain metastases in patients with malignant melanoma, the agents in this class of antitumor compounds are of particular interest because of their relatively high lipid solubility and ability to cross the blood–brain barrier. Dose regimens of BCNU alone have varied, but 80–100 mg/m^2 iv daily for three consecutive days seems to be fairly well tolerated.

That BCNU is active against malignant melanoma was substantiated by a number of clinical investigations utilizing either BCNU alone or in combination with other chemotherapeutic agents, especially DTIC with or without vincristine or hydroxyurea. Investigating its use as a single agent, DeVita et al. (1965) reported 3 responders in 20 patients (15%) with advanced malignant melanoma treated with systemic BCNU. As shown in Table III, of 140 patients treated with

TABLE III. Response Rate to BCNU Alone

Investigators	No. evaluated	No. responders	%
DeVita, et al. (1965)	20	3	15
Lessner, (1968)	3	0	0
Ramirez, et al. (1972)	99	19	19
Hill, et al. (1974)	18	3	17
Total	140	25	18

TABLE IV. Response Rate to BCNU plus Vincristine

Investigators	No. evaluated	No. responders	%
Moon (1970)	20	9	45
Johnson and Jacobs (1971)	7	1	14
Marsh et al. (1971)	11	2	18
Primack et al. (1973)	16	2	13
Stolinsky et al. (1974)	8	2	25
Moon et al. (1975)	51	8	16
Total	113	24	21

BCNU alone in four reports, 25 (18%) had objective tumor response. Since the activity of BCNU as a single agent against malignant melanoma has been confirmed, an extensive trial with different combinations of agents with BCNU has been reported. A BCNU and vincristine combination (Table IV) has been tried and reported on by many investigators between 1970 and 1975. Of 113 patients with advanced malignant melanoma treated mainly as phase II trials, 24 patients (21%) responded with a response range from 13% to 45%. Most of these trials consist of 20 or fewer patients, and this is part of the reason for the wide range of response. Moon et al. (1975) had treated 51 patients and reported 8 (16%) responses, which is no different than the response rate (18%) to BCNU alone in malignant melanoma patients.

The BCNU, vincristine, and DTIC (BVD) combination has been adequately evaluated by many investigators and cooperative groups. Some of these studies were phase II and phase III trials with many treatment arms for comparison (Table V). These reports were published between 1970 and 1979. In 587 patients treated with BVD combination, 142 (24%) had objective responses, ranging from 19% to 63%. In four large series reported, the response rate for BVD was no different than that for BCNU or DTIC alone. Einhorn et al. (1974) reported a

TABLE V. Response Rate to BCNU, Vincristine, and DTIC (BVD)

Investigators	No. evaluated	No. responders	%
Luce et al. (1970)	27	12	44
Cohen et al. (1972)	16	10	63
Beretta et al. (1973)	41	8	20
Einhorn et al. (1974)	106	20	19
Carmo-Pereira et al. (1976)	20	7	35
Carter et al. (1976)	65	15	23
Cohen et al. (1977)	24	7	29
McKelvey et al. (1977)	132	31	23
Hill et al. (1979)	156	32	21
Total	587	142	24

TABLE VI. Response Rate to BCNU Combinations

Investigators	Agents	No. evaluated	No. responders	%
Costanza et al. (1972)	BCNU, DTIC	61	12	19
Costanzi et al. (1975) Costanzi (1976)	BCNU, hydroxyurea, DTIC (BHD)	123	44	36
Carter et al. (1976)	BHD	63	8	13
Costanzi et al. (1975) Costanzi (1976)	BHD–VCR	124	39	31
Vandyk and Falkson (1975)	BVD–procarbazine	25	7	28
McKelvey et al. (1977a)	BVD–chloropromazine	111	27	24
	Total	507	137	27

19% (20/106) response rate, McKelvey et al. (1977b) reported 23% (21/132), Carter et al. (1976) 23% (15/65), and Hill et al. (1979) 21% (32/156).

Many other BCNU combinations have been reported in the treatment of patients with disseminated melanoma (Table VI). Costanza et al. (1972) reported a response rate of 19% (12/61) with the combination of BCNU and DTIC. The combination of BVD and procarbazine has been reported by VanDyk and Falkson (1975) with a response rate of 28% (7/25). McKelvey et al. (1977b) showed no advantage to adding chlorpromazine to BVD with a response rate of 24% (27/111) as compared to BVD alone (23%) in randomized trial. The combination of BCNU, hydroxyurea, and DTIC (BHD) with or without vincristine has been tried by the Southwest Oncology Group (SWOG) and reported by Costanzi and co-workers (Costanzi et al., 1975; Costanzi, 1976). The response rates were similar, 36% (44/123) to BHD and 31% (39/124) to BHD–VCR. Subsequent to this study, BHD has been utilized further in many phase III trials in the SWOG. However, Carter et al. (1976) reported a response rate of 13% (8/63) to BHD in patients with advanced malignant melanoma. Table VII gives overall response

TABLE VII. Overall Response Rates to BCNU Alone or in Combination

Agents	No. evaluated	No. responders	%	Range of response (%)
BCNU	140	25	18	0–19
BCNU, VCR	113	24	21	13–45
BCNU, VCR, DTIC	587	142	24	19–63
BVD–chlorpromazine	111	27	24	24
BCNU, DTIC	61	12	19	19
BCNU, hydroxyurea, DTIC	186	52	28	13–36
BHD–VCR	124	39	31	31
Total	1322	321	24	0–63

rates to BCNU alone and in combination. The dose schedule utilized in some of these trials are shown in Table VIII.

B. CCNU (NSC-79037)

CCNU was the first analog of BCNU selected for clinical trial because of its experimental activity in leukemia L1210 and greater lipid solubility. In comparison with BCNU, CCNU has superior antitumor activity against intracerebrally inoculated L1210 and the Lewis lung carcinoma. It is also significantly active against Walker 256 carcinosarcoma and B16 melanoma (Carter, S.K. *et al.* 1972).

As a single agent, CCNU has usually been given at a dose of 130 mg/m^2 orally every 6–8 weeks. In phase I–II trials, CCNU was found to be active in patients with metastatic malignant melanoma (Table IX). In 82 patients treated, 15 (18%) had objective tumor response with a response range of 0–33%. Most of these trials comprise a small number of patients.

Many trials of malignant melanoma treated with CCNU combinations were reported between 1971 and 1980 (Table X). The overall response rate in these trials is 29% (71/248). In three investigations in which CCNU, vincristine, and DTIC were used, the response rate was 19% (31/164) (Carter and Krementz, 1971, Carter *et al.*, 1976; Hill, G.J. *et al.*, 1979).

Of interest is the CCNU combination chemotherapy in disseminated

TABLE VIII. Response Rate and Dose in Nitrosourea Combinations

Investigators	Agents	Doses	Resp./Eval.	%
McKelvey *et al.* (1977a)	BCNU VCR DTIC	150/mg/m^2 iv/d 1 every 4 wk 1.4 mg/m^2 iv/d 1 & 5 every 4 wk 150 mg/m^2 iv/d 1–5 every 4 wk	31/132	23
Cohen *et al.* (1972; 1977)	BCNU VCR DTIC	65 mg/m^2 iv/d 3–4 every 6 wk 1.1.5 mg/m^2 iv/d 1 & 14 then wkly 250 mg/m^2 iv/d 3–4 every 6 wk	17/40	43
Costanzi *et al.* (1975); Costanzi (1976)	BCNU hydroxyurea DTIC	150 mg/m^2/d 1 every other course 1480 mg/m^2/d 1–5 every 4 wk 150 mg/m^2/d 1–5 every 4 wk	44/123	36
Einhorn and Furnas (1977)	MeCCNU VCR DTIC	175 mg/m^2 (o) every 6 wk 1.4 mg/m^2 iv weekly × 6 then every 3 wk 250 mg/m^2 iv/d 1–5 every 3 wk	9/23	39
Hill *et al.* (1979)	CCNU VCR DTIC	2–2.4 mg/kg d 2 (o) every 6 wk 0.027–0.032 mg/kg iv d 1 & 5 every 16 wk 2.7–3.2 mg/kg iv d 1–5 every 6 wk	16/86	19
Hill *et al.* (1979)	BCNU VCR DTIC	2–3.6 mg/kg iv d 2 every 6 wk 0.027–0.032/mg/kg iv d 1 & 5 every 6 wk 2.7–5.4 mg/kg iv d 1–5 every 6 wk	22/146	15

TABLE IX. Response Rate to CCNU Alone

Investigators	Evaluated	Responders	%
Hansen and Muggia (1971)	2	0	0
Ahmann et al. (1972)	19	1	5
Hoogstraten and Luce (1973)	17	3	18
Broder and Hansen (1973)	4	0	0
DeConti et al. (1973)	4	1	25
Cruz et al. (1974)	27	7	26
Perloff et al. (1974)	3	1	33
Hill, J. et al. (1974)	6	2	33
Total	82	15	18

melanoma without DTIC. DeWesch et al. (1976) reported a response rate of 49% (17/35) with CCNU, vincristine, and bleomycin; Carmo-Pereira et al. (1980) used CCNU, vincristine, and procarbazine in 30 patients; 18 (60%) had objective response, and Green et al. (1980) reported 33% (4/12) with CCNU, vincristine, procarbazine, and cyclophosphamide.

These results are interesting and may need to be confirmed in the future and compared in phase II–III trials with other known active combination chemotherapy regimens in patients with metastatic malignant melanoma. An example of the dose schedule with CCNU combination is shown in Table VIII.

C. Methyl CCNU (NSC-95449)

Because of its superior activity in the B16 melanoma system (Schabel, 1973) and its ability to be given orally, MeCCNU was investigated in patients with advanced malignant melanoma. Hill, R.P. and Stanley (1975) reported on the response of hypoxic B16 melanoma cells to *in vivo* treatment with nitrosoureas (BCNU, CCNU, MeCCNU) or cyclophosphamide. When these four agents are

TABLE X. Response Rate to CCNU Combinations

Investigators	Agents	No. evaluated	No. responders	%
Carter and Krementz (1971)	CCNU, DTIC, VCR	11	4	36
Carter et al. (1976)	CCNU, DTIC, VCR	67	11	16
Hill et al. (1979)	CCNU, DTIC, VCR	86	16	19
Einhorn et al. (1973)	CCNU, adriamycin	7	1	14
DeWesch et al. (1976)	CCNU, BLM, VCR	35	17	49
Carmo-Pereira et al. (1980)	CCNU, VCR, procarbazine	30	18	60
Green et al. (1980)	CCNU, VCR, procarbazine, cyclophosphamide	12	4	33
	Total	248	71	29

TABLE XI. Response Rate to MeCCNU Alone

Investigators		No. evaluated	No. responders	%
Young *et al.* (1973)		28	6	21
Costanza and Nathanson (1974)		23	2	9
Tranum and Haut (1974)		15	0	0
Ahmann *et al.* (1974)		19	5	26
Firat and Tekuzman (1974)		5	0	0
Costanza *et al.* (1977)		119	18	15
Cohen *et al.* (1980)		77	9	12
	Total	286	40	14

assessed in terms of the tumor kill at the lethal dose of 10% of the mice, MeCCNU was found to be the most effective, followed by CCNU and then BCNU and cyclophosphamide together. The superiority of MeCCNU is possibly related to the fact that it seems to be longer lived in the mice than are the other agents. MeCCNU doses have varied in the trials from 150–250 mg/m^2 given orally every 6–8 weeks. A single oral dose of 225 mg/m^2 was clinically well tolerated.

As a single agent, MeCCNU was found to have 14% (40/286) activity in patients with malignant melanoma. In two large series, Costanza *et al.* (1977) reported a response rate of 15% (18/119) with MeCCNU and Cohen *et al.* (1980) had a 12% (9/77) response rate (Table XI).

Combinations containing MeCCNU have been tried in patients with disseminated melanoma (Table XII). In 302 patients treated, 50 (17%) had objective responses. This is rather low as compared to the BCNU or CCNU combinations in malignant melanoma. Most of these MeCCNU combinations were with DTIC alone or with other agents, and the result is no different than the response rate to DTIC alone or MeCCNU alone.

TABLE XII. Response Rate to MeCCNU Combinations

Investigators	Agents	No. evaluated	No. responders	%
Costanza and Nathanson (1974)	MeCCNU, DTIC	26	6	23
Costanza *et al.* (1977)	MeCCNU, DTIC	122	18	14
Ahmann *et al.* (1975)	MeCCNU, cyclophosphamide	20	1	5
Livingston *et al.* (1975)	MeCCNU, cyclophosphamide, VCR, bleomycin	41	7	17
Einhorn and Furnas (1977)	MeCCNU, DTIC, VCR	23	9	39
Cohen *et al.* (1980)	MeCCNU, chlorpromazine, caffeine	70	9	13
	Total	302	50	17

D. Streptozotocin (NSC-85998)

Streptozotocin is a nitrosourea compound that is found to lack overt hematologic toxicity in the animal studies. Streptozotocin has shown antitumor activity against various tumors in man, but the clinical usefulness of this drug has been limited, mainly because of renal and gastrointestinal toxicity. It is found to be active in pancreatic islet-cell carcinoma (Broder and Carter, 1973) and in patients with malignant lymphoma (Schein et al., 1977; O'Connell and Wiernik, 1975) Schein et al. (1974) reported two responses in 16 patients (13%) with malignant melanoma treated with streptozotocin.

E. Chlorozotocin (NSC-178248)

Chlorozotocin (CZT) is the 2-chloroethyl analog of the nitrosourea streptozocin. Chlorozotocin shows good activity against L1210 leukemia. P388 leukemia, and B16 melanoma, with cures being reported in all three systems (Hoth et al., 1977).

Chlorozotocin was found to be superior to CCNU and produces no suppression of the peripheral neutrophils (Panasci et al., 1977). CZT has produced a response in the following cancers: melanoma, breast, lymphoma, colon, adenocarcinoma and squamous cell carcinoma of the lung, renal, and ovary cancers (Hoth et al., 1977, Tan et al. 1978; Kovach et al. 1978; Hoth et al., 1978). In a phase II trial with chlorozotocin in metastatic malignant melanoma, using doses of 120 mg/m^2 iv every 6 weeks, Hoth et al. (1969) reported 5 partial remissions in 34 patients (14%). Sites of response included two lung, two subcutaneous, and one node. In patients with no prior therapy, partial response was observed in 4/23 patients; in patients with prior therapy, partial response was seen in 1/11. Talley et al. (1979) reported no response in eight melanoma patients treated with chlorozotocin; all of these patients had extensive prior chemotherapy. VanAmburg et al. (1980) reported 3 complete remissions (2 with visceral metastases) and 1 partial response in 10 melanoma patients previously untreated when they received CZT 150 mg/m^2 iv every 6 weeks, whereas 1 partial response in 28 patients previously treated with chemotherapy was obtained.

In one trial at Wayne State University by Samson (1980, personal communication), chlorozotocin in combination with DTIC and actinomycin D is being tried in patients with advanced malignant melanoma; a response rate of 16% (4/25) is reported.

III. DISCUSSION

Malignant melanoma is one of the few malignant diseases in which the majority of active standard and investigational drugs have been adequately evaluated, but none has yielded a response rate of greater than 20-30%.

A variety of combination chemotherapy regimens have been employed in metastatic malignant melanoma, primarily utilizing DTIC in combination with a nitrosourea, with or without the addition of vincristine. However, most of these combination regimens have not been shown to be superior to DTIC alone or to a nitrosourea agent alone.

A few important factors may influence the response to systemic chemotherapy in patients with metastatic malignant melanoma. Performance status is one important factor. In those patients who are ambulatory, active, and with minimal symptoms, the chance of response to chemotherapy is higher and survival longer than in those patients with poor performance status (Costanza et al., 1977; Amer et al., 1979).

For patients with visceral organ involvement from malignant melanoma, especially liver or brain, the chance of response to chemotherapy is smaller, and survival is shorter than for those patients with lymph node and subcutaneous involvement (Costanza et al., 1977, Costanzi et al., 1975; Amer et al., 1979; Carter et al., 1976).

Another very important factor is previous systemic chemotherapy. The chances are very small that patients with prior chemotherapy will respond to other agents, single or combination, regardless of how effective they were as first treatment (Costanzi et al., 1977; VanAmburg et al., 1980; Ahmann 1976). In a trial comparing MeCCNU versus DTIC and vincristine (Ahmann 1976), the response rate to MeCCNU as the first agent was 26% (5/19), whereas none of the patients who received DTIC and VCR first and crossed over to MeCCNU responded (0–11).

Immunotherapy in combination with a variety of agents have been evaluated in malignant melanoma (Al-Sarraf et al., 1980a; Costanzi et al., 1979; Costanzi et al., 1980). In good randomized trials with a large number of patients, immunotherapy with BCG or levamisole did not add to the response of single-agent or combination therapy with nitrosourea agents.

Many adjuvant trials for patients with malignant melanoma stage I and II are under way utilizing single-agent or combination chemotherapy containing a nitrosourea. In those randomized studies in which a large number of patients are included, no advantage of such therapy is reported (Quagliana et al., 1980).

In conclusion, the introduction of the nitrosoureas into clinical trials has provided another class of chemotherapeutic agents possessing modest activity in patients with metastatic malignant melanoma.

ACKNOWLEDGMENTS

Thanks to Marie Schumacher and Rose Marie Mason for secretarial assistance.

REFERENCES

Ahmann, D.L. (1976). *Cancer Treat. Rep. 60*, 747–751.

Ahmann, D.L., Hahn, R.G., and Bisel, H.F. (1972). *Cancer Res. 32*, 2432–2434.

Ahmann, D.L., Hahn, R.G., and Bisel, H.F. (1974). *Cancer 33*, 615–618.

Ahmann, D.L., Hahn, R.G., Bisel, H.F., Eagen, R., and Edmonson, J.H. (1975). *Cancer Chemother. Rep. 59*, 451–453.

Al-Sarraf, M., Costanzi, J.J., and Dixon, D.O. (1980a). *In* "Tumor Progression" (Crispen, ed.), pp. 299–309. Elsevier/North Holland, New York.

Al-Sarraf, M., Costanzi, J.J., and Dixon, D.O. (1980b). *In* "Tumor Progression" (Crispen, ed.), pp. 197–209. Elsevier/North Holland, New York.

Amer, M.H., Al-Sarraf, M., and Vaitkevicius, V.K. (1979). *Surg. Gyn. Obst. 149*, 687–692.

Beretta, G., Bajetta, E., Bonadonna, G., Tancini, G., Orefice, S., and Veronesi, U. (1973). *Tumori 59*, 239–248.

Broder, L.E., and Carter, S.K. (1973). *Ann. Intern. Med. 79*, 108–118.

Broder, L.E., and Hansen, H.H. (1973). *Eur. J. Cancer 9*, 147–152.

Carmo-Pereira, J., Costa, F.O., and Pimentel, P. (1976). *Cancer Treat. Rep. 60*, 1381–1383.

Carmo-Pereira, J., Costa, F.O., Pimentel, P., and Henrigues, E. (1980). *Cancer Treat. Rep. 64*, 143–145.

Carter, R.D., and Krementz, E.T. (1971). *Proc. AACR 12*, 88.

Carter, R.D., Krementz, E.T., Hill, G.J. II, Metter, G.E., Fletcher, W.S., Golomb, F.M., Grage, T.B., Minton, J.P., and Sparks, E.C. (1976). *Cancer Treat. Rep. 60*, 601–609.

Carter, S.K., Schabel, F.M. Jr., Broder, L.S., and Johnston, T.P. (1972). *Adv. Cancer Res. 16*, 273–332.

Cohen, S.M., Greenspan, E.M., Weiner, M.J., and Kabakow, B. (1972). *Cancer 29*, 1489–1495.

Cohen, S.M., Greenspan, E.M., Ratner, L.H., and Weiner, M.H. (1977). *Cancer 39*, 41–44.

Cohen, M.H., Schoenfeld, D., and Wolter, J. (1980). *Cancer Treat. Rep. 64*, 151–153.

Comis, R.L. (1976). *Cancer Treat. Rep. 66*, 165–176.

Comis, R.L., and Carter, S.K. (1974). *Cancer Treat. Rep. 1*, 285–304.

Costanza, M.E., and Nathanson, L. (1974). *Proc. AACR and ASCO 15*, 173.

Costanza, M.E., Nathanson, L., Lenhard, R., Wolter, J., Colsky, J., Oberfield, R.A., and Schilling, A. (1972). *Cancer 30*, 1457–1461.

Costanza, M.E., Nathanson, L., Schonfeld, D., Wolter, J., Colsky, J., Regelson, W., Cunningham, T., and Sedransk, N. (1977). *Cancer 40*, 1010–1015.

Costanzi, J.J., Vaitkevicius, V.K., Qualgiana, J.M., Hoogstraten, B., Coltmon, C.A., and Delaney, F.C. (1975). *Cancer 35*, 342–396.

Costanzi. J.J. (1976). *Cancer Treat. Rep. 6*, 189–192.

Costanzi, J.J., Al-Sarraf, M., and Dixon, D.O. (1979). *Proc. ASCO 20*, 362.

Costanzi, J.J., Al-Sarraf, M., and Dixon, D.O. (1980). *Proc. ASCO 21*, 474.

Cruz, A.B., Jr., Armstrong, D.M., and Aust, J.B. (1974). *Proc. AACR and ASCO 15*, 184.

DeConti, R., Hubbard, S.P., Pinch, P. and Bertino, J.R. (1973). *Cancer Chemother. Rep. 57*, 201–207.

DeVita, V.T., Carbone, P.P., Owens, A.H., Jr., Gold, G.L., Krant, M.J., and Edmonson, J. (1965). *Cancer Res. 25*, 1876–1881.

DeWesch, G., Bernheim, J., Michel, J., Lejeune, F., and Kanis, Y. (1976). *Cancer Treat. Rep. 60*, 1273–1276.

Einhorn, L.H., and Furnas, B. (1977). *Cancer Treat. Rep. 61*, 881–883.

Einhorn, L.H., Livingston, R.B., and Gottlieb, J.A. (1973). *Cancer Chemother. Rep. 57*, 437–445.

Einhorn, H., Burgess, M.A., Vellejos, C., Bodey, G.P., Gutterman, J., Mavligit, G., Hersh, E., Luce, J., Frei, E., Freireich, E., and Gottlieb, J. (1974). *Cancer Res. 34*, 1995–2004.

Firat, D., and Tekuzman, G. (1974). *Proc. AACR and ASCO 15*, 5.

Green, M.R., Dillman, R.O., and Horton, C. (1980). *Cancer Treat. Rep. 64*, 139–142.

Hansen, H.M., and Muggia, F.M. (1971). *Cancer Chemother. Rep. 55*, 99–100.

Hill, G.J., Metter, G.E., Krementz, E.T., Fletcher, W.S., Golomb, F.M., Ramirez, G., Grage, T.B., and Moss, S.E. (1979). *Cancer Treat. Rep. 63*, 1989–1992.

Hill, J. II, Ruess, R., Berris, R., Philpott, G.W. and Parkin, P. (1974). *Ann. Surg. 180*, 167–174.

Hill, R.P., and Stanley, J.A. (1975). *Cancer Res. 35*, 1147–1153.

Hoogstraten, B., and Luce, J.K. (1973). *Proc. AACR 14*, 3.

Hoth, D., Schein, P., MacDonald, J., Buscaglia, D., and Haller, D. (1977). *Proc. AACR and ASCO 18*, 309.

Hoth, D., Butler, T. Winokur, S., Kales, A., Woolley, P., and Schein, P. (1978). *Proc. ASCO 19*, 381.

Hoth, D., Robichaud, K., Wolley, J.S., MacDonald, J.S., Price, N., Gullo, J., and Schein, P.S. (1979). *Proc. AACR and ASCO 20*, 413.

Johnson, F.D., and Jacobs, E.M. (1971). *Cancer 27, 1306–1312*.

Kovach, J.S., Moertel, C.G., Shutt, A.J., and O'Connell, M.J. (1978). *Proc. ASCO 19*, 408.

Lessner, H.E. (1968). *Cancer 22*, 451–456.

Livingston, R.B., Einhorn, L.H., Bodey, G.P., Burgess, M.A., Freireich, E.H., and Gottlieb, J.A. (1975). *Cancer 36*, 327–332.

Luce, J.K. (1972). *Cancer 30*, 1604–1615.

Luce, J.K., Thurman, W.G., Isaacs, B.L., and Talley, R.W. (1970a). *Cancer Chemother. Rep. 54*, 119–124.

Luce, J.K., Torin, L.B., and Price, H. (1970b). *Proc. AACR 11*, 50.

Marsh, J.C., DeConti, R.C., and Hubbard, S.P. (1971). *Cancer Chemother. Rep. 55*, 599–606.

McKelvey, E.M., Luce, J.K., Talley, R.H., Hersh, E.M., Hewlett, J.S., and Moon, T.E. (1977a). *Cancer 39*, 1–4.

McKelvey, E.M., Luce, J.K., Vaitkevicius, V.K., Talley, R.W., Bodey, G.P., Lane, M., and Moon, T.E. (1977b). *Cancer 39*, 5–10.

Moon, J.H. (1970). *Cancer 25*, 468–473.

Moon, J.H., Gailani, S., Cooper, M.R., Hayes, D.M., Rege, V.B., Blom, J., Falkson, G., Mourice, P., Brunner, K., Glidewill, W., and Holland, J.F. (1975). *Cancer 35*, 368–371.

O'Connell, M.J., and Wiernik, P.G. (1975). *Cancer Chemother. Rep. 59*, 443–445.

Panasci, L.C., Green, D.C., Nagourney, R., Fox, P., and Schein, P.S. (1977). *Cancer Res. 37*, 2615–2618.

Perloff, M., Muggia, F.M., and Ackerman, C. (1974). *Cancer Chemother. Rep. 58*, 421–424.

Primack, A., Dhru, D., Kiryabwire, J.W.M., and Vogel, C.L. (1973). *Cancer 31*, 337–341.

Quagliana, J. Tranum, B., Neidhardt, J., and Gagliano, R. (1980). *Proc. AACR and ASCO 21*, 399.

Ramirez, G., Wilson, W., Grage, T., and Hill, G. (1972). *Cancer Chemother. Rep. 56*, 787–790.

Rozencweig, M., Slavik, M., Muggia, F.M., and Carter, S.K. (1976). *Med. Pediat. Oncol. 2*, 417–432.

Schabel, F.M., Jr. (1973). *Cancer Chemother. Rep. 4*, 3–6.

Schein, P.S., O'Connell, M.J., Blom, J., Hubbard, S., Magrath, I.T., Bergevin, P., Wiernik, P.H., Ziegler, J.L., and DeVita, V.T. (1974). *Cancer 34*, 993–1000.

Stolinsky, D.C., Pugh, R.P., Bohannon, R.A., Bogdon, D.L., and Bateman, J.R. (1974). *Cancer Chemother. Rep. 58*, 947–950.

Talley, R.W., Brownlee, R.W., Baker, L.H., Oberhauser, N.A., and Pitts, K. (1979). *Proc. AACR and ASCO 20*, 440.

Tan, C., Gralla, R., Steinherz, P., and Young, C.W. (1978). *Proc. AACR 19*, 126.

Tranum, B.L., and Haut, A. (1974). *Proc. AACR and ASCO 15*, 171.

VanAmburg, A., Ratkin, G., Washington, U., and Presant, C. (1980). *Proc. AACR and ASCO 21*, 353.

VanDyk, J.J., and Falkson, B. (1975). *Med. Pediat. Oncol. 1*, 107–111.

Wasserman, T.M., Slavik, M., and Carter, S.K. (1974). *Cancer Treat. Rep. 1*, 131–151.

Wittes, R.E., Wittes, J.T., and Golbey, R.B. (1978). *Cancer 41*, 415–421.

Young, R.C., Canellos, G.P., Chabner, B.A., Schein, P.S., Brereton, H.D., and DeVita, V.T. (1973). *Clin. Pharmacol. Ther. 15*, 617–622.

Chapter 26

NITROSOUREAS IN THE TREATMENT OF MULTIPLE MYELOMA

O. Ross McIntyre
Thomas Pajak
Louis Leone
Robert Kyle
Gibbons G. Corwell
John Harley
Shaul Kochwa
Babette B. Weksler
Linda P. Glowienka

I. INTRODUCTION

In 1967 the Cancer and Leukemia Group B (CALGB) began studies with BCNU in multiple myeloma. Since that time, the CALGB and other cooperative groups have defined a role for this agent in the treatment of the disease.

Certain results of these studies have been previously reported (Leone, 1974; Salmon, 1976; McIntyre *et al.*, 1978, Kyle *et al.*, 1979; and Harley *et al.*, 1979). We now present information on response and survival for patients from three CALGB protocols in which previously untreated patients were randomized to regimens of BCNU with or without prednisone or to melphalan and prednisone. We also discuss results from a later CALGB protocol in which patients were randomized to receive three intravenous drugs (BCNU, melphalan, and cyclophosphamide) with oral prednisone or a standard melphalan and prednisone regimen. Finally, we present our experience with BCNU-containing drug combinations for the treatment of patients failing on, or refractory to, other treatment. A current study of the CALGB includes patients randomized to either BCNU or CCNU and, when complete, will provide further information on the usefulness of these agents in the disease.

II. METHODS

The criteria for the diagnosis of multiple myeloma utilized by the CALGB have been published elsewhere (Kyle *et al.*, 1973). Responses were evaluated according to the criteria of the Chronic Leukemia-Myeloma Task Force (1973). Briefly, the patient was considered to have had an objective response if there was a decrease of more than 50% in serum M protein or in urinary M protein or a decrease of more than 50% in measured cross-sectional area of a plasmacytoma or recalcification of bone lesions. The treatment schedules employed are described in the figures that accompany each section of the results.

III. RESULTS

A. CALGB Protocols 6701, 6803

Figure 1 shows the treatment schedules for the melphalan or BCNU-containing regimens employed in these protocols.

Regimen 1 consists of an oral loading dose of melphalan given in conjunction

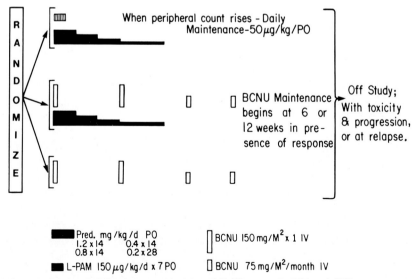

Fig. 1. Melphalan, BCNU treatment schedules, CALGB protocols 6701 and 6803.

TABLE I. Responses[a] CALGB 6701, 6803

Regimen	No. Eval.[b]	Responders No. (%)
L-PAM + Pred	44	21 (48)
BCNU + Pred	34	12 (35)
BCNU	35	11 (31)

[a] Myeloma Task Force criteria.
[b] One patient on BCNU alone and one on LPAM + Pred were excluded because of incomplete data.

with a 70-day tapering course of prednisone. This was compared with a regimen containing BCNU 150 mg/m² and the same prednisone dose. If necessary, a second dose of BCNU of 150 mg/m² was given. Responding patients on this regimen received BCNU 75 mg/m² every 28 days. The third regimen consisted of the same BCNU dose without the course of prednisone. Appropriate dose adjustments were made for drug toxicity. Only patients designated as good risk were eligible for these protocols. Good risk was defined according to the following criteria:

1. An estimated survival of greater than 2 months.
2. A white count of > 4000 and a platelet count of > 100,000, a BUN > 30 mg/dl.
3. Serum calcium of less than 12 mg/dl.
4. The absence of a significant infection.

All patients treated appear in the survival curves; no patients were excluded because they died early.

Table 1 summarizes objective responses for patients assigned to the three regimens. Although more patients receiving the melphalan and prednisone com-

TABLE II. M Protein Response, CALGB 6701, 6803

Regimen	No. Eval.	Serum responders No. (%)
L-PAM + Pred	37	14 (38)
BCNU + Pred	19	6 (32)
BCNU	30	5 (17)

Regimen	No. Eval.	Urine responders No. (%)
L-PAM + Pred	16	11 (69)
BCNU + Pred	12	6 (50)
BCNU	16	8 (50)

bination achieved a response, this trend was not significant. Table II displays serum M protein and urine myeloma protein responses. Fewer patients receiving BCNU alone achieved a serum M protein response, but this trend was not significant. Figure 2 displays the survival of these patients. The survival for patients receiving melphalan and prednisone is initially better than that achieved by patients randomized to BCNU and prednisone, but the difference is not significant ($p = 0.67$).

There were no significant differences in the pretreatment prognostic factors of patients assigned to the three regimens. Table III describes toxicity experienced by patients assigned to the three treatment schedules. There was no difference in degree of leukopenia or thrombocytopenia caused by these schedules. More severe or life-threatening infections were experienced in the two regimens containing prednisone, but this difference was not significant ($p = 0.49$).

B. CALGB Protocol 7161

Figure 3 displays the treatment schedules used in CALGB 7161. Intravenous melphalan given with prednisone was compared with a combination of intraven-

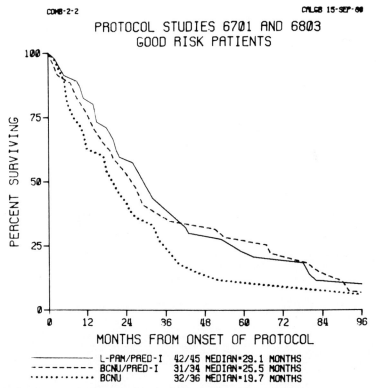

Fig. 2. Survival according to treatment, CALGB protocols 6701 and 6803.

TABLE III. Toxicity of Treatment Regimen, CALGB 6701, 6803

		Leukopenia			
Regimen	Total Eval.	% Mild	% Moderate	% Severe	% Life threatening
L-PAM + Pred	44	23	30	30	2
BCNU + Pred	34	15	26	15	9
BCNU	35	34	26	9	3
		Thrombocytopenia			
L-PAM + Pred	43	12	14	19	7
BCNU + Pred	32	19	13	9	6
BCNU	35	17	11	11	6
		Infectious Complications			
L-PAM + Pred	42	17		10	
BCNU + Pred	34	29		3	
BCNU	35	11		0	

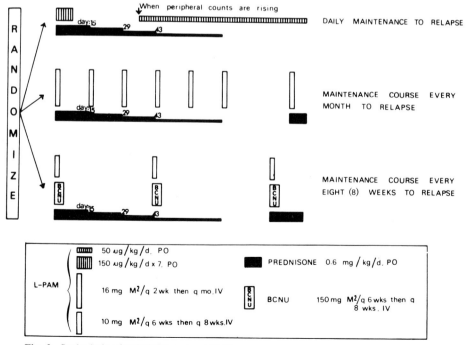

Fig. 3. Study design for CALGB protocol 7161.

ous melphalan–BCNU and prednisone or with oral melphalan and prednisone. After completions of six doses given every 2 weeks, the intravenous melphalan–oral prednisone regimen was repeated every 4 weeks. The intravenous melphalan–BCNU dose was administered twice 6 weeks apart and then every 8 weeks with prednisone. No prednisone was given after the initial 10 weeks for the patients receiving the oral melphalan.

Treatment-related deaths occurred in 5 of 48 patients receiving the intravenous melphalan prednisone combination and in 2 of 40 patients receiving intravenous melphalan, BCNU, and prednisone ($p = 0.58$). This suggests that the toxicity of the intravenous melphalan prednisone regimen may have been more intense. Our experience with intravenous melphalan treatment of myeloma will be summarized in another publication. There were no significant differences in pretreatment prognostic factors in the randomized treatment groups.

Table IV provides information concerning the response of patients randomized to the three regimens. A trend exists for an increased percentage of serum and urinary M protein responses in the group of patients receiving intravenous melphalan and BCNU with prednisone. This trend when compared with the group receiving intravenous melphalan and prednisone does not achieve

TABLE IV. Responses,[a] CALGB 7161

Regimen	No. Eval.[b]	Responders No. (%)
L-PAM + Pred iv	46	26 (57)
L-PAM + BCNU + Pred iv	38	22 (58)
L-PAM + Pred	49	20 (41)

Serum Protein Response

Regimen	No. Eval.	Responders No. (%)
L-PAM + Pred iv	33	14 (42)
L-PAM + BCNU + Pred iv	27	17 (63)
L-PAM + Pred	34	13 (38)

Urinary Protein Response

Regimen	No. Eval.	Responders No. (%)
L-PAM + Pred iv	25	14 (56)
L-PAM + BCNU + Pred iv	21	14 (67)
L-PAM + Pred	20	9 (45)

[a] Myeloma Task Force criteria.
[b] 6 patients on iv LPAM + Pred, 2 patients on iv LPAM + BCNU + Pred were excluded because of incomplete data.

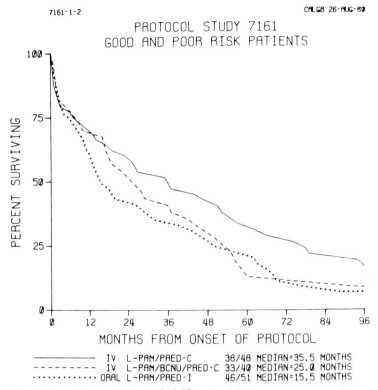

7161·1·2   CALGB 26-AUG-80

PROTOCOL STUDY 7161
GOOD AND POOR RISK PATIENTS

───────────	IV L-PAM/PRED-C	38/48	MEDIAN=35.5 MONTHS
─ ─ ─ ─ ─ ─ ─	IV L-PAM/BCNU/PRED-C	33/40	MEDIAN=25.0 MONTHS
·············	ORAL L-PAM/PRED-I	46/51	MEDIAN=15.5 MONTHS

Fig. 4. Survival according to treatment, CALGB protocol 7161.

statistical significance, however (p = 0.19). Figure 4 shows survival for the patients randomized to these three regimens. The addition of BCNU to the intravenous melphalan-prednisone regimen did not produce a survival advantage (p = 0.42).

Table V describes the toxicity of the three regimens. Life-threatening leukopenia was more common in the patients receiving intravenous melphalan and prednisone or intravenous melphalan, BCNU, and prednisone than in the group receiving oral melphalan and prednisone. Severe and life-threatening infections were also more frequent among patients randomized to receive the intravenous medications. These differences were not statistically significant, however.

C. CALGB Protocol 7261

Figure 5 displays the treatment schedule for CALGB 7261. Results of this study have been previously published (Harley *et al.*, 1979), but survival will be updated and discussed here.

TABLE V. Toxicity of Treatment Regimen, CALGB 7161

Regimen	Total Eval.	Leukopenia			% Life threatening
		% Mild	% Moderate	% Severe	
L-PAM + Pred iv	47	6	23	35	30
L-PAM + BCNU + Pred iv	40	8	28	33	25
L-PAM + Pred	50	14	24	35	12
		Thrombocytopenia			
L-PAM + Pred iv	47	13	11	28	28
L-PAM + BCNU + Pred iv	40	15	18	18	30
L-PAM + Pred	50	16	8	18	28
		Infectious Complications			
L-PAM + Pred iv	47	28		21	
L-PAM + BCNU + Pred iv	40	20		25	
L-PAM + Pred	51	36		12	

As previously reported, the frequency of objective responses for the regimen consisting of intravenous BCNU, cyclophosphamide, and melphalan with oral prednisone was significantly higher than for the oral melphalan–prednisone regimen ($p = 0.0047$). The frequency of a 50% or more reduction in abnormal serum protein was also significantly higher for the intravenous combination ($p < 0.001$). However, no difference in survival was noted for the group receiving the

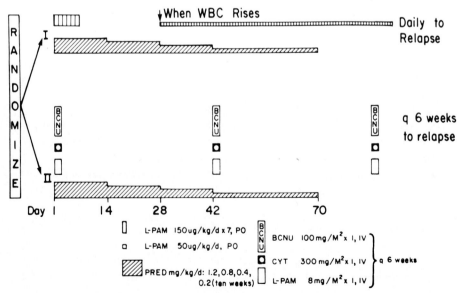

Fig. 5. Study design for CALGB protocol 7261.

intravenous combination despite the improved responses produced by this treatment. Further analysis revealed that patients in the high-risk group or high tumor load group had an improved survival if they received the intravenous combination. Patients in the standard-risk group or those with intermediate and low tumor cell burdens had a trend toward poorer survival if they received the combination therapy. We have recently reassessed these results using updated survival information. This analysis again indicates that patients defined as increased risk have a significantly increased survival if they initially receive the intravenous combination regimen ($p = 0.036$). The intravenous combination regimen also significantly prolongs the survival of patients with high tumor cell load ($p = 0.01$). In the updated analysis those patients originally classified as standard risk now have a statistically significant decrease in survival if they receive the intravenous combination ($p = 0.049$), and there is a trend toward a poorer survival for those patients in the low and intermediate tumor cell load category ($p = 0.14$).

D. CALGB Protocol 7361

Figure 6 shows the design for this study, which has been previously reported (Kyle *et al.*, 1979). Patients who failed to improve after at least 3 months of adequate therapy with melphalan or who had progressed following an initial objective response were entered in this study. Of the 87 evaluable patients, three objective responses (7%) were observed in those receiving the cyclophosphamide–prednisone regimen, and seven (17%) were noted in the cyclophosphamide–BCNU–prednisone group. The trend in favor of improved responses in the group receiving the BCNU is, however, not statistically significant ($p = 0.22$). There was no difference in the survival of patients randomized to either of the two treatments.

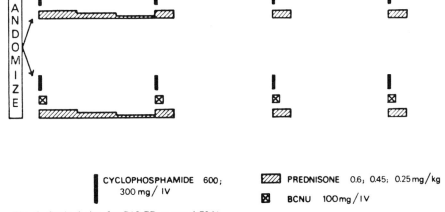

CYCLOPHOSPHAMIDE 600; 300 mg/ IV

PREDNISONE 0.6; 0.45; 0.25 mg/kg

BCNU 100 mg/ IV

Fig. 6. Study design for CALGB protocol 7361.

E. Pulmonary Toxicity of BCNU

Pulmonary fibrosis has been reported in patients receiving BCNU (Weiss and Muggia, 1980). A retrospective review of the records of the 283 evaluable patients receiving BCNU on these protocols was performed in order to ascertain the frequency of this complication. The retrospective nature of this examination and the fact that pulmonary toxicity to BCNU was not known to investigators during the early phase of these studies makes this analysis difficult. Also, given the background of bacterial pneumonias, congestive heart failure, and pulmonary emboli in this group of patients, it was frequently difficult to ascertain what the cause of deteriorating pulmonary function was. One patient had progressive pulmonary symptoms and died of complications of pulmonary fibrosis. She had received 2175 mg of BCNU over a 2-year period and died 7 months after the last dose of BCNU was given. Another patient who received 1650 mg of BCNU over a 5-month period developed progressive dyspnea. The patient was removed from protocol for other reasons, and follow-up of the pulmonary symptoms was not possible. Another patient may possibly have developed pulmonary fibrosis, but this was not sufficiently well documented to ascertain whether the pulmonary disease was BCNU related.

IV. DISCUSSION

The results from these trials indicate that BCNU alone or in combination with prednisone is an effective agent for the treatment of myeloma. When all the evidence is considered, these trials suggest that BCNU may be slightly less effective when used alone or in combination with prednisone than with melphalan-containing regimens. Our observation that patients in the increased risk or high tumor cell burden group may survive longer on regimens employing melphalan, cyclophosphamide, and BCNU, while provocative, requires further confirmation. For instance, in the trial reported by Bergsagel *et al.* (1979), improved survival was not observed for patients with high tumor loads who received the three alkylating agents regimen. Further information on this point will come from a CALGB study in progress in which the three alkylating agents regimen is being further compared with conventional therapy. In addition, a CALGB protocol currently under analysis compares the effects of oral CCNU with that of intravenously administered BCNU.

Our studies show a nonsignificant trend toward improved responses in refractory patients who receive regimens employing BCNU in combination with cyclophosphamide and prednisone. Nonrandom studies by others have suggested that combinations of BCNU with doxorubicin and cyclophosphamide or combinations of BCNU and doxorubicin when given with prednisone may be useful in refractory patients (Presant *et al.*, 1978; Alberts *et al.*, 1976).

Further investigation is required in order to define the optimal role for nitrosoureas in the therapy of myeloma as well as to establish whether particular subgroups of myeloma patients may especially benefit from treatment with nitrosoureas. The activity of these agents in advanced mouse tumors that are refractory to treatment with other antineoplastic agents suggests that these efforts will be rewarding (Schabel, 1976).

ACKNOWLEDGMENTS

The following principal investigators or investigators represent institutions contributing cases to this study. Grants are from the National Institutes of Health: Edward S. Henderson, Roswell Park Memorial Institute, Buffalo, NY (CA-02599); Charles L. Spurr, Bowman Gray School of Medicine, Winston-Salem, NC (CA-03927); Gibbons G. Cornwell III and O. Ross McIntyre, Norris Cotton Cancer Center, Hanover, NH (CA-04326); Farid Haurani, Thomas Jefferson Hospital, Philadelphia, PA (CA-05462); Robert Kyle, Mayo Clinic, Rochester, MN (CA-04646); Lorne Shapiro, Royal Victoria Hospital, Montreal, Canada; G. Watson James III, Medical College of Virginia, Richmond, VA (CA-03735); James F. Holland, Mt. Sinai Hospital, New York, NY (CA-04457); Arthur Levine, National Cancer Institute, Bethesda, MD; Richard T. Silver, New York Hospital, New York, NY (CA-07968); Louis Leone, Rhode Island Hospital, Providence RI (CA-08025); Stanley Lee, Jewish Hospital of Brooklyn, Brooklyn, NY (CA-12886); Johannes Blom, Walter Reed Army Medical Center, Washington, DC; Raymond B. Weiss, West Virginia University Medical Center, Morgantown, WV (CA-07757); Nis I Nissen, Finsen Institute, Copenhagen, Denmark; Geoffrey Falkson, H. F. Verwoerd, Pretoria, South Africa; Kurt Brunner, Inselspital, Bern, Switzerland; Arthur Sawitsky, Long Island Jewish-Hillside Medical Center, New Hyde Park, NY (CA-11028); Robert W. Carey, Massachusetts General Hospital, Boston, MA (CA-12449); Arthur Bank, Columbia-Presbyterian Hospital, New York, NY (CA-12011); Patrick Henry, University of Missouri, Columbia, MO (CA-12046); Faith Kung, University Hospital, San Diego, CA (CA11789); Rose Ruth Ellison, E. J. Meyer Memorial Hospital, Buffalo, NY (CA-16450); B. J. Kennedy, University of Minnesota, Minneapolis, MN (CA-16450).

REFERENCES

Alberts, D. S., Durie, B. G. M., and Salmon, S. E. (1976). *Lancet 1*, 926–928.
Bergsagel, D. E., Bailey, A. J., Langley, G. R., MacDonald, R. N., White, D. F., and Miller, A. B. (1979). *N. Eng. J. Med. 301*, 743–748.
Chronic Leukemia-Myeloma Task Force (1973). *Chemother. Rep. 4*, 145–158.
Harley, J. B., Pajak, T. F., McIntyre, O. R., Kochwa, S., Cooper, M. R., Coleman, M., and Cuttner, J. (1979). *Blood 54*, 13–22
Kyle, R. A., Costa, G., Cooper, M. R., Ogawa, M., Silver, R. T., Glidewell, O., and Holland, J. F. (1973). *Cancer Res. 33*, 956–960.

Kyle, R. A., Gailani, S., Seligman, B., Blom, J., McIntyre, O. R., Pajak, T. F. and Holland, J. F. (1979). *Cancer Treat. Rep. 63*, 1265–1269.

Leone, L. (1974). Oral presentation, Myeloma Symposium, Atlanta, Georgia.

McIntyre, O. R., Leone, L., and Pajak, T. F. (1978). *Blood 52 Supplement I*, 274.

Presant, C. A., and Klahr, C. (1978). *Cancer 42*, 1222–1227.

Salmon, S. (1976). *Cancer Treat. Rep. 60*, 789–794.

Schabel, F. M., Jr. (1976). *Cancer Treat. Rep. 60*, 665–698.

Weiss, R. B., and Muggia, F. M. (1980). *Am. J. Med. 68*, 259–266.

Chapter 27

CLINICAL TRIALS OF NITROSOUREAS
IN METASTATIC SARCOMAS[1]

Saul E. Rivkin

I. INTRODUCTION

The nitrosourea group of antitumor agents was developed by the Division of Cancer Treatment of the National Cancer Institute as a result of rationally based analog synthesis and delineation of structure–activity relationship. Three nitrosoureas, BCNU, CCNU, and methyl CCNU (1-3-Bis [2-chloroethyl]-1-nitrosourea, 1-[2–chloroethyl]-3-cyclohexyl-1-nitrosourea, and 1-[2-chloroethyl]-3-[4–methylcyclohexyl]-1-nitrosourea), have undergone extensive clinical trials. Two of them, BCNU and CCNU, are commercially available. The purpose of this paper is to review clinical trials of the three major nitrosoureas in the treatment of metastatic soft tissue and bony sarcomas.

The first clinical studies involving a nitrosourea (BCNU) started in 1962. Tumor activity was shown quickly and resulted in analog development.

CCNU and methyl CCNU are both lipid soluble, possess an asymmetric structure (permitting easier radiolabeling for pharmacokinetic studies), and can be administered orally. These characteristics provided impetus for their clinical usage. CCNU also was found to have greater activity than BCNU in the ic inoculated L1210 leukemia system. Methyl CCNU was found to have superior

[1]Supported in part by NCI Grant CA-20319, DHEW.

activity in Lewis lung carcinomas. Clinical trials of CCNU began in 1969 and methyl CCNU in 1971.

A. Toxicology

The three nitrosoureas produce similar qualitative toxic effects in man (Carter *et al.*, 1972; Oliverio; 1973, Wasserman *et al.*, 1974a; Wasserman *et al.*, 1974b). Toxicity is principally manifested by acute nausea and vomiting and by delayed dose-limited myelosuppression. The latter consists of moderate thrombocytopenia and leukopenia. Other infrequently observed effects include stomatitis, alopecia, anemia, anorexia, hepatotoxicity, and neurotoxicity. Hepatotoxicity, which is uncommon, appears to occur more frequently with BCNU. An added toxic effect of BCNU is occasional phlebitis and burning at the injection site due to its parenteral formulation.

II. MATERIALS AND METHOD

This review includes clinical data on all known major studies of the nitrosoureas as single agents and combinations in the treatment of advanced sarcomas. These studies were performed by either clinical cooperative groups or independent investigators. They fall into three categories: (a) broad phase II trials of the nitrosoureas as single agents against a variety of tumor types; (b) phase II–III disease-oriented trials as single agents in special tumors; and (c) phase III trials in sarcoma of use in combination with other agents.

A. Clinical Dose Formulation

The usual dose regimens for the nitrosoureas are as follows: BCNU, 200 mg/m^2 iv every 6 weeks or 100 mg/m^2 iv day 1 and 2 every 6 weeks; CCNU, 100–130 mg/m^2 orally every 6 weeks; methyl CCNU, 200–225 mg/m^2 orally every 6 weeks or 130–170 mg/m^2 iv every 6 weeks. When used in combination with adriamycin, methyl CCNU was given 150 mg/m^2 orally on day 1, adriamycin 60 mg/m^2 iv day 1, 45 mg/m^2 day 22 every 6 weeks. When combined with cytoxan, 100–125 mg/m^2 of methyl CCNU was given orally day 1, cytoxan 800–100 mg/m^2 iv on day 1 every 6 weeks.

B. Evaluation Criteria

Evaluable patients include those who received a nitrosourea on an appropriate dose schedule and could be evaluated for objective response. Most of the studies determined response and toxic effects in patients after the first 6-week course, but a few required a minimum of two courses of an adequate trial.

TABLE I. BCNU as a Single Agent and Combined with Vincristine in Sarcomas

Site	No. pts. evaluated	No. pts. responding	% Response	Remarks	Reference
BCNU					
Osteogenic sarcoma	1	0	–		Reyes et al. (1973)
	1	0	–	stable 3 wks	Lessner (1968)
Fibrosarcoma	1	0	–	stable 3 wks	DeVita et al. (1965)
Mesothelioma	1	0	–		Iriarte et al. (1966)
Rhabdomyosarcoma	1	0	–		Lessner (1968)
Chondrosarcoma	1	0	–		Iriarte et al. (1966)
	36	6	17		Slavik (1976)
Sarcoma/unspecified type	11	2	18	3 mos duration	Ramirez et al. (1972)
Subtotal	53	8	15		
BCNU + Vincristine					
Osteogenic sarcoma	6	0	–		Stolinsky et al. (1974)[a]
Fibrosarcoma	4	0	–		
Mesothelioma	2	0	–		
Rhabdomyosarcoma	3	1	33		
Anaplastic sarcoma	5	1	20		
Synovial cell	1	0	–		
Spindle cell	4	0	–		
Liposarcoma	3	0	–		
Myxosarcoma	2	0	–		
Leimyosarcoma	11	1	9		
Subtotal	41	3	7		
Total	94	11	12		

[a] All data on BCNU + Vincristine are from Stolinsky et al. (1974).

TABLE II. BCNU in the Treatment of Kaposi's and Ewing's Sarcoma

Site	No. pts. evaluated	No. pts. responding	% Response	Remarks	Reference
Kaposi's sarcoma	21	9 (4 CR)	43%	several responding	Tranum *et al.* (1975)
	1[a]	0	–	>1 yr	Stolinsky *et al.* (1974)
Total	22	9	41%		
Ewing's sarcoma	5	0	–		Sutow *et al.* (1971)
	12	5 (1 CR)	42%	one prolonged regression	Palma *et al.* (1972)
	1	1 (1 CR)	100%	CR = 63 wks	DeVita *et al.* (1965)
	2	1	50%		Iriarte *et al.* (1966)
	1[a]	0	–		Stolinsky *et al.* (1974)
Total	21	7	33%		

[a] BCNU combined with vincristine.

Complete response (CR) was defined as the disappearance of all tumor masses and other evidence of disease; partial response (PR) refers to $\geqslant 50\%$ decrease in tumor mass without the appearance of new disease. The required minimum duration of response is 1 month. No change (NC) was defined as a steady state or a response less than necessary for PR; or disease progression involving an increase of less than 50% in the lesion size with no new lesions appearing. Increasing disease was defined as an unequivocal increase of at least 50% in the size of any measurable lesions. The appearance of new lesions also constituted increasing disease.

III. RESULTS

BCNU appears to have a low order of activity in most sarcomas when used as a single agent or in combination with other agents (Table 1), with the exception of Kaposi's and Ewing's sarcoma (Table II). In Kaposi's sarcoma BCNU produced 9 responses (4 CR) in 21 patients (43%). Seventeen of these patients had extensive prior therapy and had responses similar to those without prior chemotherapy, with prolonged remissions over 1 year. BCNU as a single agent

TABLE III. CCNU in the Treatment of Sarcoma

Site	No. pts. evaluated	No. pts. responding	% Response	Reference
Sarcoma/unspecified type	50	4	8	Slavik (1976)
	5	0	–	Hoostraten *et al.* (1973)
Total	55	4	7	

TABLE IV. Methyl CCNU as a Single Agent in Sarcomas

Site	No. pts. evaluated	No. pts. responding	% Response	Remarks	Reference
Osteogenic sarcoma	4[a]	0	—		Chang et al. (1976)
Chondrosarcoma	2[a]	1	50		Chang et al. (1976)
	1	0	—		Pinedo and Kenis (1977)
Rhabdomyosarcoma	5[a]	1	20	1 stable	Chang et al. (1976)
Leiomyosarcoma	7[a]	0	—	2 stable	Chang et al. (1976)
Mesothelioma	1[a]	0	—		Chang et al. (1976)
	2	0	—		Pinedo and Kenis (1977)
Liposarcoma	2[a]	0	—		Chang et al. (1976)
Synovial sarcoma	1[a]	0	—		Chang et al. (1976)
Myxofibrosarcoma	1[a]	0	—		Chang et al. (1976)
Malignant giant cell tumor	2[a]	0	—		Chang et al. (1976)
Malignant fibrous histiocytic sarcoma	3[a]	0	—	1 stable	Chang et al. (1976)
Hemangiopericytoma	1	0	—		Pinedo and Kenis (1977)
Soft tissue	1	0	—		Pinedo and Kenis (1977)
Soft tissue & osseous sarcoma	70	5	7	no change; 11 responses leimyosarcoma, 1 synovial cell	Tranum et al. (1975)
	11 (primary)	0	—	1 osteogenic stable for 6 mos	Creagan et al. (1976)
	4 (secondary)	0	—		Creagan et al. (1976)
	5	0	—		Young et al. (1973)
	15	3	20		Wasserman et al. (1974)
Total	138	10	7		

[a] Iv methyl CCNU.

TABLE V. Methyl CCNU in Combination with Adriamycin and Vincristine in Adults with Sarcomas[a]

Site	No. pts. evaluated	No. pts. responding	% Response
Synovial	1	1	100%
Alveolar soft part	2	0	–
Chondrosarcoma	–	–	–
Fibrosarcoma	3	0	–
Leiomyosarcoma	15	3	20%
Liposarcoma	3	2	67%
Mesothelioma	1	0	–
Rhabdomyosarcoma	6	2	33%
Spindle cell	8	1	12%
Ewing's	1	0	–
Osteogenic	6	1	17%
Total	46	10	22%

[a] Data shown are from Krakoff (1975).

also showed definite palliative effects in metastatic Ewing's sarcoma with 7 remissions in 21 patients (33% response rate).

The clinical data on BCNU combined with vincristine (Table 1) yields a response rate of 7%. When combined with the results of BCNU used as a single agent (omitting Kaposi's and Ewing's sarcoma), the response rate is only 12% (11/94 responses).

CCNU has only a 7% response rate in sarcoma (Table III).

TABLE VI. Methyl CCNU and Adriamycin in Sarcomas[a]

Site	No. pts. evaluated	No. pts. responding		% Response
		CR	PR	
Angiosarcoma/hemangiosarcoma	6	0	4	66%
Ewing's sarcoma	2	1	0	50%
Fibrosarcoma	6	0	3	50%
Fibrous histiocytoma	1	0	1	50%
Kaposi's sarcoma	4	0	2	50%
Leiomyosarcoma	13	1	5	46%
Liposarcoma	6	1	0	17%
Mesothelioma	5	1	0	20%
Osteogenic	1	0	0	–
Rhabdomyosarcoma	3	0	3	100%
Undifferentiated	2	0	1	50%
Others	4	1	0	25%
Total	53	5	19	44%

[a] Data are from Rivkin et al. (1980).

TABLE VII. Methyl CCNU and Cytoxan in Sarcoma

Site	No. pts. evaluated	No. pts. responding	Reference
Sarcoma/unspecified type	6	0	Pinedo and Kenis (1977)
	5	0	Wasserman et al. (1974)
Gastric leiomyosarcoma	1	0	Bedikian et al. (1979)
Total	12	0	

The summary of methyl CCNU as a single agent (Table IV) shows a response rate of 7%. Most of these patients had failed adriamycin therapy prior to methyl CCNU. Intravenous methyl CCNU also had a 7% response rate.

Promising results with methyl CCNU were obtained when combined with adriamycin. A total of 10 responses in 46 patients were obtained from the combination of adriamycin, vincristine, and methyl CCNU (Table V). The combination of methyl CCNU and adriamycin (Table VI) produced 5 complete remissions and 19 partial remissions in 54 patients (44%). Cytoxan added to methyl CCNU resulted in no responses in 12 patients (Table VII).

IV. DISCUSSION

The nitrosoureas appear to have low activity in metastatic soft tissue sarcomas when used as a single agent. The response rate to BCNU, CCNU, and methyl CCNU used as a single agent in metastatic sarcoma is less than 13% (omitting Kaposi's and Ewing's sarcoma). Most patients were treated in phase II protocols and had received previous therapy. However, the small number of patients without previous therapy did not respond any better than did those with prior therapy.

The nitrosoureas have been shown to be very effective in Kaposi's sarcoma when used as a single agent. (Tranum et al., 1975). The four complete responders to BCNU had complete remissions lasting over 12 months, and all were unmaintained. Similar response rates occurred with the combination methyl CCNU and adriamycin with long durations of remission (Rivkin et al., 1980). The data indicates that nitrosoureas in combination with other agents would be indicated in Kaposi's sarcoma.

Ewing's sarcoma had a 33% response rate to BCNU (De Vita et al., 1965; Iriarte et al., 1966; Palma et al., 1972; Sutow et al., 1971). A similar response rate was found with methyl CCNU and adriamycin (Rivkin et al., 1980). Thorough data analysis shows that BCNU has a definite role in the palliation of metastatic Ewing's sarcoma, even in patients with previous therapy. When methyl CCNU was combined with adriamycin, one of two patients responded with a complete remission (Rivkin et al., 1980). Therefore, the nitrosoureas have a definite role in the palliation of metastatic Ewing's sarcoma. They probably should be used in combination with other agents that have been effective in Ewing's sarcoma, such as adriamycin, cytoxan, actinomycin D, and vincristine.

The rationale for combining methyl CCNU and adriamycin is that adriamycin has been determined to be the single most effective agent in the treatment of metastatic sarcoma (O'Bryan *et al.*, 1973). In two series of patients, a combination of methyl CCNU and adriamycin was well tolerated (Gottlieb *et al.*, 1972b; Skarin *et al.*, 1973). Krakoff (1975) had a response rate of 22% in a variety of sarcomas when methyl CCNU was combined with adriamycin and vincristine, which in only slightly better than adriamycin alone. Rivkin *et al.* (1980) obtained a response rate of 45% with methyl CCNU and adriamycin (MAD) in sarcoma. This response rate was similar to that of those with no prior chemotherapy with an overall median survival of 10.2 months (Fig. 1). The median length of all remissions was 4.2 months. Of the 24 achieving CR and PR, those less than 50 years of age had a significantly longer remission period ($p < 0.01$) (Fig. 2).

As reported by Rivkin *et al.* (1980), there was no response relationship to patient sex, prior surgery, leukopenia/granulocytopenia, or previous chemotherapy. The response relationship to marrow status and prior radiotherapy was nearly significant ($p = 0.06$). No response and increasing disease was found in 29 patients. The median survival time of patients achieving CR was 18 months. The median survival time of those achieving PR was 12 months, 9 months for those manifesting a stable state, and 8.5 months for those with increasing disease.

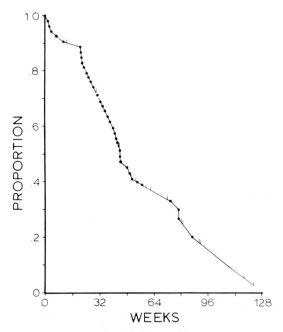

Fig. 1. Kaplan–Meier curve for survival for all 53 patients. The proportion is percentage surviving at stated time. Living patients are indicated by small vertical lines on the survival curve (Rivkin *et al.*, 1980).

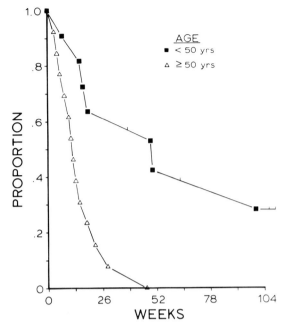

Fig. 2. Kaplan–Meier curve for patient remission length by age of patient; 11 patients < age 50; 13 > age of 50 (Rivkin *et al.*, 1980).

The response rate and duration of survival seem to be better with the combination of methyl CCNU and adriamycin than with adriamycin alone.

The toxicity (Carter *et al.*, 1972; Di Marco *et al.*, 1969; Gottlieb *et al.*, 1972a; Mhatre *et al.*, 1971) and mechanism of action (Calendi *et al.*, 1965; Carter *et al.*, 1972) of adriamycin and methyl CCNU are different. The drugs do not appear to demonstrate cross-reactivity with each other, or with the commonly used agents in metastatic sarcoma treatment (e.g., alkylating agents and antimetabolites). Thus, combination therapy with these two agents seems reasonable, especially since both have some activity in patients with advanced metastatic sarcoma.

When compared to the results of the Southwest Oncology Group study using cytoxan, DIC, vincristine, and adriamycin (CYVADIC), the response rate was the same and the median survival time for all eligible patients was 10 months ($p = 0.31$) (Fig. 3) (Benjamin *et al.*, 1977). However, the patients that responded to CYVADIC had a median survival time of 24 months as compared to 14.5 months with adriamycin and methyl CCNU ($p = 0.15$) (Fig. 4). The 23 patients less than 50 years of age had significantly longer remission periods and survival time (median 18 months) as compared to the 30 patients over 50 ($p < 0.01$). The patient–age variable may account for the difference in survival between MAD and CYVADIC. Responder survival was not analyzed by age in the CYVADIC study.

The response rate of MAD is similar to that of adriamycin and DTIC (Gottlieb

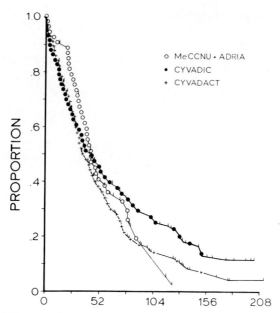

Fig. 3. Kaplan–Meier curve for patient survival comparing *patient* survival in three drug regimens (Rivkin *et al.*, 1980).

	Total	Fail
Methyl CCNU	53	37
CYVADIC	230	178
CYVADACT	224	186

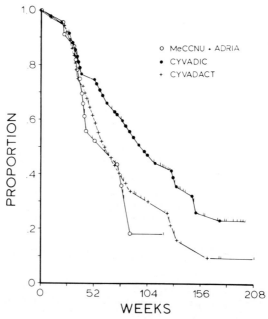

Fig. 4. Kaplan–Meier curve for survival of responding patients in three drug regimens (Rivkin *et al.*, 1980).

and Wharton, personal communication). The mean survival for adriamycin and DTIC was 10 months for all patients and 15 months for all responders. Although no significant survival difference was found in eligible patients for any of these regimens, the combination of methyl CCNU and adriamycin is more convenient and necessitates fewer clinic visits and significantly fewer injections than other combinations.

In conclusion, as single agents the nitrosoureas have a low order of activity in metastatic sarcoma. However, combination trials such as adriamycin and methyl CCNU, or adriamycin combined with other nitrosoureas, would be indicated in patients with previous chemotherapy. There is rationale for the combinations of the nitrosoureas with adriamycin, and one study has shown that the combination of methyl CCNU and adriamycin is just as effective as any other combination previously reported.

REFERENCES

Bedikian, A. Y., Valdivieso, M., Khankahanian, N., Benjamin, R. S., and Bodey, G. P. (1979). *Cancer Treat. Rep. 63*, 411–414.

Benjamin, R. S., Baker, L. H., Rodriquez, V., Moon, T. E., O'Bryan, R. M., Stephens, R. L., Sinkovics, J. G., Thigpen, T., King, G. W., Bottomley, R., Groppe, C. W., Bodey, G. P., and Gottlieb, J. A. (1977). *In* "Management of Primary Bone and Soft Tissue Tumors." (The University of Texas System Cancer Center, M.D. Anderson Hospital and Tumor Institute, 21st Annual Clinical Conference on Cancer), pp. 309–315. Yearbook Medical Publishers, Chicago.

Calendi, E., Di Marco, A., Reggiani, M., Scarpinato, B., and Valentino, L. (1965). *Biochim. Biophys. Acta 103*, 25–49.

Carter, S. K., Shabel, F. M. Jr., Broder, L. E., and Johnston, T. P. (1972). 1,3-Bis (2-chloroethyl)-1-nitrosourea and other nitrosoureas in cancer treatment: a review *Adv. Cancer Res. 16*, 273–332.

Chang, P., Levine, M. A., Wiernik, P. H., and Walker, M. D. (1976). *Cancer 37*, 615–619.

Creagan, E. T., Hahn, R. G., Ahmann, D. D., Edmonson, J. H., Bisel, H. F., and Eagan, R. T. (1976). *Cancer Treat. Rep. 60*, 1385–87.

De Vita, V. T., Carbone, P. P., Owens, A. H. Jr., Gold, G. L., Krant, M. J., and Edmonson, J. (1965). *Cancer Res. 25*, 1876–1881.

Di Marco, A., Gaetani, M., and Scarpinato, B. (1969). *Cancer Chemother. Rep. 53*, 33–37.

Gottlieb, J. A., Baker, L. H., Quagliana, J. M., Luce, J. K., Whitecare, J. P., Sinkovics, J. G., Rivkin, S. E., Brownlee, R., and Frei, E. III (1972a). *Cancer 30*, 1632–1638.

Gottlieb, J. A., McCredie, K. B., Hersh, E. M., and Frei, E. III (1972b). *Proc. AACR 13*, 79.

Hoogstraten, B., Gottlieb, J. A., Caoili, E., Tucker, W. G., Talley, R. W., and Haut, A (1973). *Cancer 32*, 38–43.

Iriarte, P. V., Hananian, J., and Cortner, J. A. (1966). *Cancer 19*, 1187–1194.

Krakoff, I. H. (1975). *Cancer Chemother. Rep. 6*, 253–257.

Lessner, H. E. (1968). *Cancer 22*, 451–456.

Mhatre, R., Herman, E., Huidobro, A., and Waravdekar, V. (1971). *J. Pharmacol. Exper. Ther. 178*, 216–222.

O'Bryan, R. M., Luce, J. K., Talley, R. W., Gottlieb, J. A., Baker, L. H., and Bonadonna, G. (1973). *Cancer 32*, 1–8.

Oliverio, V. T. (1973). *Cancer Chemother, Rep. 4*, 13–20.

Palma, J., Gailani, S., Freeman, A., Sinks, L., and Holland, J. F. (1972). *Cancer 30*, 909–913.

Pinedo, H. M., and Kenis, Y. (1977). *Cancer Treat. Rev. 4*, 67–86.

Ramirez, G., Wilson, W., Grage, T., and Hill, G. (1972). *Cancer Chemother. Rep. 56*, 787–790.

Reyes, E. S., Talley, R. W., O'Bryan, R. M., and Gastesi, R. A. (1973). *Cancer Chemother. Rep. 57*, 225–230.

Rivkin, S. E., Gottlieb, J. A., Thigpen, T., Gad El Mawla, Nazli, Saiki, J., and Dixon, D. O. (1980). *Cancer 46*, 446–451.

Skarin, A., Lokich, J., Biano, G., Chawla, P., and Frei, E. III (1973). *Cancer Chemother. Rep. 57*, 104–105.

Slavik, M. (1976). *Cancer Treat. Rep. 60*, 795–800.

Stolinsky, D. C., Pugh, R. P., Bohannon, R. A., Bogdon, D. L., and Bateman, J. R. (1974). *Cancer Chemother. Rep. 58*, 947–950.

Sutow, W. W., Vietti, T. J., Fernbach, D. J., Lane, D. M., Donaldson, M. H., and Lonsdale, D. (1971). *Cancer Chemother. Rep. 55*, 67–78.

Tranum, B. L., Haut, A., Rivkin, S., Weber, W. Quagliana, J. M., Shaw, M., Tucker, W. G., Smith, F. E., Samson, M., and Gottlieb, J. (1975). *Cancer 35*, 1148–1153.

Vogel, C. L., Clements, D., Wanume, A. K., Toya, T., Primack, A., and Kyalwazi, S. (1973). Phase II clinical trials of BCNU (NSC)409962 and bleomycin (NSC-125066) in treatment of Kaposi's sarcoma.

Wasserman, T. H. Slavik, M., and Carter, S. K. (1974a). *Cancer Treat. Rev. 1*, 131–151.

Wasserman, T. H., Slavik, M., and Carter, S. K. (1974b). *Cancer Treat. Rev. 1*, 251–269.

Young, R. C., Walker, M. D., Canellos, G. P., Schein, P. S., Chabner, B. A., and DeVita, V. T. (1973). *Cancer 31*, 1164–1169.

Chapter 28

HIGH-DOSE NITROSOUREA (BCNU) AND AUTOLOGOUS BONE MARROW TRANSPLANTATION: A PHASE I STUDY

Geoffrey P. Herzig
Gordon L. Phillips
Roger H. Herzig
Joseph W. Fay
R. L. Weiner
Steve N. Wolff
Hillard M. Lazarus

I. INTRODUCTION

Clinical trials with BCNU (1,3-bis(2-chloroethyl)-1-nitrosourea, NSC-409962) have documented substantial antitumor activity against lymphomas, small-cell lung carcinoma, and central nervous system tumors, and to a lesser degree against melanoma and gastrointestinal cancer (Wasserman et al., 1975). Studies of BCNU in vitro, and against murine tumors in vivo, suggest that a steep dose-response curve exists such that higher doses of BCNU result in greater tumor-cell killing (Bruce et al., 1967). However, bone marrow toxicity that is prolonged and cumulative limits the maximum dose of BCNU that can be given safely. The usual dose of BCNU used as a single agent is 200 mg/m² repeated at 6–8 week intervals, and rarely have doses in excess of 300 mg/m² been employed (De Vita et al., 1965). Bone marrow transplantation provides one means by which the hematologic toxicity of cytotoxic drugs and irradiation can be reduced. Transplantation with a patient's own (autologous) bone marrow which is removed and preserved by freezing prior to therapy has been successful in preventing lethal myelosuppression following high-dose total body irradiation (1000 rads) (Gale *et al.*, 1980).

In the present study we have used autologous marrow transplantation to ameliorate the bone marrow toxicity of BCNU in order to determine the maximum tolerable BCNU dose when myelosuppression is no longer the limiting toxicity.

II. MATERIALS AND METHODS

A. Patients

Patients aged 70 years or less with refractory malignancies and a minimum 8-week anticipated survival were eligible for study if the bone marrow was free of tumor and if serious cardiac, pulmonary, hepatic, and renal dysfunction was not present.

B. Marrow Cryopreservation

Marrow was aspirated from the iliac crests under general anesthesia. A total volume of approximately 10 ml/Kg is obtained. The heparinized marrow is centrifuged to remove red blood cells and is mixed with tissue culture medium and dimethylsulfoxide (final concentration of 10%). Freezing is accomplished by cooling at a rate of 1–2°C/min, and the frozen marrow is stored in a liquid nitrogen freezer ($\leq -160°C$). For transplantation, marrow is thawed at the patient's bedside and infused immediately intravenously.

C. Study Design

BCNU is given by intravenous infusion over 2 hr daily for 3 consecutive days. Autologous marrow is infused 7 days after the last dose of BCNU. BCNU dose is increased by 25% in successive groups of four patients until further dose escalation is prevented by toxicity. Total doses of 600 mg/m^2 to 2850 mg/m^2 were employed.

III. RESULTS

A. Patient Population

A total of 68 patients were studied; their median age was 45 years with a range of 1–68 years. The majority had refractory solid tumors, chiefly lung, melanoma, or glioblastoma multiforme.

B. Toxicity

1. Hematopoietic Toxicity

Virtually all patients developed marked pancytopenia. The median time to recovery of neutrophils ($>500/\mu$l) was 2 weeks (range 8–25 days) with platelet recovery ($>10,000/\mu$l) approximately 1 week later. Only 5 of 46 evaluable patients had persistent severe cytopenia beyond day 28 (day 0 is the day of marrow infusion); one of these patients died of pneumonia while neutropenic, 40 days posttransplant. This represents the only fatal complication due to myelosuppression beyond day 28. None of the 24 patients who survived 8 or more weeks experienced late myelosuppression. Hematopoietic recovery had no obvious relation to BCNU dose, marrow cell dose, or marrow storage interval.

2. Hepatic Toxicity

Thirty-one patients were evaluable for liver toxicity (see Table I). Severe toxicity (defined as liver chemistries $>$ 5 times normal or clinical signs of liver failure) developed in 9 patients and was fatal in 7. Of 19 patients who received 1200 mg/m^2 of BCNU or less, 2 developed transient elevations in transaminases without permanent sequellae. At doses of 1500 mg/m^2 or more, fatal hepatotoxicity occurred. The onset of liver abnormalities was delayed approximately 2 months posttransplantation with a rapidly progressive course terminating in hepatic coma.

3. Pulmonary Toxicity

Of 27 patients, 3 developed pulmonary toxicity characterized by the delayed occurrence of diffuse interstitial infiltrates with biopsy-proven fibrosis in 2 of the cases. Resolution of the process occurred without specific treatment in 2 patients (at BCNU doses of 1050/m^2 and 1200/m^2), the third patient died of complications following lung biopsy (BCNU dose of 1500 mg/m^2). There was no apparent relationship between BCNU dose and the incidence of pulmonary toxicity.

4. Neurotoxicity

Central nervous system toxicity not previously reported due to BCNU was observed in 3 of 4 evaluable patients at the highest BCNU dose levels (2250–

TABLE I. Hepatic Toxicity

BCNU dose (mg/m^2)	No. evaluable patients	No. with toxicity	
		Severe	Fatal
600–1200	19	2	0
1500–1800	8	3	3
2250–2850	4	4	4

2850 mg/m^2), and was fatal in 2. In contrast, no neurologic toxicity occurred among 26 patients who received BCNU doses of 600–1800 mg/m^2. CNS toxicity was manifest by the onset of confusion and personality changes 1–6 weeks posttransplantation. Symptoms progressed rapidly and terminated with death in deep coma in 2 weeks.

C. Tumor Response

Although it was not a primary goal of this study, tumor response could be evaluated in 51 patients with measurable disease. Six complete and 21 partial responses were observed. The greatest activity was manifest in melanoma (5/10 responses), glioblastoma (5/10 responses), and hematologic malignancy (2/5 responses). Also of note is the finding of 10 responses among 17 patients with metastatic disease of the central nervous system. Most responses were of short duration, however, two patients remain free of disease more than 2 years after marrow transplant (one glioblastoma, one meningeal leukemia).

IV. DISCUSSION

The present study clearly demonstrates the ability of cryopreserved autologous marrow transplantation to limit the severity of myelosuppression produced by high doses of BCNU. Restoration of normal blood counts occurred within 4 weeks of transplantation in most patients, with BCNU doses as high as 2850 mg/m^2; this is approximately 10 times the usual dose of BCNU and nearly 5 times the highest dose reported without marrow transplantation (De Vita *et al.*, 1965).

Unfortunately, dose-related nonhematopoietic toxicity was encountered which limits the maximum BCNU dose to 1200 mg/m^2. Fatal hepatic toxicity occurred in 7/12 (58%) of patients treated with BCNU doses of 1500 mg/m^2 or more. Characteristically, this toxicity was delayed in onset, beginning 6–8 weeks after marrow transplantation. Delayed liver toxicity has been reported previously in animal studies (Thompson *et al.*, 1969). Clinical trials of BCNU have revealed a low incidence of reversible liver function abnormalities, but fatal toxicity due to BCNU has not been reported (Wasserman *et al.*, 1975; DeVita *et al.*, 1965).

At the highest BCNU doses (2250–2850 mg/m^2) fatal CNS toxicity not previously attributed to BCNU was observed in 3 of 4 evaluable patients. Presumably this toxicity is related to good penetration of systemically administered BCNU into the CNS, a property of potential value in treating primary or metastatic tumors in that location.

Finally, three cases of pulmonary toxicity attributed to BCNU occurred. Similar lung toxicity has recently been documented in patients treated with BCNU in conventional regimens after cumulative doses similar to those given in the present study (Aronin *et al.*, 1980; Weiss *et al.*, 1980).

Based on these studies we have established a dose of 1200 mg/m^2 for further evaluation of toxicity and antitumor activity. Preliminary results obtained in 30 patients reveal an incidence of severe pulmonary toxicity of 13% (4/30), one fatal. Severe liver toxicity occurred in the one patient who died of pulmonary toxicity. A full analysis of the spectrum of BCNU antitumor activity will require treatment of additional patients; however significant responses have been seen in patients with melanoma, glioblastoma multiforme, and hematologic malignancy.

REFERENCES

Aronin, P. A., Mahaley, M. S., Jr., Rudnick, S. A. (1980). *N. Eng. J. Med. 303*, 183–188.

Bruce, W. R., Valeriote, F. A., and Meeker, B. E. (1967). *J. Nat. Cancer Inst. 39*, 257–266.

DeVita, V. T., Carbone, P. P., Owens, A. H. Jr., Gold, G. L., Krant, M. J., and Edmonson, J. (1965). *Cancer Res. 25*, 1876–1881.

Gale, R. P. (1980). *JAMA 243*, 540–542.

Thompson, G. R., and Larson, R. E. (1969). *J. Pharmacol. Exp. Ther. 166*, 104–112.

Wasserman, T. H., Slavik, M., and Carter, S. K. (1975). *Cancer 36*, 1258–1268.

Weiss, R. B., and Muggia, F. M. (1980). *Am. J. Med. 68*, 259–266.

Chapter 29

AMPHOTERICIN B: INTERACTIONS WITH NITROSOUREAS AND OTHER ANTINEOPLASTIC DRUGS[1]

Cary A. Presant
Frederick Valeriote
Richard Proffitt
Gerald Metter

I. INTRODUCTION

Amphotericin B (Am B) is the most widely used systemic antifungal antiobiotic. It is a polyene molecule, and is cytotoxic to fungi based on its binding to sterol components of the cell membrane. Since the binding to ergosterol in fungi is greater than the binding to cholesterol in mammalian cells, it is more cytotoxic to fungi than to normal cells. However, since Am B does bind to cholesterol in mammalian cells, it modifies cell membranes, permitting the uptake of various molecules to be enhanced. Increased uptake of antibiotics (Medoff, 1972), and even DNA fragments (Kumar, 1974) have been observed.

Since it seemed possible that Am B could also increase the uptake of anti-

[1]This summarizes some experiments of the Amphotericin B project of Washington University, St. Louis, Missouri, and the City of Hope National Medical Center, Duarte, California. Supported in part by USPHS Grant CA 15665.

cancer drugs into neoplastic cells, a series of experiments was performed to determine in experimental tumors whether Am B could augment antitumor effects. As a result of these positive studies, described below, clinical trials were initiated. Both the laboratory and the clinical investigations have indicated a definite biological effect of Am B that enhances the therapeutic effects of certain drugs in certain circumstances. This report will summarize selected preclinical studies that have led to the development of the clinical treatment programs and will update the clinical data previously published.

II. PRECLINICAL STUDIES

A. Immunological Effects of Amphotericin B

The initial trials of Am B were performed in transplantable AKR leukemia. In these experiments, mice injected with AKR leukemia cells were treated with Am B alone, BCNU alone, or AM B plus BCNU (Medoff *et al.*, 1974). These results, summarized in Table I, indicated a quite remarkable effect of the BCNU. Whereas there were no cures from any single-drug treatment, including treatment with a high dose of BCNU (which was capable of reducing the leukemic cell population from 6.4×10^5 cells down to only 7 cells), treatment with the lower dose of BCNU plus Am B resulted in cures in 60–100% of animals. This occurred despite the fact that the degree of reduction in leukemia cells, measured as leukemic colony forming units (LCFU), in the mouse femur 24 hr after drug treatment, which was reduced to 3.6×10^3 by BCNU alone, was reduced only fourfold further by the addition of Am B to BCNU. This was a surprising result since Am B alone had no effect in reducing LCFU survival, and Am B enhancement of BCNU cytotoxicity was less marked than the effect of increasing the BCNU dosage alone. Yet survival of the animals was very dramatically changed.

A likely explanation of the results of these experiments was that the Am B was acting as an immunoadjuvant. In addition to Am B's very minimally enhancing BCNU-induced cytotoxicity, the resulting augmented immune response was ca-

TABLE I. Am B and BCNU in Transplantable AKR Leukemia

Treatment *in vivo*	Survival at 28 days	LCFU/femur 24 hr after treatment
None	0	6.4×10^5
Am B (0.5 mg)	0	5.0×10^5
BCNU (0.2 mg)	0	3.6×10^3
Am B + BCNU (0.2 mg) 24 hours later	60–100%	8.0×10^2
BCNU (0.4 mg)	0	7.0×10^0

pable of killing the residual viable tumor cells. Subsequent experiments confirmed that this was so. If immunosuppressive drugs were used, the increase in life span was only proportional to the degree of increase in immediate leukemic cell killing by the addition of Am B. Since BCNU was not immunosuppressive, augmentation of the existing immune response by Am B was possible. Furthermore, using immunodeficient animals did not result in a marked increase in survival. Lastly, immunity could be adoptively transferred from animals that had been treated with Am B plus BCNU (unpublished data).

This highly impressive result in the transplantable AKR leukemia was extended to a series of experiments attempting to protect AKR mice against the emergence of spontaneous AKR lymphoma (Valeriote et al., 1976). In these experiments, mice were given injections of either Am B or dextrose every 2 weeks, beginning at 8 weeks of age. Mice were examined weekly for the emergence of spontaneous lymphoma, which was confirmed by autopsy. The results, summarized in Table II, indicated that while 100% of control animals had developed spontaneous lymphoma by 56 weeks of age, only 61% of Am B treated animals had developed lymphoma. Furthermore, the time to development of lymphoma in these mice had been delayed from 35 weeks to 46 weeks. These results indicated that the immunoadjuvant effects observed in the transplantable leukemia experiments could also be relevant to protection against the spontaneous lymphoma from which the transplantable leukemia was derived.

Additional experiments to define the nature of the Am B–BCNU interaction in transplantable AKR leukemia indicated a dose–response relationship: The maximal effects of the interaction between Am B, BCNU, the host, and the leukemia occurred when greater amounts of Am B were administered (Medoff et al., 1977). Furthermore, if Am B was administered 18 to 24 hr prior to BCNU, the effects were maximal, whereas the administration of Am B simultaneously with or after BCNU was less effective. These results were subsequently used in design of the clinical trials to be presented below.

In order to confirm that Am B was an immunopotentiator, a series of experiments was performed in non–tumor–bearing mice. Experiments to test whether Am B was able to enhance the humoral immune reaction were performed in mice immunized with the antigen trinitrophenyl-human serum albumin (TNP-HSA) along with Am B or control injections. Spleens were harvested, and the ability of

TABLE II. Am B Protection Against *In Vivo* Spontaneous AKR Lymphoma

Treatment	Incidence of spontaneous lymphoma[a]	Time to lymphoma in 50% of mice
Control	100%	35 wk
Am B	61%	46 wk

[a] At 56 weeks.

spleen cells to produce immunoglobulins was measured in a plaque-forming assay (Blanke *et al.*, 1977). In order to test cellular immune responses resulting in delayed type hypersensitivity, mice were immunized with dinitrofluoro-benzene (DNFB) by cutaneous painting. In addition, injections of Am B or control materials (saline or sodium deoxycholate) were given. Mice were challenged with DNFB skin painting on the ear, and thickness was measured (Shirley and Little, 1979). These results, summarized in Table III, indicate a highly meaningful and statistically significant enchancement of both humoral and cellular immune responses.

The mechanism of enhancement of immune response remains undefined. Preliminary experiments suggested that Am B is cytotoxic to a subpopulation of T cells. It is possible that these represent immunoregulatory T cells with suppressor activity. Furthermore, Am B has been shown to stimulate macrophages (Lin *et al.*, 1977; Chapman and Hibbs, 1978). It is also possible that macrophage activation results in enhanced T or B cell activity by increasing macrophage–effector cell interaction, or by release of lymphokines.

As yet, there have been no clinical studies of immune response in Am B-treated normal individuals. Preliminary experiments were performed in patients with bronchogenic carcinoma treated with BCNU plus Am B (Presant *et al.*, 1980a). Nine of twelve patients tested for delayed type hypersensitivity, mitogen-induced lymphocyte transformation, and anti-DLH antibody synthesis demonstrated a significant increase in at least one test of immune response. This was usually delayed-type hypersensitivity skin test, and/or mitogen-induced lymphocyte transformation. However, a controlled prospective trial is necessary in nonimmunosuppressed patients to determine if Am B possesses immunoadjuvant effects in man.

B. Pharmacological Effects of Am B

Following the investigations of the potentiation of BCNU effect in transplantable AKR leukemia, which indicated a small (fourfold) increase in BCNU toxic-

TABLE III. Am B Immunoadjuvant Effects in Mice

Parameter measured	Antigen	Treatment *in vivo*	Immune response
Splenic plaque-forming cells	TNP-HSA	Control	21[a]
		Am B	470
Delayed type hypersensitivity	DNFB	Control	11[b]
		Am B	17

[a] Number of IgG-synthesizing cells per 10^6 spleen cells.
[b] Inches ($\times$ 10^{-3}) of skin thickness at antigen application site.

ity measured with 24 hr of treatment, it was of interest to extend these observations to other drugs and to determine in cell suspensions whether uptake of other drugs was enhanced with treatment of Am B. Experiments with actinomycin D and Am B were performed in HeLa cells (Medoff *et al.*, 1975). HeLa cells in which resistance to actinomycin D had been induced were unable to transport actinomycin D effectively. The pretreatment of such cells with Am B restored the ability to transport actinomycin D, and partially restored cytotoxic effects of actinomycin D in this cell culture. Additional studies were performed with HeLa cells that had not become drug resistant (Presant and Carr, unpublished data). In studies with 5-fluorouracil and methotrexate, Am B did not enhance uptake. In studies with PALA (phospho-N-acetyl-l-asparatic acid), uptake was increased by 8.3%. In studies with adriamycin, uptake was increased by 20%.

In vivo studies of the uptake of CCNU by a murine ependymoblastoma were performed by Laurent *et al.* (1980). The uptake of CCNU cyclohexyl degradation products into tumor cells was significantly enhanced by the intraperitoneal injection of Am B. These degradation products have been associated with the carbamoylating effects of CCNU. Furthermore, the inhibition of RNA synthesis produced by CCNU was also enhanced by the addition of Am B.

The effect of Am B on uptake of nitrogen mustard by human tumor cells was studied in cell lines including HT29 human colon carcinoma cells, and SKMES epidermoid carcinoma cells (Presant and Carr, 1980). Enhancement of nitrogen mustard uptake was produced by Am B. This was characterized by an increase in the V_{max} of transport without a change in the K_m, suggesting a true change in the choline carrier transport mechanism, rather than the nonspecific induction of cell membrane "pores." Furthermore, the uptake of nitrogen mustard by fresh ovarian carcinoma cells was also enhanced.

If Am B is able to enhance the cellular uptake of many antitumor drugs, it is logical to expect increased cytotoxicity in *in vivo* treatments. An extensive series of experiments has been performed by our group (Medoff *et al.*, 1980), as indicated in Table IV. In these studies, mice with transplantable AKR leukemia were given injections of drug alone, or injections of drug plus Am B. Twenty-four hours later, the survival of leukemic cells was determined by evaluating the LCFU in a spleen cell clonogenic assay *in vivo*. The use of this assay eliminates the effects of immune potentiation. Single doses of anticancer agent were chosen which reduced leukemic cell survival to between 10^{-2} and 10^{-3} of control. For certain drugs (cytosine arabinoside, BCNU, hydroxyurea, and methotrexate), enhancement of cytotoxicity was less than 10-fold. For vincristine, the enhancement was approximately 10-fold. However, for the alkylating agents, including L-phenylalanine mustard, cyclophosphamide, and nitrogen mustard, enhancement was between 100 and 200-fold. For adriamycin, enhancement was over 300-fold.

Surprisingly, a marked difference was seen between BCNU and CCNU. Although BCNU drug enhancement by Am B was only approximately 3-fold, enhancement of lethality for CCNU was greater than 1000-fold. This effect also

TABLE IV. Am B Enhancement of Drug Lethality in AKR Leukemia

Drug	Potentiation of drug cytotoxicity *in vivo* by Am B[a]
Cytosine arabinoside	2.3
BCNU	2.8
Hydroxyurea	4.4
Methotrexate	7.9
Vincristine	12
L-phenylalanine mustard	147
Cyclophosphamide	180
Nitrogen mustard	209
Adriamycin	330
CCNU	>1000

[a] Measured as survival of leukemic colony-forming units in spleen-cell clonogenic assay *in vivo* 24 hr after drug alone compared to survival after drug plus Am B. Each number represents the *n*-fold increase in leukemic cell killing induced by Am B.

suggests that Am B was not simply producing holes or "pores" in the cell membrane. If that were the case, the enhancement of BCNU and CCNU would have been expected to be similar since their size is similar. However, the marked difference suggests that other factors such as lipid solubility or charge were important, and therefore the Am B effect on the cell membrane is more complex.

In order to determine if these effects could be seen with other types of tumors, preliminary experiments have also been performed in L1210 leukemia and EMT-6 mammary carcinoma. In each of these circumstances, BCNU cytotoxicity has been potentiated by the addition of Am B.

The enhancement of drug cytotoxicity by Am B has been further characterized. By increasing the number of courses of Am B administered, the potentiation of drug cytotoxicity increases. Furthermore, the effects of an infusion of Am B on potentiation are greater than the effects of a bolus injection of Am B.

These results have been confirmed in other tumors as well. Laurent *et al.* (1976) and Muller and Tator (1978) have shown that Am B potentiates the effects of CCNU on ependymoblastoma injected either subcutaneously or intracerebrally. In addition, Block *et al.* (1979) have indicated that a cell line derived from a human malignant melanoma when tested with DTIC alone or in combination with Am B showed potentiation of cytotoxicity by the administration of the combination.

Additional experiments were performed by Valeriote *et al.* (1979) to determine whether preexisting resistance to actinomycin D might be overcome by the administration of Am B (Table V). AKR leukemia cells were transplanted into mice and were initially sensitive to the lethal effects of intravenous administration of actinomycin D. If mice were treated with actinomycin D once and the resulting leukemia cells harvested (one passage), those cells when injected into

TABLE V. Am B Reversal of Resistance to Actinomycin D in AKR Leukemia *In Vivo*

AKR leukemic cells	Leukemic cell survival *in vivo*[a]		Potentiation of cell kill by Am B
	Actinomycin D (0.5 μg)	Actinomycin D plus Am B	
Sensitive to actinomycin D	0.02	0.002	10
Resistance to actinomycin D			
4 prior exposures (passages)	0.2	0.015	13
10 prior exposures	0.9	0.07	13
20 prior exposures	0.8	0.4	2

[a] Leukemic cell survival was measured as survival of LCFU in a spleen colony assay 24 hr after drug treatment *in vivo*. The number represents the fraction of LCFU surviving the drug treatment.

mice showed less killing by actinomycin D. This process was repeated serially, and eventually a highly resistant population of AKR leaukemia cells resulted in which actinomycin D administration *in vivo* resulted in practically no leukemic cell killing. If actinomycin D was injected with Am B, the killing of initially sensitive AKR leukemia cells was potentiated by a factor of 10. If four or ten passages had occurred, the same degree of potentiation by Am B was observed (although the sensitivity to actinomycin D alone had decreased to no apparent cytotoxicity after ten prior exposures). However, if 20 passages had been performed, not only was there no cytotoxicity from the administration of actinomycin D alone, but the cytotoxicity following the administration of actinomycin D plus Am B was minimal. The potentiation of cell kill by a factor of two was not considered to be a meaningful one.

This experiment is important in its possible extrapolation to the clinical situation. It indicates that for certain drugs and certain tumors, initially sensitive tumors may show enhancement of cytotoxicity by the addition of Am B. Furthermore, even when cells become sufficiently resistant to a drug so that further administration of the drug produces no apparent cytotoxicity, addition of Am B may still potentiate cytotoxicity (as with 10 prior exposures to actinomycin D). However, very marked resistance (as with 20 prior exposures) may result in a situation in which not only does the chemotherapeutic drug not induce cytotoxicity, but even the administration of that drug with Am B produces no cytotoxicity.

Preliminary experiments have begun to determine the potentiation of cytotoxic effects on normal cells in an experimental situation. Experiments performed with normal hematopoietic colony-forming units in marrow following the *in vivo* injection of actinomycin D alone or with Am B indicated no significant potentiation of cytotoxicity (unpublished observations). An extensive series of observations is accumulating from the human trials that indicates that normal cell toxicity is not consistently enhanced by the addition of Am B. However, further experimental studies quantitating the interaction of chemotherapeutic drugs and Am B in normal tissues is needed.

C. Effects of Am B on Tumor Cell Clonogenicity

In order to predict which types of treatment programs might be effective in the clinical setting, *in vivo* clonogenic assays of cell suspensions from primary human tumor biopsies have been developed. The most notable of those is that of Hamburger *et al.* (1978). Over the last 4 years, we have developed a nearly identical technique for the purposes of testing whether the cytotoxicity of chemotherapeutic drugs against human colony-forming tumor cells could be potentiated by Am B. We have modified our techniques to make the results directly comparable with those laboratories utilizing a previously published system.

Tumor specimens have been obtained from patients with a variety of tumors. Single-cell suspensions were prepared from these tumors by sequentially mincing, screening, and passing the suspension through needles of decreasing size to 23 gauge. Neoplastic effusions were processed by centrifugation over a Ficoll-Hypaque gradient to eliminate tissue clumps. To the cell suspension test drugs or control medium without drugs were added. Chemotherapeutic drugs were tested at one to three concentrations either alone or with the addition of Am B at a final concentration of 2 μg/ml. Additional control cultures included Am B without chemotherapeutic drugs.

Cells were then placed in soft agar culture. A feeder layer consisted of 0.5% agar with enriched McCoy's 5A medium plus 15% fetal calf serum with penicillin–streptomycin. The upper layer containing the cell suspension (5×10^5 cells per dish) consisted of 0.3% agar, CMRL 1066 medium with 20% horse serum, and $5 \times 10^{-5}M$ 2-mercaptoethanol. Cells were incubated at 37° in a humidified chamber containing 95% oxygen–5% CO_2 for 14 to 21 days. Colony growth was assessed at 3-day intervals by inverted phase microscopy. Plating efficiency of control cells incubated in medium alone was compared to the plating efficiency of cells incubated with drug alone, cells incubated with Am B alone, and cells incubated with the combination.

The frequency of successful cultures was approximately 50% (Table VI). Fifteen trials with Am B were evaluable. Of those, only four trials indicated minimal cytotoxicity induced by Am B (26% to 46% reduction in colonies), suggesting that Am B is unlikely to have a significant cytotoxic effect in human malignancy when used alone. This is consistent with a lack of prior observations of antitumor effects in patients harboring malignant neoplasms who develop systemic fungal infections and are treated with Am B.

Am B potentiated drug-induced cytotoxicity in four trials, or 38%. Of the drugs tested, the following showed potentiation *in vivo:* chlorozotocin, vincristine, and L-phenylalanine mustard. (In each case where Am B alone produced a reduction in colonies, potentiation was defined as a reduction in clones greater than the sum of reductions by Am B alone plus drug alone.) The following drugs have not yet demonstrated potentiation by Am B: adriamycin,

TABLE VI. Am B Effects on Human Colony-Forming Tumor Cells *In Vitro*[a]

	No. of Samples		No. of Samples
Total cultures	12	Drugs potentiated	
Successful cultures	6	Chlorozotocin	2/3
Am B trials evaluable	15	L-phenylalanine mustard	1/2
Am B induced		Vindesine	1/2
cytotoxicity[b]	4	Vincristine	0/1
Am B potentiation		Adriamycin	0/4
of drug-induced		BCNU	0/1
cytotoxicity[c]	4	DTIC	0/1
Tumor cells showing		Methotrexate	0/1
Am B-induced potentiation			
Melanoma	2/6		
Breast carcinoma	1/3		
Ovarian carcinoma	1/3		
Bronchogenic carcinoma	0/3		

[a] 5×10^5 human tumor cells in single-cell suspensions were placed into soft agar cell culture. A technique similar to that of Hamburger *et al.* (1978) was used, except that drugs were present in the agar continuously. Colonies were enumerated after 14 to 21 days of culture.

[b] Greater than 30% decrease in colony formation from Am B alone (no chemotherapeutic drug added).

[c] Greater than 20% decrease in colony formation from Am B plus drug, compared to drug alone.

BCNU, vincristine, DTIC, and methotrexate. The types of tumors in which Am B has shown potentiation have included ovarian carcinoma (1/3), carcinoma of the breast (1/3), and malignant melanoma (2/6). Two cases of bronchogenic carcinoma showed no potentiation.

Additional observations have been made using the clonogenic assay. First, in tumor cell suspensions in which Am B induced potentiation is observed, it is usually observed with only one of several drugs tested. Therefore, enhancement of cytotoxicity may be highly selective for particular drugs. Second, in at least one circumstance, potentiation of cytotoxicity was only observed at high concentrations of L-phenylalanine mustard, suggesting that clinical trials of Am B potentiation should be performed at maximally tolerated chemotherapeutic drug dosages. Third, it is apparent that drugs not previously appreciated to be potentiated by Am B can be evaluated in the clonogenic assay. The potentiation of chlorozotocin cytotoxicity was first observed in the clonogenic assay, and is currently undergoing clinical trials.

This approach to evaluation of Am B–chemotherapeutic drug effects is also being performed in other laboratories (Wilson *et al.*, 1980). Preliminary evidence indicates no potentiation by Am B of adriamycin, melphalan, or BCNU in ovarian carcinoma cell cultures. The number of patients studied and the exact details of the tests for resistance or potentiation have not yet been published.

III. CLINICAL STUDIES

A. Am B Reversal of Drug Resistance

Based on the work of Valeriote *et al.* (1976) demonstrating enhancement of the cytotoxicity of BCNU and other drugs (Table IV), and the reversal of resistance to actinomycin D (Table V), clinical studies of the addition of Am B to chemotherapy were initiated in patients with a variety of human neoplasms. Preliminary results have been published (Presant *et al.*, 1977; Presant and Garrett, 1980). Patients were eligible for this treatment program if they had become resistant to a chemotherapeutic program administered over at least two cycles of therapy. Progression was indicated by a greater than 50% increase in the measurable tumor parameter. As indicated in Table VII, any of a variety of treatments were allowed. Patients were then treated with the same chemotherapy in the same dosages and schedules but following pretreatment with Am B for 2 days (7.5 mg/m^2 day 1 and 30 mg/m^2 day 2, followed by chemotherapy day 2) or 4 days (7.5 mg/m^2 day 1, 15 mg/m^2 day 2, and 30 mg/m^2 days 3 and 4, followed by chemotherapy day 4). Am B was given intravenously over 6 hr (for details, see

TABLE VII. Reversal of Resistance to Anticancer Drugs Using Am B

Treatment	No. trials	Responses[a]		
Adriamycin, BCNU, cyclophosphamide	13	(1 CR	3 PR	1 OR)
Vinblastine, vincristine	1	(1 OR)		
Actinomycin D, cyclophosphamide, vincristine	1	(1 PR)		
Actinomycin D	1			
CCNU	1	(1 OR)		
Chlorozotocin	2			
Platinum	1			
Adriamycin	5	(1 OR)		
Adriamycin, cyclophosphamide, DTIC	1			
Adriamycin, DTIC	1			
Adriamycin, cyclophosphamide	3			
Adriamycin, cyclophosphamide, methotrexate	1			
	31			

[a] CR = complete response; PR = partial response; OR = objective response.

Presant *et al.*, 1977). Patients remained on study until there was evidence of tumor progression. Responses were defined as complete, partial (greater than 50% reduction in tumor measurements), or objective (25–50% reduction in tumor measurements), as conventionally accepted.

Thirty-eight patients have entered 40 trials. Four are inevaluable (less than 2-week survival). For five it is too early for evaluation (less than 2 months on trial) at the present time. Of the remaining 31 trials, one complete response, four partial responses, and four objective regressions were obtained. Reversal of previously established resistance was observed in patients treated with the combination of adriamycin, BCNU plus cyclophosphamide; or actinomycin D, cyclophosphamide, and vincristine. In addition, single-drug treatments previously associated with tumor progression showed evidence of measurable tumor response when combined with Am B. These included vinblastine and vincristine, adriamycin, and CCNU.

A complete response of 3 months was obtained on two occasions in a patient with acute myelomonocytic leukemia treated with adriamycin, BCNU plus cyclophosphamide, and Am B. Progression was observed when the chemotherapeutic drugs were used alone. Of three patients with multiple myeloma, there was one partial response for 25 months following treatment with adriamycin, BCNU, cyclophosphamide, and prednisone; one objective response of 3 months with the same regimen; and one patient with progression of disease. Of three patients with carcinoma of the breast, treated with the same chemotherapeutic program, there were two partial responses of 1 and 2 months each, and one patient with progressive disease. A patient in blastic transformation of chronic myelomonocytic leukemia had an objective regression of 1 month following treatment with vincristine and vinblastine and Am B. Of three patients with carcinoma of the thyroid, one patient had an objective response of 3 months following adriamycin plus Am B, and two other patients showed progression of disease. Of two patients with renal cell carcinoma, one patient had an objective response of 1 month's duration following CCNU plus Am B, and another patient treated with adriamycin plus Am B showed progression of disease.

Of four patients with soft-tissue sarcomas, three treated with adriamycin-containing combination treatment programs plus Am B showed progression of disease. One patient showed a partial response of 3 month's duration after treatment with actinomycin D, cyclophosphamide plus vincristine and Am B. Three patients with carcinoma of the ovary, two patients with acinic cell carcinoma, one patient with testicular carcinoma, six patients with bronchogenic carcinoma, and one patient with epidermoid carcinoma of the head and neck all failed to show evidence of response.

The response in the patient with soft-tissue sarcoma is notable. That patient had previously received adriamycin, cyclophosphamide, and methotrexate. The patient demonstrated resistance to this regimen, and when Am B was added to that treatment program, further progression of the disease was noted. The patient was then treated with actinomycin D, cyclophosphamide plus vincristine. Further progression of disease was noted, and the patient was semicomatose with hypoxia from extensive pulmonary metastases. When Am B was given in addition to actinomycin-D, cyclophosphamide plus vincristine, a nearly complete remission was achieved, with restoration of a Karnofsky performance status of 80%. This patient demonstrates that when Am B produces potentiation of drug-induced cytotoxicity, it may do so in a setting of no prior response to the chemotherapeutic drugs, and may do so only with selected chemotherapeutic drugs.

Dose-limiting toxicity with this treatment program was the same as that observed with the chemotherapeutic agent alone. In approximately one-third of patients, the dose-limiting myelosuppression previously observed with the chemotherapeutic drugs alone was enhanced by the addition of Am B. In another third of patients, the degree of myelosuppression was unchanged, and in another third, the myelosuppression following the addition of Am B was less than that observed with the chemotherapeutic drugs alone. Clearly, accentuation of toxicity can be observed, and therefore toxicity must be carefully monitored during the addition of Am B to other chemotherapeutic drugs. Overall, during the Am B trials, severe granulocytopenia ($< 1000/mm^3$) was observed in 40% of cases, and severe thrombocytopenia ($< 25000/mm^3$) in 15%. In addition, fever, chills, occasional hypotension, and occasional nephrotoxicity were observed with Am B. These side effects are further detailed below.

In an attempt to confirm our preliminary results, Krutchik *et al.* (1978) used an identical treatment schema with 11 consecutive evaluable breast carcinoma patients and three evaluable sarcoma patients. Patients were treated with a variety of adriamycin-containing treatment programs. One patient with breast carcinoma achieved a partial response for 4 months. The remaining patients failed to show measurable responses. It is of note that the degree of previous treatment with chemotherapy was greater in this study than that which we have performed and described above. If the same relationship between degree of previous treatment and Am B response obtains clinically as that previously described in experimental systems (Table V), one would expect less frequent potentiation of cytotoxicity in the more heavily drug treated patient population. It is apparent that additional studies are needed of the addition of Am B to chemotherapy to define the frequency of reversal of resistance and to identify the clinical situations in which this may be important.

B. Nitrosourea–Am B Interactions

Based on the early trials in experimental animals performed by Medoff and Valeriote *et al.* (1979) (Table I), the initial clinical trials of Am B in combination with chemotherapeutic agents studied BCNU (Table VIII). The initial trial was a phase I–phase II study to determine if the addition of four doses of Am B to BCNU produced enhanced toxicity (Present *et al.*, 1976). The results of this trial

TABLE VIII. BCNU plus Am B Trials

Response	BCNU alone	Am B plus BCNU
No. patients evaluable	137[a]	18
Response >50%	7% (0%–21%)	33%
Response >25%	8%	44%
Median survival	2.5 mo[b]	3 mo
Toxicity		
No. patients evaluable	13[c]	24
Thrombocytopenia:		
Nadir <100,000/mm³	33%	38%
Nadir < 50,000/mm³	17%	13%
Leukopenia:		
WBC nadir <2000/mm³	15%	
PMN nadir <1500/mm³		25%
Abnormal liver function tests	31%	16%
Elevated BUN	10%	13%
Fever, chills	—	100%
Hypotension	—	25%

[a] Includes historical trials of one dose only, daily × 3, daily × 5, and weekly therapy.
[b] From Ramirez et al. (1972).
[c] From DeVita et al. (1965) at a BCNU dose of 250 mg/m² every 6–8 weeks.

indicated that there was no enhancement of BCNU toxicity. In Table VIII, we have compared the results of that trial and our subsequent phase II study (Presant et al., 1980a) with the data on BCNU alone in identical doses and schedules (as described by DeVita et al., 1965). The frequency and severity of thrombocytopenia and leukopenia were similar. Approximately 35% of patients had thrombocytopenia and approximately 20% of patients had leukopenia, either with or without Am B.

Furthermore, less common side effects of nitrosoureas, including abnormal liver function tests and transient mild azotemia, were not exacerbated by the addition of Am B. However, toxicity related to Am B was naturally observed. Fever and chills of at least minimal degree were universal. Although these symptoms were minimized by the administration of acetaminophen to prevent fever and diphenhydramine to prevent chills, at least mild reactions were observed. In situations where fever and chills were still observed despite premedication, 100 mg of hydrocortisone was added to each Am B 6-hour infusion. It was also found that meperidine, 25 mg intravenously, could abrogate the chills once they were observed (Burkes et al., 1980). A transient mild fall in blood pressure was observed in approximately 25% of patients receiving Am B, and this was usually associated with fever and vasodilatation. Increased fluid administration or cessation of Am B reversed the toxic effect.

The addition of Am B to BCNU appeared to enhance the therapeutic effects (Table VIII). When the results of the phase II trial (Presant *et al.,* 1980a) in patients with bronchogenic carcinoma were compared to the historical results of BCNU used alone in any of four different schedules, the partial response rate (greater than 50% reduction in tumor area) was 33% with Am B versus 7% without. Furthermore, if one accepted the criterion for response of a 25 % or greater reduction in tumor area, then the response rate to Am B plus BCNU was 44%, compared to 8% with BCNU alone. Nevertheless, median survival was short, either with or without Am B. Based on the favorable results of this pilot study, Am B is currently being combined with CCNU as well as adriamycin, methotrexate, and hexamethamelamine in patients with non–small-cell carcinoma of the lung.

One other clinical study has been performed utilizing Am B in combination with CCNU (Vogel, 1977). One of 7 patients with bronchogenic carcinoma treated with the combination obtained a partial response in that study. However, patients in the study had extensive previous treatment with other drugs (Vogel, personal communication).

In an attempt to extend the results of this study to the management of patients with malignant gliomas, Opfell *et al.* (1980) treated such patients with surgery, radiation therapy, plus Am B and BCNU as used in the previous studies (Presant *et al.,* 1976; Presant *et al.,* 1980a). The median survival of 57 weeks was quite comparable to that obtained with the same therapy without Am B. However, the actuarial survival of 49% at 24 months was suggestively better than would be expected from the same therapy without Am B. However, only a small number of patients (12) were studied.

C. Ongoing Investigations

There are five ongoing trials of Am B that should help to elucidate the role of Am B in patients with malignant neoplasms (Table IX). Since three of these

TABLE IX. Am B Trials in Progress

Disease	Phase	Schema
Non–small-cell lung cancer	III	Adriamycin, CCNU, MTX & hexamethylmelanine ± Am B
Soft tissue sarcoma	III	Adriamycin, cyclophosphamide & MTX ± Am B
Malignant lymphoma	III	CCNU & vinblastine ± Am B
Renal cell carcinoma	II	CCNU + Am B
Miscellaneous	II	Drugs (previously ineffective) + Am B

studies are randomized studies comparing chemotherapy with and without Am B, results in a variety of tissue types (bronchogenic carcinoma, soft-tissue sarcoma, and lymphoma) should be definitive. Furthermore, since the Am B reversal experiments (Table VII) are continuing, and since that study provides a unique opportunity to use each patient as his own control, the role of Am B in other malignant neoplasms with other drugs should be further defined with the coming years.

Thirty-five patients with extensive non–small-cell carcinoma of the lung, previously untreated with chemotherapy, have been randomized to receive either adriamycin, CCNU, methotrexate, and hexamethylmelamine alone, or following 2 days of Am B (Presant *et al.*, 1980b). In the first 24 evaluable patients, response rates were low in each group, and Am B did not appear to increase the response rate or survival. The degree of thrombocytopenia was marginally greater in patients receiving Am B compared to those who did not. The early results from this trial do not suggest that Am B can potentiate effects of the four drugs chosen in this trial. These data are consistent with those of the study of Sarna and co-workers (personal communication), who observed in a small number of patients no beneficial effect by adding Am B to a combination of methotrexate, vincristine, cyclophosphamide, and adriamycin. However, since the schedule of administration and doses of the chemotherapeutic agents were changed, and since the chemotherapy alone control arm was an historical one, the conclusions regarding Am B are only suggestive.

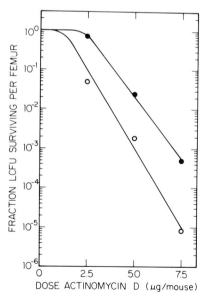

Fig. 1. Fractional survival of femoral marrow LCFU in AKR leukemia following various doses of actinomycin D *in vivo*. Solid circles, mice injected with actinomycin D alone. Open circles, mice injected with antinomycin D plus Am B. The enhanced killing by addition of Am B is 15-fold at drug doses of 2.5 and 5.0 μg per mouse, but 50-fold at 7.5 μg per mouse.

However, our experimental data may indicate why little effect was seen with this 4-drug combination, even though potentiation is noted with all these agents. As shown in Fig. 1, the extent of potentiation is dependent on the dose of the anticancer agent—the higher the dose of the anticancer drug, the greater the extent of potentiation by Am B. When multidrug regimens are employed clinically, it is common practice to decrease the dose of each of the drugs compared to each drug used as a single agent. Thus, suboptimal doses of each agent (in terms of Am B potentiation) may be a consequence of multidrug therapy.

Patients with metastatic soft-tissue sarcomas previously untreated with chemotherapy are currently being randomized to receive adriamycin, cyclophosphamide, and methotrexate alone or with Am B (Presant *et al.*, 1980c). Patients who achieve responses are randomized to receive either actinomycin D alone or with Am B as maintenance therapy. The results of this study are too preliminary to be conclusive at this time.

CCNU is being used in two trials at the City of Hope National Medical Center. Patients with malignant lymphoma refractory to adriamycin and cyclophosphamide are being randomized to receive either CCNU or vinblastine alone or with Am B. In a smaller trial of patients who have metastatic renal cell carcinoma, a phase II trial of CCNU plus Am B is being performed.

IV. CONCLUSIONS

The role of Am B, if any, in management of patients with neoplastic diseases is still undefined. However, from the preclinical studies, it is clear that the polyene antibiotics, and Am B in particular, have immunoadjuvant properties equal to or greater than those of previously well known immunoadjuvants such as BCG or *C. parvum*. The mechanism of this immunoadjuvant effect undoubtedly involves cell–cell interaction, and will be defined in the coming years.

Of perhaps greater interest in antineoplastic therapy is the well-documented enhancement of cellular uptake of other drugs, especially alkylating agents, and actinomycin D. Associated with this is a marked increase in the cytotoxic effect of some other antineoplastic agents, especially nitrosoureas, alkylating agents, and adriamycin. This enhancement of cytotoxic effect is present in a number of different experimental systems, and is seen in the tumor cells of some patients in *in vitro* clonogenic assays.

In clinical studies, it has been proven that Am B can be safely given with certain other drugs without a consistent increase in dose-limiting toxicity of those drugs. The toxic effects of Am B are not severe (probably because of the short duration of exposure to Am B). The studies in which Am B has been added to drugs known to be ineffective in patients leads to the conclusion that sensitivity to some of those drugs can be induced in tumor cells previously resistant. The frequency of this effect is not high, and appears to be drug dependent. Preliminary evidence suggests that Am B will not enhance the therapeutic response to a polychemotherapy regimen in non-small cell bronchogenic carcinoma, but the results in other ongoing studies are too preliminary to be conclusive.

In order to further define the role of Am B in anticancer therapy, an optimal schedule of Am B therapy must be defined. Furthermore, since the clinical circumstances in which Am B can be of benefit are undoubtedly limited, one must define diseases and drugs best potentiated by Am B. Clinical investigations must be initiated to determine whether the immunoadjuvant effects observed in preclinical studies are also observed in the clinical setting. Furthermore, there is a suggestion from preclinical studies that other polyene antibiotics may be more effective than Am B in laboratory trials. If this is the case, the preliminary conclusions which can be made at the present time may warrant clinical trials of potentially more effective polyenes.

REFERENCES

Blanke, T. J., Little, J. R., Shirley, S. F., and Lynch, R. G. (1977). *Cell Immunol. 33*, 180–190.

Block, J. B., Tabbarah, H., Isacoff, W., and Drakes, T. P. (1979). *J. Dermatol. Surg. Oncol. 5*, 118–123.

Burks, L. C., Aisner, J., Fortner, C. L., and Wiernik, P. H. (1980). *Arch. Int. Med. 140*, 483–484.

Chapman, J. A., and Hibbs, Jr., J. B. (1978). *Proc. Nat. Acad. Sci. 75*, 4349–4353.

DeVita, V., Carbone, P. P., Owens, A. H. Jr., Gold, G. L., Krant, M. J., and Edmonson, J. (1965). *Cancer Res. 25*, 1876–1881.

Hamburger, A. W., Salmon, S. E., Kim, M. B., Trent, J. F., Soehlen, B. J., Alberts, D. S., and Schmidt, J. H. (1978). *Cancer Res. 38*, 3438–3444.

Krutchik, A. N., Buzdar, A. U., Blumenschein, G. R., and Sinkovics, J. G. (1978). *Cancer Treat. Rep. 62*, 1565–1567.

Kumar, V. K., Medoff, G., Kobayashi, G. S., and Schlessinger, D. (1974). *Nature 250*, 323–325.

Laurent, G., Atassi, G., and Hildebrand, J. (1976). *Cancer Res. 36*, 4069–4073.

Laurent, G., Dewerie-Vanhouche, J., Machin, D., and Hildebrand, J. (1980). *Cancer Res. 40*, 939–942.

Lin, H.-S., Medoff, G., and Kobayashi, G. S. (1977). *Antimicrob. Agents Chemother. 11*, 154–160.

Medoff, G., Kobayashi, G. S., Kwan, C. B., Schlessinger, D., and Veukov, P. (1972). *Proc. Nat. Acad. Sci. 69*, 196–199.

Medoff, G., Valeriote, F., Lynch, R. G., Schlessinger, D., and Kobayashi, G. S. (1974). *Cancer Res. 34*, 974–978.

Medoff, G., Goldstein, M., Schlessinger, D., and Kobayashi, G. S. (1975). *Cancer Res. 35*, 2548–2552.

Medoff, G., Valeriote, F., Ryan, J., and Tolen, S. (1977). *J. Nat. Cancer Inst. 58*, 949–953.

Medoff, G., Valeriote, F., and Dieckman, J. (1980). *J. Nat. Cancer Inst.* Submitted for publication.

Muller, P. J., and Tator, C. H. (1978). *J. Neurosurg. 49*, 579–588.

Opfell, R. W., Grahn, E., and Baughen, M. (1980). *Proc. Amer. Soc. Clin. Oncol. 21*, 477.

Presant, C. A., and Carr, D. (1980). *Biochem. Biophys. Res. Commun. 93*, 1067–1073.

Presant, C. A., and Garrett, S. (1980). *Proc. Amer. Assoc. Cancer Res. 21*, 138.

Presant, C. A., Klahr, C., Olander, J., and Gatewood, D. (1976). *Cancer 38*, 1917–1921.

Presant, C. A., Klahr, C., and Santala, R. (1977). *Ann. Intern. Med. 86*, 47–51.

Presant, C. A., Hillinger, S., and Klahr, C. (1980a). *Cancer 45*, 6–10.

Presant, C. A., Metter, G. E., Bertrand, M., Chang, F. L., Farbstein, M., Forman, S. J., Rappaport, D., Sarauw, A., Kendregan, B., Mackie, A., and Garrett, S. (1980b). *In* "Immunotherapy of Cancer; Current Results of Trials in Man" (W. Terry, ed). Elsevier, New York. In press.

Presant, C. A., Bartolucci, A. A., Lowenbraun, S., and the Southeastern Cancer Study Group (1980c) *In* "Immunotherapy of Cancer: Current Results of Trials in Man" (W. Terry, ed.). Elsevier, New York. In press.

Ramirez, G., Wilson, W., Grage T., and Hill, G. (1972). *Cancer Chemother. Rep. 56,* 787–790.

Shirley, S. F., and Little, J. R. (1979). *J. Immunol. 123,* 2878–2882.

Valeriote, F., Lynch, R., Medoff, G., and Kumar, B. V. (1976). *J. Nat. Cancer Inst. 56,* 557–560.

Valeriote, F., Medoff, G., and Dieckman, J. (1979). *Cancer Res. 39,* 2041–2045.

Vogel, S. E. (1977). *Proc. Am. Assoc. Cancer Res. 18,* 169.

Wilson, J. K. V., Ozols, R. F., Grotzinger, K. R., and Young, R. C. (1980). *Proc. Am. Assoc. Cancer Res. 21,* 175.

Chapter 30

ACTIVITY OF NITROSOUREAS ON HUMAN TUMORS *IN VITRO*

Brian F. Issell
Claude Tihon
M. Elaine Curry

I. INTRODUCTION

The nitrosoureas are a group of anticancer compounds with a long clinical history. BCNU was first clinically evaluated in the early 1960s and was followed by CCNU in 1969 and methyl CCNU in 1971. Despite this long period of evaluation, it is still uncertain whether one of these compounds has any superior clinical utility over the others or whether they may be clinically interchangeable.

Against mice with intracerebrally implanted L1210 cells, CCNU appeared superior to BCNU, which in turn was more active than methyl CCNU (Schabel, 1976). However, methyl CCNU (MeCCNU) was the most active of the three compounds against established transplanted Lewis lung carcinoma (Schabel, 1976), and was selected for further clinical study mainly on this basis (Wasserman *et al.*, 1975).

Relatively few clinical studies have addressed a direct comparison between the three compounds. The analysis of composite data and sequential studies to identify compound differences is almost impossible due to the inability to obtain and determine patient and therapy comparability. Many of the sequential studies have compared the nitrosoureas in patients who had significant differences in the extent of their prior treatments (Carter and Wasserman, 1976). Two randomized studies by the Cancer and Leukemia Group B in patients with Hodgkin's disease who had failed previous chemotherapy suggested a superiority of CCNU over BCNU (Selawry and Hanson, 1972) and CCNU over MeCCNU (Maurice *et al.*, 1978). A superiority for CCNU over BCNU when given as adjuvant chemotherapy following surgery plus radiotherapy was also suggested for patients with glioblastoma multiforme in a prospective randomized study (Solero *et al.*, 1979). Other direct comparisons between CCNU and MeCCNU have been examined in breast cancer by the Southwest Oncology Group (Gottlieb *et al.*, 1974), lung cancer by the Mayo Clinic (Eagan *et al.*, 1974), and in gynecological malignancies by the Gynecologic Oncology Group (Omura *et al.*, 1977; and Omura *et al.*, 1978). No definite conclusions could be reached since both compounds were relatively inactive in these settings. However, there was a suggestion that CCNU may be more active than MeCCNU in breast cancer.

In view of these largely inconclusive clinical data regarding the relative activities of BCNU, CCNU, and MeCCNU, we decided to undertake studies comparing nitrosourea activities *in vitro* using the soft agar colonogenic assay. This assay has been shown to be highly predictive for *in vivo* patient drug response when tested by two independent institutions (Alberts *et al.*, 1980; and VonHoff, 1980). This chapter outlines the interim results of an ongoing study.

II. MATERIALS AND METHODS

The soft agar clonogenic assay has been previously described in detail elsewhere (Hamburger and Salmon, 1977; Hamburger *et al.*, 1978; and Salmon *et al.*, 1978). Briefly, tissue samples for culture were obtained from malignant effusions or from surgically resected solid tumor tissue. Effusions were collected in preservative-free heparinized containers. Single-cell suspensions from solid tissue were obtained by a mechanical disaggregating procedure that was modified from the previously described methods and is as follows. The specimen was cut into cubes of 2 mm or less with scapels, suspended in nutrient medium, and further disaggregated by being forced hydrostatically through a series of "syringe screens" ranging from 20-mesh to 400-mesh in size. A syringe screen consisted of a 50 ml syringe barrel with the needle end cut off and replaced by a stainless steel screen.

Tumor cell suspensions were incubated in serum-free nutrient medium alone or in medium containing 0.2 μg/ml of BCNU, CCNU, or MeCCNU at 37°C for 1 hr. Drug concentrations of 0.2 μg/ml were chosen based on average serum BCNU concentrations obtainable during the 60 min following drug administration (Levin *et al.*, 1978).

After drug incubation, the cells were washed twice in medium. Either 1×10^6 or 2×10^6 cells, depending on the viability previously determined by the trypan blue exclusion method, were mixed in 1.0 ml of enriched medium containing 0.3% Low Melting Point Agarose™ (Bethesda Research Laboratories) and placed on a 2 ml 0.5% agarcontaining nutrient feeder layer in 35 mm plastic petri dishes. The 35 mm assay dishes were placed in 150 mm humidified petri dishes and incubated in 5% CO_2 at 37°C. They were left undisturbed until counting at 15 and 21 days later.

Colonies were defined as tight aggregations of > 30 cells. Assay dishes and controls were run in triplicates, a procedure which we have generally found gives individual dish variability of less than 15%. The assay was considered evaluable if 30 or more colonies had grown in the control dishes. A positive drug effect was defined as a 70% or more decrease in the number of colonies of the drug-treated plates compared to control plates. This percentage decrease had been previously found to predict for a positive patient effect (Salmon and VonHoff, 1980).

III. RESULTS

Over 100 separate tumor specimens from either solid tumor excision, malignant effusion aspiration, or malignant bone marrow sampling have been received for assay by our laboratory. Seventy-four percent of the samples received grew colonies. In 57% of the total samples, there were multiple plates of sufficient numbers of colonies for satisfactory drug testing. Virtually all the major histological tumor types have been successfully cultured. Comparative nitrosourea testings on 14 separate human tumors are completed to date.

The tumor types tested, sources of tumor cells, number of control colonies,

TABLE I. Nitrosourea Comparative Activity

Tumor type	Source	No. control colonies	% decrease of control		
			BCNU	CCNU	MeCCNU
Colorectal adeno.	Solid	179	3	2E	18
Colorectal adeno.	Solid	55	11	7E	42
Colorectal adeno.	Solid	214	21	41	17
Colorectal adeno.	Solid	278	28	35	33
Lung adeno.	Pleural effusion	84	25	24	23
Ovary	Ascites	7200	59	–	43
Ovary	Ascites	345	–	10E[a]	34
Prostate	Ascites	8000	17	6	–
Pancreatic	Ascites	93	36	33E	–
Lymphoma DHL	Solid	67	48	63	63
Unknown primary	Solid	61	33	20	45
Rhabdomyosarcoma	Solid	37	41	38	62
Renal cell	Solid	1400	27	22	37
Breast	Solid	31	39	22E	42E

[a] E = Enhancement of growth compared to control.

TABLE II. Compound Ranking (25% or Greater Difference)

Ranking	Frequency	Tumor types
MeCCNU > CCNU	4	2 Adeno. rectum, 1 ovary, 1 unknown primary
MeCCNU > BCNU	1	Adeno. rectum
CCNU > MeCCNU	1	Adeno. rectum
CCNU > BCNU	0	
BCNU > CCNU	1	Breast
BCNU > MeCCNU	1	Breast

and decrease in colony numbers compared to controls for each of the three nitrosoureas tested are outlined in Table I. None of the tumors tested to date appeared to show significant sensitivity to any of the nitrosoureas as measured by the criterium of 70% or greater colony number decrease.

Table II outlines the comparative ranking of the three nitrosoureas based on a 25% or more difference in drug sensitivity for each tumor type. MeCCNU was superior to CCNU in five tumors and to BCNU in two tumors. CCNU was superior to MeCCNU in one tumor, whereas BCNU was superior to both CCNU and MeCCNU in one assayed breast cancer specimen. There is no evidence from our experience or from other reported data that these *in vitro* differences would be reflected by different *in vivo* patient responses to the nitrosoureas tested.

IV. DISCUSSION

The inability of transplanted murine tumors to predict relative activities of anticancer drugs in patients is clearly demonstrated in the case of the nitrosoureas and is a major problem of cancer drug development. The introduction of the soft agar clonogenic assay with its high *in vitro–in vivo* correlation now provides a technology with considerable potential for selecting drugs relevant to the treatment of patients. It is true that further understandings of the biology of the assay and additional technological refinements are required before its results can be considered with utmost confidence. Nevertheless, in comparison with our previously available methods, this system could well constitute a major breakthrough.

In addition to its utilization in helping to select appropriate drugs for individual cancer patients, the assay is also being used in our laboratory to aid the selection of analogs and novel compounds for development at the preclinical and phase I clinical stages. It is intended to test compounds head-to-head against at least ten individual tumors of several signal histological types. To aid the supply of a sufficient number of tumors of each histological type, the validity of using thawed frozen specimens is also under investigation.

As is the case with the nitrosoureas, there are instances where analogs have undergone substantial clinical development without established evidence of any superiority for one over another. This is mainly because the compounds were developed sequentially and historical trends had established the utilization of each compound in separate tumor types. Head-to-head clinical comparisons were rarely carried out in a single tumor type. An important question often asked by clinicians is, Can one of the analogs with a more convenient form of administration (oral) substitute for its solely parenterally formulated analog in all patients? The hope that the clonogenic assay may help answer these questions form the rationale for this study.

Because of the small number of tumors tested so far, the interim results presented in this study are inconclusive. What can be said is that the lack of *in vitro* sensitivity thus far shown for BCNU, CCNU, and MeCCNU is not inconsistent with the drug response expected for patients with these tumors.

ACKNOWLEDGMENTS

The authors wish to thank Mrs. Audrey Farwell for her technical assistance and Ms. Judi Brinck for preparation of this manuscript.

REFERENCES

Alberts, D. S., Chen, H. S. G., Soehnlen, B., Salmon, S. E., Surwit, E. A., Young, L., and Moon, T. E. (1980). *Lancet 2,* 340–342.

Carter, S. K., and Wasserman, T. H. (1976). *Cancer Treat. Rep. 60,* 807–811.

Eagan, R. T., Carr, D. T., Coles, D. T., Dines, D. E., and Ritts, R. E. (1974). *Cancer Chemother. Rep. 58,* 913–918.

Gottlieb, J. A., Rivkin, S. E., Spiegel, S. C., Hoogstraten, B., O'Bryan, R. M., Delaney, F. C., and Singhakowinta, A. (1974). *Cancer 33,* 519–526.

Hamburger, A. W., and Salmon, S. E. (1977). *Science 197,* 461–463.

Hamburger, A. W., Salmon, S. E., Kim, M. B., Trent, J. M., Soehnlen, B. J., Alberts, D. S., and Schmidt, H. J. (1978). *Cancer Res. 38,* 3438–3444.

Levin, V. A., Hoffman, W., and Weinkam, R. J. (1978). *Cancer Treat. Rep. 62,* 1305–1312.

Maurice, P., Glidewell, O., Jacquillat, C., Silver, R. T., Carey, R., TenPas, A., Cornell, C. J., Burningham, R. A., Nissen, N. I., and Holland, J. F. (1978). *Cancer 41,* 1658–1663.

Omura, G., DiSala, P., Blessing, J., Boronow, R., Hreshchyshyn, M., and Park, R. (1977). *Cancer Treat. Rep. 61,* 1533–1535.

Omura, G. A., Shingleton, H. M., Creasman, W. T., Blessing, J. A., and Boronow, R. C. (1978). *Cancer Treat. Rep. 62,* 833–835.

Salmon, S. E., Hamburger, A. W., Soehnlen, B., Durie, B. G. M., Alberts, D. S., and Moon, T. E. (1978). *New Eng. J. of Med. 298,* 1321–1327.

Salmon, S. E., and VonHoff, D. D. (1980). *Semin. Oncol.* In Press.

Schabel, F. M. (1976). *Cancer Treat. Rep. 60,* 665–698.

Selawry, O. S., and Hansen, H. H. (1972). *Proc. Am. Assoc. Cancer Res. 13,* 46.

Solero, C. L., Monfardini, S., Brambilla, C., Vaghi, A., Valagussa, P., Morello, G., and Bonadonna, G. (1979). *Cancer Clin. Trials 2,* 43–48.

VonHoff, D. D. (1980). *Proc. AACR 21,* 134.

Wasserman, T. H., Slavik, M., and Carter, S. K. (1975). *Cancer 36,* 1258–1268.

Chapter 31

STREPTOZOTOCIN

Patrick J. Byrne
Philip S. Schein

I. INTRODUCTION

Streptozotocin is a naturally occurring product of the fermentation of *Streptomyces achromogenes* (Wiggans *et al.*, 1978). It was initially developed as an antibiotic, with antibacterial activity having been demonstrated against both gram-positive and gram-negative bacteria (Levi *et al.*, 1977). The drug was subsequently screened for antitumor activity in rodent tumor models; and following evidence of antineoplastic properties, the subsequent development of this drug was transferred to the National Cancer Institute. In further testing, streptozotocin demonstrated reproducible activity against the intraperitoneally implanted L1210 murine leukemia. It was during the preclinical toxicologic evaluation in rodents and dogs that streptozotocin's diabetogenic properties became evident. The chemical structure of streptozotocin is composed of methylnitrosourea (MNU) attached to the C-2 position of glucose (Herr *et al.*, 1967). As such, it represents a unique naturally occurring nitrosourea. Like its methylnitrosourea cytotoxic group, streptozotocin possesses mutagenic and carcinogenic properties. The two unique pharmacologic properties of streptozotocin are its potent diabetogenicity, mediated through the destruction of the pancreatic islet beta cell, and antitumor activity without the necessity for bone marrow toxicity. The former has been exploited in the treatment of human islet cell carcinoma (Schein *et al.*, 1973a).

II. RESULTS

A. Metabolism

Under physiological conditions streptozotocin decomposes spontaneously with a chemical half-life of 247 min in 0.1 M phosphate-buffered saline pH 7.4 (Colvin *et al.*, 1976; Heal *et al.*, 1979). In the process of degradation a series of alkylating moieties are formed of which an alkyldiazehydroxide precursor and a carbonium ion are considered the most important. The alkylating activity of streptozotocin measured with nitrosbenzyl pyridine is 7% of that of chlorozotocin. While it is assumed that an isocyanate is also formed, streptozotocin possesses relatively little carbamoylating activity as a result of internal carbamoylation (Loranger, 1980).

Heal *et al.* (1979) have examined the importance of nitrosourea carbamoylating activity by comparing the clinical and biological properties of streptozotocin to a second analog with the MNU cytotoxic group attached to carbon 1 of glucose (GNU). While the latter compound has high carbamoylating activity, the alkylating activity of the two drugs is comparable. Production of single strand breaks (SSB) produced in L1210 DNA by streptozotocin was comparable to that produced by GNU at equimolar drug concentrations. However, repair of these breaks was complete by 8 hr in the streptozotocin-treated cells compared to 12 hr required for GNU-treated cells, thus demonstrating a delay in DNA repair with the compound possessing high carbamoylating activity. Nevertheless, there was no difference in either *in vitro* or *in vivo* cytotoxicity for the L1210 leukemia, suggesting a minor, if any, therapeutic role for carbamoylation. In detailed structure–activity analyses, a role for methylnitrosourea carbamoylating activity could not be identified for either bone marrow toxicity or lethal toxicity (Panasci *et al.*, 1977), whereas alkylation is the principal mediator of both the therapeutic and the nondiabetogenic toxic properties.

B. Mechanisms of Streptozotocin Activity at the Cellular Level

Similarities were noted between the diabetogenic effect of streptozotocin and alloxan found in rodents and dogs. Both agents caused selective destruction of pancreatic islet cells with a resultant diabetic state. Glutathione, cysteine, nicotinic acid, and nicotinamide administered in pharmacologic doses prior to alloxan were known to prevent this diabetogenic action, whereas only nicotinamide protects against the beta-cell toxicity of streptozotocin (Schein *et al.*, 1967). This protection has been demonstrated in mice and rats and subsequently was found to cross species barriers in monkeys and dogs (Schein *et al.*, 1973). Hepatic oxidized and reduced nicotinamide adenine dinucleotide (NAD, NADH) levels in mice are acutely depressed after a single injection of streptozotocin (Schein and Loftus, 1968); this condition persists for 24 hr, with a gradual return to normal concentration. If a single pharmacologic dose of nicotinamide is given 10 min prior to the streptozotocin, liver NAD levels are protected for 16 hr, and this depression can be totally prevented if a repeat injection of nicotinamide is given at 8 hr. Alloxan has no effect on liver NAD levels, while methylnitrosourea (MNU) and any compound with an N-nitroso methyl or ethyl end group can also reduce the concentration of pyridine nucleosides (Schein and Loftus, 1968). It is likely that alkylation of DNA is the principal means of antileukemic activity for streptozotocin, whereas depression of NAD is responsible for diabetogenicity. Prevention of the lowering of NAD with nicotinamide protects against diabetogenicity but does not alter its alkylation of DNA. This represents a unique instance where the mechanisms of antitumor activity of a drug can be separated from the mechanism of the principal toxicity.

The mechanism by which streptozotocin lowers NAD levels has been investigated. The major cytotoxic action of streptozotocin is considered to be alkylation of cellular DNA. This alkylation results in depurination which, in turn, may lead to alkali-induced chemical or enzymatic single-strand breaks in DNA. Within the nucleus the enzyme poly (ADP-ribose) polymerase utilizes NAD as a substrate for the generation of ADP-ribose (Smulson et al., 1977). This enzyme has been implicated in the repair of lesions in DNA by increasing sites of DNA polymerase activation (Smulson et al., 1975). This effect may be due to the reduced affinity of ADP-ribosylated histones for DNA, allowing localized regions of relaxation of the chromatin architecture and thus facilitating the action of the DNA polymerase. The pretreatment of streptozotocin or MNU on HeLa cells has been shown to markedly stimulate poly (ADP-ribose) polymerase with a concomitant release of template restriction for DNA polymerase. Further work (Sudhakar et al., 1979) has shown that DNA fragmented by endonucleases or by incubation with MNU, the active component of streptozotocin, is particularly effective in stimulating poly (ADP-ribose) polymerase. Activation of poly (ADP-ribose) polymerase was found following MNU treatment of cells in both nuclei and isolated nucleosomes (Sudhakar et al., 1979b). This suggests that the increase in enzyme activity is not a result of a nonspecific effect of MNU on the nuclear membrane or on other structural elements of nuclei unrelated to the enzymes. Thus this stimulated poly (ADP-ribose) polymerase activity leads to the increased utilization of cellular NAD.

The structure of chromatin is composed of spherical nucleosomes joined to each other by "linker" DNA. Poly (ADP-ribose) polymerase has been shown to be associated with the histone proteins that are found in the linker region of chromatin (Smulson et al., 1977). It has also been shown that MNU preferentially alkylates this linker region of chromatin and carbamoylates the histone proteins associated with the linker DNA (Sudhakar et al., 1979b). It has been demonstrated that all compounds that possess a N-nitrosomethyl or ethyl end group depress liver NAD concentrations and that the methyl compounds are tenfold more active based on molar dose (Anderson et al., 1974). The placement of a halide on the alkyl end of many nitrosoureas, as in the case of BCNU or

CCNU, completely removes the ability to depress NAD concentrations. This correlates well with the evaluation of CCNU in HeLa cell chromatin; it failed to demonstrate a stimulatory effect on poly (ADP-ribose) polymerase (Sudhakar *et al.*, 1979b). The two major antitumor actions of streptozotocin are thus related to direct cytotoxicity from alkylation of DNA and the lowering of cellular NAD by the stimulation of poly (ADP-ribose) polymerase.

C. Diabetogenic Effect of Streptozotocin

The diabetogenic effect of streptozotocin was found during preclinical toxicity testing in many animal models. This was likened to the alloxan effect, with both having selective beta cell toxicity leading to degranulation and beta cell lysis. Both alloxan- and streptozotocin-induced diabetes can be prevented by nicotinamide, although alloxan requires pretreatment whereas the streptozotocin effects can be successfully avoided by nicotinamide given up to 60 min after treatment (Ho and Hashim, 1972). Alloxan diabetes may be inhibited by a wide range of chemicals such as nicotinamide, nicotinic acid, glutathione, and cysteine, which probably work by interacting directly with the alloxan molecule causing the inactivation in blood. In contrast to streptozotocin, alloxan does not lower pancreatic islet cell or hepatic NAD concentrations (Anderson *et al.*, 1974). It was subsequently found that the cytotoxic part of the streptozotocin molecule, MNU, was diabetogenic when administered at high doses. The differences seen in the doses required for diabetogenicity by streptozotocin and MNU have been related to their relative *in vivo* uptake in islet cells. Streptozotocin had a 3.8-fold increased concentration in islet cells compared to MNU, and the molar dose of MNU for diabetogenicity is 3.5 times the streptozotocin dose required to produce equivalent beta cell destruction and reduction in islet NAD.

Following the administration of streptozotocin, there is a triphasic response of blood glucose concentrations. The initial hyperglycemia at 3 hr post injection is accompanied by an increase in plasma free fatty acids and decreased insulin levels (Schein *et al.*, 1971). This is followed by severe hypoglycemia at 7–12 hr, corresponding to the degranulation and disruption of beta cells; then the persistent hyperglycemic state begins at 1–2 days after treatment. This diabetic state is permanent with alloxan and streptozotocin at high dose (150–200 mg/kg in mice, but may be reversible 12–14 months following low-dose streptozotocin (50mg/kg) in rats (Rakieten *et al.*, 1976b). With long-term follow-up of 1–2 years in monkeys and dogs, no evidence of vascular disease or ketoacidosis develops in animals demonstrating a sustained diabetic state (Schein *et al.*, 1973b).

The diabetogenic effect of streptozotocin can be prevented by administration of pharmacologic doses of nicotinamide, as mentioned previously, but there is also evidence that certain carbohydrates also can protect the islet in rats. These include 2-deoxy-D-glucose and 3-0-methyl-D-glucose. The mechanism of the protective action of 3-0-methyl-D-glucose is proposed to be mediated by an interaction with the beta cell surface because it is a nonmetabolized analog of glucose and does not stimulate insulin secretion. The protective action of 2-deoxyglucose is additive to that of 3-0-methyl-glucose and may be effected by blocking streptozotocin entry into the cell (Franza *et al.*, 1980). Also, mannoheptulose given with 3-0-methyl glucose to rats prior to streptozotocin did not negate this protective effect as it does when this combination is given prior to alloxan. There are clear differences between the protective effects of carbohydrates with alloxan and with streptozotocin diabetogenicity.

There is an interesting cross-species variation in the diabetogenic effect of streptozotocin; dogs, monkeys, hamsters, rats, and mice all show islet-cell destruction after treatment, whereas the beta cells of rabbits and guinea pigs are resistant to this effect of streptozotocin (Kushner *et al.*, 1969). The resistance of the latter two species is not due to an unusually rapid rate of inactivation of streptozotocin, but may be related to a difference in the preferential synthetic pathway of NAD in these animals. Dogs and rats tend to aminate most of their nicotinic acid to form nicotinamide, whereas rabbits and guinea pigs tend to deaminate nicotinamide to form nicotinic acid which is then utilized to form NAD. Of particular clinical importance is the relative resistance of the human beta cell to the diabetogenic effect of streptozotocin. At the doses used in clinical phase II trials in 106 patients with advanced malignancies, only one patient was noted to have sustained hyperglycemia following streptozotocin treatment (Schein *et al.*, 1974).

D. The Carcinogenic Effects of Streptozotocin

During observations of the long-term effects of streptozotocin-induced diabetes in rats, an increased incidence of renal adenomas in rats was noted (Rakieten *et al.*, 1968). It is not surprising that streptozotocin has carcinogenic activity because of the known carcinogenicity of the MNU cytotoxic groups. The tumorigenic properties depend on the N-nitrosoalkane end of the molecule and on the *in vivo* release of this group as diazomethane. Attempts to protect against tumorgenesis with nicotinamide was effective in reducing the incidence of renal adenomas, but a surprising outcome was the development of pancreatic islet cell tumors (nesidioblastomas) in a high percentage of those rats receiving the combined treatment (Smulson *et al.*, 1977). These tumors first became apparent 7–8 months into the study, but most were detected 1–1 1/2 years after treatment. Only a single pancreatic islet-cell tumor was observed in 26 animals treated with streptozotocin alone. These islet-cell tumors were insulin secreting and were first found in animals presenting with seizures from hypoglycemia. Interestingly, streptozotocin without nicotinamide was used to treat these islet cell tumors and was successful in increasing the blood glucose, reducing circulating insulin, and showing necrosis of these tumors at postmortem exam (Rakieten *et al.*, 1976b). It is proposed that islet cells take up the drug but, having been protected against destruction by nicotinamide, remain subject to methylation of DNA by the 1-methyl-1-nitrosourea moiety of the streptozotocin molecule. This transformation is expressed after a long latent period of 1 to 1½ years with the proliferation of neoplastic beta cells.

The nucleotide cyclic 3':5'-GMP has been implicated in carcinogenesis by recent investigations (Vavra *et al.,* 1960). The enzyme guanylate cyclase catalyzes the conversion of GTP to cyclic 3':5'-GMP and has been shown to have marked *in vitro* activation by both streptozotocin and MNU (Hadden *et al.,* 1972) in the rat liver, heart, kidney, brain, lung, and pancreas.

In studies of mutagenicity using the Ames assay (Franza *et al.,* 1980). streptozotocin was much more potent than any other nitrosourea in actual clinical use. Of particular interest was the markedly increased mutagenic activity compared to the MNU cytotoxic group, which possesses greater intrinsic alkylating activity.

E. Clinical Trials with Streptozotocin

1. Islet Cell Carcinoma

Murray-Lyon *et al.* (1968) reported the successful treatment of a patient with a metastatic islet cell carcinoma that was secreting insulin, gastrin, and glucagon. By 1973, the Drug Evaluation Branch of the National Cancer Institute had on record a total of 52 patients with islet cell carcinoma treated with streptozotocin (Broder and Carter, 1973). Measurable disease responses were seen in 14 of 38 patients (37%) with functioning tumors. Of the responding cases, all patients with hypoglycemia had amelioration of hypoglycemic symptoms with fasting insulin levels returning toward normal; the median time to onset of response was 17 days, and median time to maximum response was 35 days. In 8 patients with nonfunctioning islet cell tumors, 5 had a measurable tumor response. A significant increase in 1-year survival rate and a doubling of median survival were shown for responders as compared with the nonresponders: 1268 days versus 518 days respectively. Complete and partial responders did equally well when 1-year survival rates were measured. Interestingly, there was also evidence that biochemical responders without measurable tumor reduction had a survival advantage over nonresponding patients. Our personal experience has been consistent with this result. Of 12 patients with islet cell carcinoma, 6 evidenced an objective response including 3 patients who achieved complete remissions; these included 1 patient with malignant insulinoma (Schein *et al.,* 1974) and 2 patients with the pancreatic cholera syndrome (Kahn *et al.,* 1975). Nephrotoxicity has been the most common treatment-limiting drug effect, expressed as proteinuria, decreased creatinine clearance, or proximal tubular injury manifested by renal glycosuria, aminoaciduria, phosphaturia, and renal tubular acidosis. Renal toxicity usually reversed within 4 weeks after cessation of drug treatment. Hepatotoxic reactions occur in the form of transient elevations in serum transaminase levels. Progressive leukopenia and thrombocytopenia was noted in a patient treated with a single dose of 7.5gm/m^2. This was the first reported example of severe bone marrow suppression reported with streptozotocin.

In 1975 the Mayo Clinic reported results in a small series of patients with meta-static islet cell carcinoma who were treated with streptozotocin alone or in com-bination with 5-fluorouracil (5-FU) (Moertel, 1975). In this nonrandomized comparison, six objective responses were seen in eight patients treated with 5-FU and streptozotocin compared to three of six patients treated with streptozotocin alone. These promising early results led to a prospective randomized trial by the Eastern Cooperative Oncology Group which has evaluated 84 patients with meta-static islet cell carcinoma treated with either streptozotocin alone or the regimen of 5-FU and streptozotocin (Moertel *et al.*, 1980). The combination showed a significant therapeutic advantage over streptozotocin alone in overall response rate, 63% versus 36%, and in complete response rate, 33% versus 12%. The median duration of all responses was 17 months, and 24 months for complete responses. Response rates were comparable in both functioning and non-function-ing tumors and there was no evidence of a preferentially favorable or unfavorable response among specific functional tumor variants. Survival for patients treated in the two groups was not statistically different with a projected median survival of 25.5 months for the combined therapy group and 16.5 months for the patients receiving streptozotocin alone. Only 25% of tumors remained in remission for periods lasting more than 12 months. Toxicity was primarily gastrointestinal, renal and hepatic in nature. One third of the patients evidenced renal toxicity, but this necessitated discontinuation of streptozotocin in only three patients. One patient developed acute liver toxicity immediately following therapy and died in hepatic coma, and one patient died from profound leukopenia and sepsis.

2. Malignant Carcinoid Tumors

There is considerable histologic and functional similarity between malignant carcinoid tumors and malignant insulinomas, both are presumed to be of neural crest origin and have been grouped with the APUD (amine precursor uptake and decarboxylation) tumors. The majority of early reports of streptozotocin treat-ment of carcinoid tumors in the literature have involved patients with very late stages of this disease characterized by massive hepatomegaly. In our initial trial (Schein *et al.*, 1973a), no response was observed in eight patients with advanced tumor originating from the ileum. In a phase II study with streptozotocin at the Mayo Clinic (Moertel, 1975), objective responses were found in three of six patients treated with streptozotocin and six of nine patients treated with the com-bination 5-FU and streptozotocin. Because of the usually indolent course of malig-nant carcinoid, these patients were treated only after there was evidence of far advanced symptomatic disease. The findings of a very high (150 mg/24 hr) urinary excretion of 5-hydroindoleacetic acid (5-HIAA) or early signs of carcinoid heart disease are poor prognostic indicators and were used to justify initiation of chemo-therapy. Considerable care was taken to avoid precipitation of an acute carcinoid crisis by rapid lysis of large bulk tumor cells by chemotherapy. These crises are characterized by a constant intense flush, rapidly rising urinary levels of 5-HIAA, and then coma. In patients with severe carcinoid syndromes, a 50% reduction of the usual therapeutic dose of chemotherapy may be required to avoid this crisis. The Western Cooperative Cancer Chemotherapy Group (Stolinsky *et al.*, 1972) reported that one of four patients with malignant carcinoid experienced an objective response to streptozotocin, with decreased liver size and reduction of the symptoms of the carcinoid syndrome.

The first prospectively randomized multiinstitutional study was performed by the Eastern Cooperative Oncology Group (ECOG), a program involving 89 evaluable patients with advanced malignant carcinoid (Moertel and Hanley, 1979). The patients received streptozotocin combined with either cyclophosphamide or 5-FU. Tumor measurement showed objective response in 8 of 38 patients (21%) treated with streptozotocin plus 5-FU, whereas 10 of 42 patients (24%) treated with streptozotocin and cyclophosphamide responded. If reduction in 24-hr urinary excretion of 5HIAA by 50% was also considered an objective parameter of response, 14 of 42 patients (33%) responded to streptozotocin and 5-FU compared to 12 of 47 patients (26%) with streptozotocin and cyclophosphamide. In 65% of patients there was concordant response of tumor size and 5-HIAA levels, and no patient had completely discordant responses of these two parameters with treatment. The median duration of response was 7 months. The objective response rate was related to the primary site of the carcinoid tumor, those tumors originating in the small intestine showed a significantly greater response rate (37%) than primaries in the lung (12%) or an unknown primary site (17%). The median corrected survival for the cyclophosphamide combination was 12.5 months compared to 11.2 months for the 5-FU combination.

3. Adenocarcinoma of the Pancreas

Several limited phase II studies of streptozotocin in patients with advanced adenocarcinoma of the pancreas demonstrated a relatively high number of objective responses (Moertel, 1975; Stolinsky *et al.*, 1972; Awrich *et al.*, 1979). The Eastern Oncology Cooperative Group treated 93 evaluable patients with advanced adenocarcinoma of the pancreas with either streptozotocin plus 5-FU or streptozotocin plus cyclophosphamide (Moertal *et al.*, 1977). Patients were stratified according to the site of primary pancreatic lesion, grade of anaplasia, and performance status. Both regimens were found to produce similar poor response rates, with 6 of 51 (12%) patients responding in the cyclophosphamide combination and 5 of 42 patients (12%) responding to the 5-FU combination. The median duration of objective response measured from onset of therapy was 5.5 months and 9 months respectively. Nine of 11 responses were complete, and 6 of the 11 responses were maintained for 6 months or longer. However, the toxicity of both combinations was significant, with severe nausea and vomiting in the majority of patients. Hematologic toxicity (mean WBC 4000 or platelets 150,000) was evidenced by 68% of patients treated with the 5-FU combination and 90% of patients treated with the cyclophosphamide regimen.

Twenty-three previously untreated patients with advanced adenocarcinoma of the pancreas (Vesely *et al.*, 1977) were evaluated. The combination chemotherapy regimen included streptozotocin, mitomycin C, and 5-FU (SMF) given in repeated doses in a 9-week cycle. All patients had objectively measurable tumor for evaluation of response. Ten of 23 patients (43%) achieved an objective response with a median duration of response in excess of 7 months. The patients who showed an objective response had a median survival of 10 months as compared to 3 months in nonresponders with one 5-year disease-free survival. Gastrointestinal toxicity was noted in all patients but required withdrawal from treatment in only one patient. Hematologic toxicity was moderate in degree with WBC median nadir for all cases of 2.9×10^3 cells/mm^3 (range 0.7–6.3×10^3 cells/mm^3) and median platelet count nadir of 73×10^3/mm^3 (range 20–207×10^3). There were three episodes of sepsis and two thrombocytopenic GI hemorrhages. Nephrotoxicity, evidenced by proteinuria, was produced in 30% of patients; this was attributable to streptozotocin, although mitomycin has been reported to cause glomerular damage at higher doses (20–40mg/m^2 iv) than used in this regimen. Two of 23 patients (9%) developed renal impairment that stabilized at the reduced level of function after streptozotocin alone was omitted from the regimen. Similar therapeutic results have been reported by Abderhalden *et al.* (1977) with 5-FU and streptozotocin combination compared to 5-FU, streptozotocin, and mitomycin C. The response rate was slightly higher with the three-drug regimen, 5 of 16 patients (31%), than for 5-FU and streptozotocin alone, 4 of 10 patients (21%).

4. Malignant Lymphomas

In a phase II study (Schein *et al.*, 1974), sixteen patients with advanced stages of Hodgkin's disease received streptozotocin at a time when the disease was refractory to more conventional chemotherapy and when further myelosuppression therapy was prohibited by severe bone marrow hypoplasia caused by extensive prior therapy. Seven patients (44%) demonstrated objective response, and in one case there was complete disappearance of disease. In four patients response was documented after demonstrated therapeutic resistance to the chloroethylnitrosourea BCNU; a partial response to CCNU was observed in one instance after initial partial remission and subsequent failure with streptozotocin. However, the response durations were short, with a median of 2 months. In two patients, white blood cell counts returned to normal while on treatment, allowing placement on more conventional myelosuppressive therapy while in partial remission.

Three of 11 patients with poorly differentiated lymphocytic lymphoma demonstrated objective remission lasting 1 to 4 months. Very transient or no objective response were found in five cases of cyclophosphamide-resistant Burkitts tumor and three patients with diffuse histiocytic lymphoma.

Because of these promising results in this phase II trial, a protocol for patients with advanced previously treated Hodgkins disease was initiated with streptozotocin, CCNU, adriamycin, and bleomycin (SCAB) (Levi *et al.*, 1977). The overall response rate was 10 of 17 patients (59%) with 6 of 17 patients (35%) obtaining a complete remission. The median duration of unmaintained complete response was greater than 8 months. The median duration of remission for partial responders was only 2 months. The toxicity of this combination of drugs was significant. Myelosuppression was severe after 25% of the courses, and five patients had documented bacterial infections during the period of leukopenia. Renal tubular dysfunction occurred in 18% of patients, and bleomycin lung toxicity was noted in three patients. Because of these toxicities, the SCAB regimen could not be generally recommended for use in previously treated Hodgkins disease patients.

5. Colorectal Adenorcarcinoma

In phase II trials, streptozotocin has shown limited activity as a single agent in colorectal adenocarcinoma with a response rate of 15% in previously treated patients (Horton *et al.*, 1975). When combined with methyl CCNU, 5-FU, and vincristine (the MOF–strep regimen), Kemeny *et al.* have found a 32% response rate in 24 of 74 patients with measurable metastatic colorectal carcinoma (Kemeny *et al.*, 1980). The median duration of remission was 8 months. The median white blood cell (WBC) nadir was $2.6 \times 10^3/mm^3$, with four patients (5%) having WBC counts below $2.0 \times 10^3/mm^3$, and the median platelet nadir was $63 \times 10^3/mm^3$. The MOF-strep protocol produced minimal nephrotoxicity, whereas nausea and vomiting were present in essentially all cases.

III. CONCLUSION

Streptozotocin has been shown to be an important antitumor agent for islet cell carcinomas and malignant carcinoids. The role of streptozotocin in the management of adenocarcinoma of the pancreas and the colon, as well as in Hodgkin's disease, is now being actively evaluated. Nephrotoxicity and gastrointestinal toxicity have been the principal treatment-limiting drug effects but can be controlled by careful clinical evaluation during therapy and dose modification when necessary. The primary role of streptozotocin in the future lies in its inclusion as a nonmyelosuppressive component to regimens of combination chemotherapy.

REFERENCES

Abderhalden, R., Bukowski, R., Groppe, C., Hewlett, J. and Weich, J. (1977). *Proc. Amer. Soc. Clin. Oncol. 18*, 301.

Anderson, T., Schein, P. S., McMenarmin, M., and Cooney, D. (1974). *J. Clin. Invest. 54*, 672–677.

Awrich, A., Fletcher, W., Klotz, J., Minton, J., Hill, G., Aust, J., Grage, T., and Multhauf, P. (1979). *J. Surg. Oncol. 12*, 267–273.

Broder, L., and Carter, S. (1973). *Ann. Int. Med. 79*, 109–118.

Colvin, M., Brundhett, R. B., Cowens, W., Jardin, I., and Ludlum, D. B. (1976). *Biochem. Pharmacol.* 25, 695–699.

Franza, B. R., Oeschger, N., Oeschger, M., and Schein, P. S. (1980). *JNCI* 65, 149–154.

Hadden, J., Hadden, E., Haddox, M., and Goldberg, M. (1972). *Proc. Nat. Acad. Sci.* 69, 3024–3027.

Heal, J. M., Fox, P. A., and Schein, P. S. (1979). *Cancer Res.* 39, 82–89.

Herr, R. R., Johnke, H., and Argondelis, A. (1967). *J. Amer. Chem. Soc.* 89, 4808–4809.

Ho, C. K., and Hashim, S. A. (1972). *Diabetes 21*, 789–793.

Horton, J., Mittleman, A., and Taylor, S. (1975). *Cancer Chem. Rep.* 59, 330–340.

Kahn, C. R., Levy, A., Gardner, J., Miller, J. V., Gorden, P., and Schein, P. S. (1975). *New Eng. J. Med.* 292, 941–945.

Kemeny, N., Yagoda, A., Braun, D., and Golbey, R. (1980). *Cancer 45*, 876–881.

Kushner, B. S., Lazar, M., Furman, M., Lieberman, T. W., and Leopold, I. H. (1969). *Diabetes 18*, 542–544.

Levi, J., Wiernik, P., and Diggs, C. (1977). *Med. Ped. Oncol. 3*, 33–40.

Lewis, C., and Barbiers, A. R. (1960). *In* "Antibiotics Annual 1959–1960," p. 247. Antibiotica, New York.

Loranger, R. A. (1980). PhD. Thesis, Georgetown University.

Moertel, C. (1975). *Cancer 36*, 675–682.

Moertel, C., Hanley, J., and Johnson, L. (1980). Submitted for publication.

Moertel, C., and Hanley, J. (1979). *Cancer Clin. Trials 2*, 327–334.

Moertel, C., Douglas, H., Hanley, P., and Carbone, P. (1977). *Cancer 40*, 605–608.

Murray-Lyon, I. M., Eddleston, A., Williams, R., Brown, M., Hogbur, B., Bennett, A., Edwards, J., and Taylor, K. (1968). *Lancet 2*, 895–898.

Panasci, L. C., Fox, P. A., and Schein, P. S. (1977). *Cancer Res. 37*, 3321–3328.

Rakieten, N., Gordon, B., Cooney, D., Davis, R., and Schein, P. (1968): *Ca. Chem. Rep. 52:* 563–567.

Rakieten, N., Gordon, B., Beaty, A., Cooney, D., and Schein, P. (1976a). *Proc. Soc. Exp. Biol. Med. 151*, 356–361.

Rakieten, N., Gordon, B., Beaty, A., Bates, R., and Schein, P. (1976b). *Proc. Soc. Exp. Biol. Med. 151*, 632–635.

Schein, P., Kahn, R., Gordon, P., Wells, S., and DeVita, V. (1973a). *Arch. Int. Med. 132*, 555–561.

Schein, P. S., Cooney, D. A., and Vernon, M. L. (1967). *Cancer Res. 27*, 2324–2332.

Schein, P. S., Rakieten, N., Cooney, D. A., Davis, R., and Vernon, M. L. (1973b). *Proc. Soc. Exp. Biol. Med. 143*, 514–518.

Schein, P. S., and Loftus, S. (1968). *Cancer Res. 28*, 1501–1506.

Schein, P. S., Alberti, K. G., and Williamson, D. H. (1971). *Endo. 89*, 827–834.

Schein, P. S., Rakieten, N., Cooney, D., Davis, R., and Vernon, M. L. (1973c). *Proc. Soc. Exp. Biol. Med. 143*, 514–518.

Schein, P. S., O'Connell, M., Blom, J., Hubbard, S., Magrath, I., Bergevin, P., Wiernik, P., Ziegler, J., and DeVita, V. (1974). *Cancer 34*, 993–1000.

Smulson, M., Stark, P., Gazzoli, M., and Roberts, J. (1975). *Exp. Cell Res. 90*, 175–182.

Smulson, M., E., Schein, P., Mullins, D. W. Jr., and Sudhakar, S. (1977). *Cancer Res. 37*, 3006–3012.

Stolinsky, D., Sadoff, L., Braunwald, J., and Bateman, J. (1972). *Cancer 30*, 61–67.

Sudhakar, S., Tew, K. D., and Smulson, M. E. (1979a). *Cancer Res. 39*, 1405–1410.

Sudhakar, S., Tes, K., Schein, P., Woolley, P., Smulson, M. (1979b). *Cancer Res. 39*, 1411–1417.

Vavra, J. J., DeBoer, C., Dietz, A., Hanka, L. J., and Sokolski, W. T. (1960). *In* "Antibiotics Annual 1959–1960," p. 230. Antibiotica, New York.

Vesely, D., Rovere, L., and Levey, G. (1977). *Cancer Res. 37*, 28–31.

Wiggins, R. G., Woolley, P., Macdonald, J., Smythe, T., Ueno, W., and Schein, P. S. (1978). *Cancer 41*, 387–391.

Chapter 32

PHASE I AND II STUDIES OF PCNU

Michael A. Friedman

I. INTRODUCTION

The nitrosoureas, a burgeoning family of antitumor agents, have been the object of intensive laboratory and clinical investigation for more than a decade. In man, these drugs have confirmed activity against brain, lung, and enteric cancers, as well as against melanoma, myeloma, and Hodgkin's disease. Nevertheless, the efficacy of these agents has been modest at best, and considerable efforts have been expended in the hope of developing a more potent analog—one with a wider clinical spectrum or a greater therapeutic index, or both (Wasserman *et al.,* 1975). Currently a dozen or more nitrosoureas are being tested in clinical trials worldwide.

Each new nitrosourea has been selected for clinical trial on the basis of its activity in animal tumor screening experiments or its pharmacologic or physical–chemical properties. Leukemia L1210 has been the usual initial tumor screen employed; it was the activity against intracerebrally (ic) inoculated L1210 that presaged similar activity in human brain tumors. In addition to intraperitoneal (ip) and ic L1210, more resistant rodent tumors (e.g., Lewis lung, B16 melanoma) have been used in this expanded screening process. It has been commonly held that a drug's effectiveness in animal tumors would correlate positively with that in human tumors. This "rational" process helped lead to the selection of CCNU, and then methyl CCNU for clinical trials.

Pharmacologic and physical–chemical features of the nitrosoureas have likewise been studied in attempts to identify those properties that confer clinical utility and efficacy. The three principal nitrosoureas in current clinical use, BCNU, CCNU, and Methyl CCNU, were so studied. They are all relatively lipophilic, a characteristic thought to relate to both the (ic) antitumor effectiveness in animals and the central nervous system (CNS) antitumor effectiveness in tumors in man. Initially, these drugs were selected to possess increasing lipophilicity in the hope of improving the delivery of drug to the CNS sanctuary.

Although it has been recognized that the nitrosoureas possess both alkylating and carbomoylating abilities, the appreciation of these functions has only more recently been employed in analog selection. Wheeler *et al.* (1974) suggested that anti-L1210 tumor activity correlated well with alkylating activity and that effectiveness increased as lipid solubility decreased. Carbamoylating activity, on the other hand, does not clearly relate to antitumor effectiveness. Because of the likely relationship between form and function, some nitrosoureas have been selected for trial both for their effectiveness in animal tumor screens and their relative alkylating/carbamoylating activity.

In spite of the efforts of "rational" drug selection, our understanding is flawed and our success limited. Not only have we not been able to design a truly superior nitrosourea, but there is little to recommend one analog over another (Wasserman *et al.*, 1975).

The subject of this chapter is the phase I and II evaluation of 1-(2-chloroethyl)-3-(2,6-dioxo-1-piperidyl)-1-nitrosourea (PCNU). Interest in this compound was generated by workers at the Brain Tumor Research Center at the University of California, San Francisco. A nationwide effort in Phase I-II testing of PCNU is ongoing, and sufficient data are now available to allow an interim report.

II. BACKGROUND

PCNU is a N-(2-chloroethyl)-N-nitrosourea similar in structure to the other clinically useful nitrosoureas, BCNU, CCNU, and methyl CCNU. Because there are many nitrosoureas currently available, it is necessary to formulate persuasive

TABLE I. Antitumor Activity Relative to Alkylating and Carbamoylating Activity

Drug	Antitumor activity	Alkylating activity	Carbamoylating activity
PCNU	100[a]	100	100
BCNU	48	75	285
CCNU	21	28	417
Methyl CCNU	15	28	379

[a] An Activity of 100 was arbitrarily assigned to PCNU for comparisons with the other drugs in this study.

TABLE II. Nitrosourea Partition Coefficients

Drug	Log P
PCNU	0.37
BCNU	1.53
CCNU	2.83
Methyl CCNU	3.30

reasons for selecting a new nitrosourea to test clinically. Briefly, the rationale for studying PCNU is as follows:

1. The National Cancer Institute tested PCNU against a battery of ic, ip, and im tumors (including L1210, 9L rat sarcoma, murine ependimoblastoma, P388 lymphoid leukemia, colon 26, B16 melanocarcinoma, CD8F mammary, and colon 38 tumors). PCNU was generally superior to the other nitrosoureas tested.
2. The *in vitro* alkylating activity of PCNU is higher and the carbamoylating activity lower than that of several other nitrosoureas (Levin and Kabra, 1974). These data are summarized in Table I.
3. Finally, Levin and Kabra (1974) attempted to describe mathematically the chemical characteristics of the nitrosourea most effective for CNS tumors. They found that the log P value's (octanol/water partition coefficient) relation to antitumor effectiveness was a parabolic function. They predicted that the ideal log P would be between -0.20 and 1.34. Comparison of some nitrosoureas is shown in Table II.

Levin and Kabra (1974) hypothesized that to be effective against solid tumors in the brain, the optimal log P would be 0.4—a value close to that for PCNU (0.37). Therefore, there seemed to be persuasive reasons to study PCNU in the clinical setting.

III. PHASE I TRIALS OF PCNU

Based on animal toxicology data from the National Cancer Institute, it was predicted that PCNU would have human organ system damage qualitatively similar to the other chloroethylnitrosoureas. In animals the principal toxicity was hematologic; less frequent liver and kidney dysfunction was also noted. Moreover, since much is now known about the toxicity of nitrosoureas, phase I testing could be initiated with a confidence uncommon in clinical oncology.

There have been five concurrent phase I trials of PCNU in the United States. All have employed intermittent schedules. The institutions testing the drug and the administration schedules included M.D. Anderson Hospital (d 1 q 6 weeks), Memorial Sloan-Kettering (d 1+2+3+4+5 q 6 weeks), University of Wisconsin (d 1+2+3+4+5 q 6 weeks), Georgetown University (d 1+2+3 q 6 weeks), and Northern California Oncology Group (d 1+2 q 6 weeks).

The initial dosage for these studies was 10–15 mg/m^2, and the dose was escalated to 100–125 mg/m^2 iv every 6 weeks. Table III describes the number of patients studied at each institution and the final recommended dosages for patients previously receiving chemotherapy and those previously untreated.

The phase I study performed at the M.D. Anderson Hospital by Stewart and co-workers (1980) is the largest of these trials reported to date. In this study, 73 patients received a total of 150 courses. For those patients who had previously received chemotherapy, a single dose of 60 mg/m^2 every 6–8 weeks was tolerable. Thrombocytopenia was detected in 75% of these patients, with a median platelet nadir of 49,000 cells/mm^3. By comparison, patients who had not received previous chemotherapy tolerated a dose of 110 mg/m^2 with a median platelet nadir of 74,000 cells/mm^3. For both groups the platelet nadir occurred on day 28 (median), and usually 1–2 weeks was required for recovery to normal levels. These investigations suggested that the bone marrow toxicity was probably cumulative and certainly dose related. Other toxicities described included granulocytopenia, which was uncommon, dose related, and occasionally delayed; nausea (12%), vomiting (17%), mild hypotension (42%), malaise (5%), stomatitis (3%), minimal transaminase elevation (2%), diarrhea (2%), rash (2%), and myalgias (1%).

The phase I study from Memorial Sloan-Kettering Cancer Center was reported by Gralla et al. (1980). Twenty-eight patients received doses ranging from 25–150 mg/m^2 administered as a divided daily dose for 5 days every 6 weeks. These investigators also noted that thrombocytopenia was more frequent than leukopenia and that the median platelet nadir day was 32 with recovery by day 42–50. They also noted that at a higher dose, leukopenia was observed with some regularity.

The phase I study conducted at the University of Wisconsin (Earhart et al., 1980) described 24 patients treated with doses escalating from 25–125 mg/m^2. This total dose was divided and administered daily × 5 every 6–8 weeks. At the higher doses the median platelet nadir was 64,000 cells/mm^3 and usually oc-

TABLE III. Phase I Trials of BCNU

Institution	No. of patients	Recommended dosage (mg/m^2)	
		No previous therapy	Previous therapy[a]
M. D. Anderson	73	110	60
Memorial Sloan-Kettering		100–125	100–125
Wisconsin	24	125	75
Georgetown	19	90–100	75
NCOG	69	100	80

[a] Chemotherapy or radiation that would damage the bone marrow.

curred on day 30 with a recovery time of approximately 2 weeks. Leukopenia was less frequently encountered; when it did occur the median count was 2900 and the nadir was day 44. Other toxicities described included nausea and vomiting in one-third of patients and anorexia in one-half. No important renal or hepatic toxicity was observed.

Georgetown University (Woolley *et al.*, 1980) reported a study of 19 patients who received escalating doses of 30–105 mg/m^2 iv, divided over 3 days every 6–8 weeks. They also detected thrombocytopenia at their higher doses occurring on day 22–30. Leukopenia was less frequent and occurred approximately 1 week to 10 days later. Again, no notable hepatic or renal toxicity was noted.

The Northern California Oncology Group (NCOG) study was initiated in December of 1978 and to date includes 69 patients. The median age of these patients was 54, with a range of 18–79 years, and there were 36 male and 33 female patients. The histological types of primary tumors included a wide variety of adenocarcinomas, sarcomas, and squamous cell carcinomas.

In Table IV, the Karnofsky performance status of the treated patients is presented. Of the 69 patients registered on this study, complete data are available for 59. There were 12 early deaths and 3 patients refused further therapy, allowing 44 patients to be fully evaluated.

The initial dosage was 12 mg/m^2 administered on two consecutive days, and at 75–80 mg/m^2 reproducible toxicity was noted. When it became clear that there was minimal nausea and vomiting and that the divided dosage schedule was cumbersome, the chemotherapy was confined to a 1-day treatment every 6 weeks. A total of 21 patients received this dose, and only 5 suffered any notable hematologic toxicity. Ten patients had white blood cell (WBC) counts of less than 4400 cells/mm^3, but only 2 patients had WBC between 1900 and 1000 cells/mm^3. Thirteen patients had some thrombocytopenia (less than 129,000 cells/mm^3). Nine patients had platelet counts of $< 25,000$/mm^3. There was, however, no bleeding or infection noted in any of these patients. At a higher dosage (100 mg/m^2) 37 patients were treated, 10 of whom had hematologic toxicity. There were 19 patients with leukopenia (< 4400 cells/mm^3), but only 1 patient had WBC of less than 1000. There were 26 patients with throm-

TABLE IV. Karnofsky Performance Status of
Patients Entered on the NCOG Study

Performance (%)	No. of patients
50	7
60	15
70	17
80	12
90	14
100	4

bocytopenia ($<$ 129,000), but there were 5 patients with platelet counts of less than 25,000. There was one episode of infection, but no hemorrhage. Twenty-seven of 58 patients had mild to moderate nausea and vomiting. There was no hepatic, kidney, or neurologic toxicity noted.

IV. CLINICAL RESPONSES

Although these studies have focused on toxicity, some clinical information has been obtained. Most of this information comes from the M.D. Anderson Hospital study (Stewart *et al.*, 1980) where a total of 52 patients could be evaluated for objective response. Ten patients (19%) evidenced such a response. Responses were noted in 7/14 patients with primary central nervous system tumors, 1/15 with melanoma, 1/4 with colon cancer, and 1/1 with Hodgkin's disease. Unfortunately, no responses were seen in 7 patients with breast cancer, 3 with lung cancer, or 2 with pancreatic cancer. In the NCOG study there were only 2 patients who had an objective response; 1 patient with colon cancer metastatic to liver and 1 patient with large-cell carcinoma of the lung. Both patients responded at doses of 100 mg/m^2 every 6 weeks, and the duration of response was 216 days and 119+ days.

V. PHARMACOLOGICAL INFORMATION

As part of the phase I evaluation of PCNU, an attempt at defining its distribution and metabolism was made. Some data are available from the Memorial Sloan-Kettering and Georgetown University studies. Unfortunately, both of these institutions used ^{14}C-labeled PCNU instead of assaying the parent compound directly. The inherent difficulties of pharmacologic studies with labeled material are known. Nevertheless, given this limitation, the Memorial Sloan Kettering study indicated that there was a terminal half-life of 35.9 hr for ^{14}C-labeled PCNU. Fifty percent of this material was recovered in the urine at 24 hr and 65% at 60 hr. Further, the peak cerebral spinal fluid level occurred at 60 min (a level that was 45% of the concurrent serum value). Three patients were studied at Georgetown University using ^{14}C-labeled PCNU. Patients receiving doses of 10–30 mg/m^2 had peak plasma levels of 6.68 to 8.95 micromolar. The initial half-life was 15 min and the terminal half-life 10 hr. The percentage of PCNU bound to plasma protein was 30–35, and 29% of the plasma level was found in the CNS at 1.5 hr and 49% of the plasma level at 2.5 hr. Direct measurement of serum levels for PCNU has been carried out by Dr. Levin of the University of California, San Francisco (see Levin, Chapter 13, this volume).

VI. CONCLUSION

There have been more than 200 patients treated with PCNU during these phase I and II evaluations, and some conclusions seem warranted:

1. The safe dose for patients who have had previous myelosuppressive therapy is 60–80 mg/m^2. For those patients not previously exposed to hematotoxic therapy, the dose is 90–125 mg/m^2 every 6 weeks.
2. The toxicity of this drug is qualitatively and quantitatively similar to other chloroethylnitrosoureas; specifically, dose-limiting thrombocytopenia occurs at approximately 30 days after the administered dose and requires 1–2 weeks for recovery. Leukopenia is somewhat less frequent and less severe and occurs 1 week to 10 days later. There does seem to be cumulative hematologic toxicity.
3. The less common nitrosourea toxicities such as pulmonary fibrosis or renal parenchymal loss have not yet been detected. Patients in these studies have not received treatment for a sufficiently long period of time at high enough doses to enable detection of these toxicities. Further studies are necessary.
4. Although the phase II data are preliminary at this time, there does seem to be a scattering of responses in precisely those tumors thought to be nitrosourea-sensitive (e.g., central nervous system, melanoma, gastrointestinal, and lung). The data are preliminary; much further evaluation is necessary to determine definitely the efficacy of PCNU.
5. Despite arguments based on physical, chemical, and *in vivo* animal tumor testing, PCNU does not yet appear to be a dramatically superior drug to other related nitrosourea compounds.

In preclinical testing, rodents could tolerate more moles of PCNU than BCNU, which resulted in better antitumor activity for the former drug. In man, however, the opposite occurs. PCNU is relatively more toxic than BCNU, and therefore only about one-half as much can be administered. The reasons for PCNU's greater hematotoxicity in man are not known.

A more complete phase II evaluation of PCNU is needed. However, unless more activity or broader clinical efficacy is observed, it is unlikely that this drug will replace the currently available nitrosoureas.

REFERENCES

Earhart, R. H., Koeller, J. M., and Davis, H. L. (1980). *Amer. Soc. Clin. Oncol. CO148*, 356.

Gralla, R. J., Young, C. W., Tan, C. T., and Sykes, M. P. (1980). *Amer. Assoc. Cancer Res. 743*, 185.

Levin, V. A., and Kabra, P. (1974). *Cancer Chemother. Rep. 1*, 787–792.

Stewart, D. J., Benjamin, R. S., Leavens, M., Valdivieso, M., Burgess, M. A., McKelvey, E., and Bodey, G. P. (1980). *Amer. Assoc. Cancer. Res. 674*, 168.

Wasserman, T. H., Slavik, M., and Carter, S. K. (1975). *Cancer 36*, 1258–1268.

Wheeler, G. P., Bowdon, B. J., and Grimsley, J. A. (1974). *Cancer Res. 34*, 194–200.

Woolley, P., Luc, V., Smythe, T., Rahman, A., Hoth, D., Smith, F., and Schein, P. (1980. *Amer. Soc. Clin. Oncol. C-72*, 336.

Chapter 33

CHLOROZOTOCIN: CLINICAL TRIALS

Daniel F. Hoth
Luz Duque-Hammershaimb

I. INTRODUCTION

Although the chloroethylnitrosoureas BCNU, CCNU, and methyl CCNU have established activity in a spectrum of human malignancies, these compounds produce severe, delayed, and cumulative bone marrow toxicity that seriously limits their clinical usefulness. Structure–activity studies with the nonmyelosuppressive methylnitrosourea streptozotocin have suggested that bone marrow toxicity could be reduced by attaching the cytotoxic group to the carbon-2 position of glucose (Wheeler *et al.*, 1974). Chlorozotocin is a newly synthesized chloroethylnitrosourea designed to modify the dose-limiting myelotoxicity while retaining antitumor activity. Its structure differs from BCNU by the substitution of a glucose carrier for the chloroethylnitrosourea cytotoxic group.

In vivo studies in animal tumor systems demonstrated that chlorozotocin was active against L1210, P388, and B16 melanocarcinoma. Schedule dependency studies of the drug show it is inactive when administered po or when used against intracranially implanted L1210 (Helman *et al.*, 1976).

During preclinical investigations, the most important feature distinguishing chlorozotocin from other chloroethylnitrosoureas was its maximum activity in the L1210 leukemia model at a dose that produced no decrease in the peripheral leukocyte count. This stood in contrast to BCNU, which at a therapeutically equivalent maximum antitumor dose produced a significant reduction in circulat-

ISBN 0-12-565060-4

ing leukocytes (Anderson *et al.*, 1975). This reduced myelosuppressive activity was further confirmed by *in vitro* cytotoxicity studies employing bone marrow cells committed to granulocyte–macrophage differentiation (CFUC). The *in vitro* exposure of bone marrow cells to BCNU at a concentration of 1×10^{-4} M for 2 hr eliminated all colony formation, whereas an equimolar concentration of chlorozotocin resulted in only a 25% decrease in CFUC. At a BCNU concentration of 5×10^{-5} M, a 55% reduction in CFUC was observed; similar concentrations of chlorozotocin produced no decrease (Schein *et al.*, 1979). It must be emphasized that this bone marrow sparing effect is relative, since maximal nonlethal doses of chlorozotocin do cause significant reduction in leukocyte counts (Anderson *et al.*, 1975).

In the large-animal toxicity studies, the most prominent toxicity in rhesus monkeys and beagle dogs was dose-related renal tubular necrosis. Bone marrow hypoplasia (including thrombocytopenia), hepatotoxicity, and gastroenteritis were also noted, but only at the highest dose levels (Helman *et al.*, 1976).

II. PHASE I TRIALS

Five phase I trials have explored dose schedules varying from single-dose administration every 6 weeks to a daily $\times$ 5 schedule repeated every 6 weeks. The qualitative toxicities in all of the studies were similar (Anderson *et al.*, 1976; Kovach, *et al.*, 1979, Gralla *et al.* 1979a; Hoth *et al.*, 1979; Taylor *et al.*, 1980). The dose-limiting toxicity was thrombocytopenia, which appeared 3 to 5 weeks after treatment and was reversible. Leukopenia was less severe at any given dose level and was rarely dose limiting. There was no evidence of cumulative thrombocytopenia. Nausea was observed in 10–30% of patients, and vomiting in less than 10%. A mild elevation of hepatic transaminases was observed sporadically; this effect appeared 3 to 4 weeks after drug administration and was usually reversible.

In confirmation of the animal toxicology studies, renal toxicity was observed during phase I trials but was confined to mild elevations of BUN and creatinine and was not dose limiting (Kovach *et al.*, 1979). Subsequently, severe renal insufficiency was reported in a woman with bronchoalveolar carcinoma who received 13 courses of chlorozotocin over 15 months (Baker *et al.*, 1979). In addition, two cases of chronic interstitial pneumonitis were observed in chlorozotocin-treated patients (Ahlgren *et al.*, 1980).

III. CLINICAL PHARMACOLOGY

The pharmacokinetic properties of chlorozotocin in humans were studied at Georgetown University and Mayo Clinic, using a spectrophotometric method for measuring plasma concentrations of the intact N-nitroso group. Following rapid intravenous administration, three distinct half-lives of 3, 8, and 30 min were obtained. The concentration of N-nitroso intact chlorozotocin at one hr was 10% of the initial peak level (Hoth *et al.*, 1978; Kovach *et al.*, 1979).

Studies with ^{14}C-radiolabeled chlorozotocin demonstrated similar plasma pharmacokinetics. Excretion was primarily renal, as 60% of the radiolabel was recovered in the urine in 48 hr. Assay of protein binding, as determined by a filtration method, reveals that approximately 90% of the drug is protein bound. There is no significant CSF penetration since at 30 min and 3 hr the CSF concentration of radiolabel was less than 5% of the corresponding plasma level (Hoth *et al.*, 1978).

IV. PHASE II TRIALS

Thirty-one disease-oriented protocols in different tumor sites were sponsored by the National Cancer Institute following the phase I trials. Fourteen studies are completed.

A. Melanoma

Georgetown University (Hoth *et al.*, 1980) reported 1 CR and 4 PRs among 35 patients with advanced metastatic melanoma, for an overall response rate of 14% (Table I). The median duration of response was 18 weeks with a range of 12 to 42+ weeks. The study showed no correlation between prior therapy and response; 15% of patients without previous therapy responded versus 12% of those with previous therapy. The Eastern Cooperative Oncology Group (ECOG) reported 2 CRs and 1 PR out of 41 evaluable patients in a randomized phase II comparison of methyl CCNU versus chlorozotocin versus neocarzinostatin (ECOG Minutes, 1980). Neither the response rate nor the median survival of responders to chlorozotocin differed significantly from the methyl CCNU arm of the study. Both of the above studies used a dose of 120 mg/m^2 every 6 weeks.

The Southeastern Cancer Study Group (SEG) discovered a marked difference in the response to chlorozotocin exhibited by patients with and without prior therapy. Using a dose of 150 mg/m^2 every 6 weeks, one PR and one stable disease (SD) were seen in 20 previously treated patients, while three CRs, one PR, and two SDs in ten patients without prior therapy yielded a 40% overall response rate. This investigation is continuing in patients without prior therapy (Van Amburg *et al.*, 1980). The Cancer and Leukemia Group B (CALGB), at a dose of 120 mg/m^2 every 6 weeks, noted one CR and one SD among nine previously untreated patients, and one CR and one SD in 23 previously treated patients (CALGB Minutes, 1980). Three institutions (Mayo Clinic, Memorial Sloan Kettering Institute, and Wayne State University), using dose escalations up to 200 mg/m^2, found one responder in a total of 57 patients. (Ahmann *et al.*, 1980; New Drug Liaison Minutes, 1978; Talley *et al.*, 1979). The combined

TABLE I. Melanoma

Institution	Regimen	No. patients evaluated	Response CR	PR	Reference
ECOG	120 mg/m^2 q 6 wk	41	2	1	ECOG Minutes (1980)
MSKI	150–200 mg/m^2 5 d q 6 wk	21	0	0	New Drug Liaison Minutes (1978)
WSU	120–150 mg/m^2 q 6 wk	10	0	0	Talley *et al.* (1979)
Mayo	150 mg/m^2 q 6 wk	26	0	1	Ahmann *et al.* (1980)
Georgetown	120 mg/m^2 q 6 wk	35 20[a]	1 0	4 3	Hoth *et al.* (1980)
SECG	150 mg/m^2 q 6 wk	20 10[a]	0 3	1 1	Van Amburg *et al.* (1980)
CALGB	120 mg/m^2 q 6 wk	23 9[a]	1 1	0 0	CALGB Minutes (1980)
WSU	CZN 90–100 mg/m^2 q 6 wk + DTIC + ACT D				

[a] No prior therapy.

results of the previous studies in advanced melanoma show an overall response rate of 8% among 181 patients. Considering previously untreated patients, this rate was 20% of a total of 39 patients. These response rates compare favorably with those seen for BCNU (18%), CCNU (13%), and methyl CCNU (15%), and were obtained with less toxicity than was the case with the other chloroethylnitrosoureas (Costanza *et al.*, 1977; Wasserman *et al.*, 1974). This suggests that combination of chlorozotocin with other agents may be possible with minimum dose alteration. One such trial combining DTIC, actinomycin D, and chlorozotocin was recently initiated at Wayne State University.

B. Hematologic Malignancies

The activity of chlorozotocin in hematologic malignancies was studied by five institutions. In collaboration with Georgetown, investigators in Argentina noted no major responses in 20 patients of whom 11 had ALL; 8 had AML, and 3 had CLL (Woolley *et al.*, 1980).

In Hodgkin's disease, chlorozotocin yielded better results (Table II). Memorial Sloan Kettering, in their broad phase II study, treated 17 patients with refractory Hodgkins disease, 13 of whom were adequately treated. Three partial responses of 1, 5, and 7 months duration were observed (Sklaroff *et al.*, 1979).

TABLE II. Lymphomas

	Institution	Regimen	No. patients evaluated	Response CR	Response PR	Reference
Hodgkin's	MSKI	150–200 mg/m^2 q 6 wk	13	0	3	Sklaroff *et al.* (1979)
	CALGB	120 mg/m^2 q 6 wk	16	0	1	CALGB Minutes (1980)
	Georgetown	120 mg/m^2 q 6 wk	4	0	2	Woolley *et al.* (1980)
	ECOG	60–80 mg/m^2 q 6 wk		too early		
Non-Hodgkin's	CALGB	120 mg/m^2 q 6 wk	31	0	4	CALGB Minutes (1980)
	WSU	90–120 mg/m^2 q 6 wk	12	1	3	Talley *et al.* (1979)
	MSKI	30–40 mg/m^2 5 d q 6 wk	5	0	1	New Drug Liaison Minutes (1978)
	Georgetown	120 mg/m^2 q 6 wk	8	0	2	Woolley *et al.* (1980)
	ECOG	60–80 mg/m^2 q 6 wk	9	too early		

Of 16 patients with Hodgkins disease, CALGB noted 1 PR (CALGB Minutes, 1980), whereas the Georgetown study noted 2 PRs out of four patients. The total response rate for refractory Hodgkins disease is 20%.

Among patients with non-Hodgkins lymphoma, Wayne State University, CALGB, and Georgetown (at doses of 90 to 120 mg/m^2) and MSKI (at 150 to 200 mg/m^2) noted one CR and ten PRs out of a total of 56 patients for a response rate of 20% (Talley *et al.*, 1979; CALGB Minutes, 1980; Woolley *et al.*, 1980). Five of the responders were of the diffuse histiocytic subtype.

C. Sarcoma

Two studies in osteosarcoma are in progress, neither of which has produced evaluable results. The Southwest Oncology Group (SWOG) is using 120 mg/m^2, while the ECOG is randomizing between chlorozotocin and bleomycin (Table III).

In soft tissue sarcoma, one patient with fibrosarcoma had a PR in the Wayne State trial (Talley *et al.*, 1979). The SECG has an ongoing study in sarcoma. MSKI evaluated three single agents (DVA, AMSA, and chlorozotocin) in 89 adults with sarcoma. One partial response was seen in 19 patients treated with chlorozotocin (Magill *et al.*, 1980). The activity of chlorozotocin is insignificant in this disease.

TABLE III. Sarcomas

| Institution | Dose | No. patients evaluated | Response | | Reference |
			CR	PR	
Osteosarcoma					
ECOG	120–150 mg/m² q 5 wk		too early		
SWOG	120 mg/m² q 6 wk		too early		
Soft tissue					
MSKI	150–200 mg/m² 5 d q 6 wk	19	0	1	Magill *et al.* (1980)
WSU	90–120 mg/m² q 6 wk	10	0	1	Talley *et al.* (1979)
SECG	150 mg/m² q 6 wk		too early		

D. Lung

The activity of chlorozotocin in advanced non–small-cell lung cancer is summarized in Table IV. Minimal activity was seen in all histologic subtypes. Among a combined total of 101 patients with adenocarcinoma of the lung evaluated by five institutions, four PRs were seen for a response rate of 4%. One PR at 200 mg/m² every 6 weeks was obtained among 21 patients with large-cell anaplastic lung cancer, for a response rate of 5%; one PR occurred in a total of 66 patients with epidermoid cancer, for a response rate of 1.5%.

In small-cell lung cancer, two PRs were seen out of 39 patients, for a response rate of 2.8% (Table V). Chlorozotocin activity in advanced lung cancer is inferior to the currently used nitrosoureas (Comis, 1976).

E. Colon

Table VI shows the results of the five reported studies in advanced adenocarcinoma of the colon. The Georgetown study, using a regimen of 120 mg/m² every 6 weeks, treated 55 evaluable patients; 45 had an ECOG performance status (PS) of 0 or 1, and 11 had a PS of 2 or 3. Six partial responses were observed, for an overall response rate of 11%; 4 of these were among the 29 patients with no prior therapy, and 2 were among the 26 previously treated. The median duration of response was 18 weeks (Hoth *et al.*, 1979).

In a broad phase II study by Wayne State, using 90 mg/m² every 6 weeks for poor risk patients and 120 mg/m² every 6 weeks for good risk patients, there were no responses among 46 evaluable patients (Talley *et al.*, 1979).

The SWOG employed the highest dose, 200 mg/m² every 6 weeks. They treated 83 patients, including 13 with no prior therapy, and observed one PR in

TABLE IV. Non–Small-Cell Lung Cancer

Histology	Institution	Regimen	No. patients evaluated	Response CR	Response PR	Reference
Adenocarcinoma	MSKI	150–200 mg/m² over 5 d q 6 wk	18	0	1	Gralla and Yagoda (1979b)
	WSU	90–180 mg/m² q 6 wk	8	0	0	Talley et al. (1979)
	SWOG	100–225 mg/m² q 6 wk	20	0	1	SWOG Minutes (1980)
	MAYO	125 mg/m²	28	0	0	Creagan et al. (1979)
		q 5 wk	8[a]	0	1	
	CALGB	120 mg/m² q 6 wk	27	0	1	Cornell et al. (1981)
Anaplastic	CALGB	120 mg/m² q 6 wk	4	0	0	Cornell et al. (1981)
	MSKI	175–200 mg/m² q 6 wk	1	0	0	Casper et al. (1979)
	SWOG	100–225 mg/m² q 6 wk	13	0	1	SWOG Minutes (1980)
	MAYO	125 mg/m²	3	0	0	Creagan et al. (1979)
Epidermoid	MSKI	150–200 mg/m² over 5 d q 6 wk	11	0	1	Casper et al. (1979)
	SWOG	100–225 mg/m² q 6 wk	28	0	0	SWOG Minutes (1980)
	CALGB	120 mg/m² q 6 wk	27	0	0	Cornell et al. (1981)

[a] No prior therapy.

TABLE V. Small-Cell Lung Cancer

Institution	Regimen	No. patients evaluated	Response CR	Response PR	Reference
MSKI	130–200 mg/m² q 6 wk	8	0	1	New Drug Liaison Minutes (1978)
WSU	90–120 mg/m² q 6 wk	11	0	1	Talley et al. (1979)
SWOG	100–225 mg/m² q 6 wk	1	0	0	SWOG Minutes (1979)
CALGB	120 mg/m² q 6 wk	2	0	0	Cornell et al. (1981)
EST	120 mg/m² q 6 wk vs cisplatin vs Maytansine	17	0	0	ECOG Minutes (1980)

TABLE VI. Colon Cancer

Institution	Regimen	No. patients evaluated	Response CR	Response PR	Reference
Georgetown	120 mg/m²	29[a]	0	4	Hoth *et al.* (1980)
	q 6 wk	26	0	2	
WSU	90–120 mg/m²	16[a]	0	0	Talley *et al.* (1979)
	q 6 wk	30	0	0	
SWOG	200 mg/m² q 6 wk	83	0	1	SWOG Minutes (1980)
CALGB	120 mg/m²	33[a]	0	0	CALGB Minutes (1980)
	q 6 wk	84	0	0	
Mayo and Georgetown	100–120 mg/m²	28[a]	0	1	
	vs	11	0	1	
	175–200 mg/m²	24[a]	1	2	
		15	0	1	

[a] No prior therapy.

the total group (SWOG Minutes, 1980). CALGB noted one PR, and six stable disease in 33 patients with no previous therapy, and four stable disease in 84 patients with prior treatment (CALGB Minutes, 1980).

In a randomized comparison of low (100 to 120 mg/m²) versus high (175 to 200 mg/m²) doses of chlorozotocin, Mayo Clinic and Georgetown demonstrated an overall response rate of 87% (6/78). In 39 patients in the low-dose arm, including 28 with no prior therapy, objective responses were observed in 2 patients (65%). This was not significantly different from the 10% response rate seen in the high-dose group (one CR and two PR) in 39 patients, including 24 with no prior therapy. The median duration of survival was 136 days for both treatment groups (Schutt *et al.*, 1981).

In summary, of 379 patients with advanced colon cancer, there was a total of 14 CRs and PRs, for an overall response rate of 4%. Of 143 patients who were previously untreated, a response rate of 2% was observed. Hence, in comparison to BCNU, CCNU, and methyl CCNU, chlorozotocin has inferior activity in colon cancer and does not merit further trials (Comis, 1976).

F. Pancreas

Based on the known activity of streptozotocin (STZ) in islet-cell carcinoma of the pancreas, chlorozotocin is also being evaluated in this disease (Table VII).

SWOG recently initiated a single agent trial, while ECOG in a phase II–III design is comparing streptozotocin plus 5-FU versus streptozotocin plus adriamycin versus chlorozotocin at a dose of 150 mg/m² every 6 weeks. It is too early for analysis of either study.

In six patients with adenocarcinoma of the pancreas, Wayne State noted no activity (Talley *et al.*, 1979). CALGB treated 17 patients with no prior therapy;

TABLE VII. Pancreas and Gastric Cancer

	Institution	Regimen	No. patients evaluated	Response CR	Response PR	Reference
Pancreas						
Islet cell	SWOG	100–200 mg/m^2 q 6 wk	9	0	1	SWOG Minutes (1980)
	ECOG	150 mg/m^2 q 6 wk vs STZ + 5-FU		too early		
Adenocarcinoma	WSU	90–120 mg/m^2 q 6 wk	6	0	0	Talley *et al.* (1979)
	GITSG	120 mg/m^2 q 6 wk	20	0	1	GITSG Minutes (1980)
	CALGB	120 mg/m^2 q 6 wk	17[a] 9	0 0	0 0	CALGB Minutes (1980)
Stomach	CALGB	120 mg/m^2 q 6 wk	9	0	0	CALGB Minutes (1980)
	WSU	90–120 mg/m^2 q 6 wk	4	0	0	Talley *et al.* (1979)
	SWOG	100–200 mg/mm^2 vs FAM vs FAM-V	8	0	0	SWOG Minutes (1980)

[a] No prior therapy.

in five, disease stabilization occurred, whereas in nine previously treated patients no responses occurred (CALGB Minutes, 1980). The Gastrointestinal Tumor Study Group (GITSG) is currently evaluating high-dose (175 mg/m^2) and low-dose (120 mg/m^2) chlorozotocin in advanced pancreatic cancer, and has noted one PR among 20 patients (GITSG Minutes, 1980).

G. Gastric

Three studies were initiated in advanced gastric cancer (Table VII). The CALGB and WSU studies observed no responses in 9 and 4 patients respectively (CALGB Minutes, 1980; Talley *et al.*, 1979). An ongoing study by SWOG randomizes between chlorozotocin versus 5-FU, adriamycin, mitomycin C (FAM) versus FAM plus vincristine.

H. Breast

Three institutions evaluated chlorozotocin in breast cancer (Table VIII). One PR of 28 patients was seen at Wayne State University at a dose of 90–180 mg/m^2,

TABLE VIII. Breast Cancer

Institution	Regimen	No. patients evaluated	Response		Reference
			CR	PR	
WSU	90–180 mg/m^2 q 6 wk	28	0	1	Talley *et al.* (1979)
MAYO	150 mg/m^2 q 6 wk	17	0	0	New Drug Liaison Minutes (1979)
CALGB	120 mg/m^2 q 6 wk	48	0	0	CALGB Minutes (1980)
		4[a]	0	0	

[a] No prior therapy.

whereas no responses were seen in 20 patients at Mayo Clinic (Talley *et al.*, 1979; New Drug Liaison Minutes, 1979). In an ongoing study by CALGB no responses were seen in 48 patients. In summary, no responses were observed in 97 patients with breast cancer.

I. Miscellaneous

In advanced renal cell carcinoma, independent studies at WSU and MSKI observed no responses among a combined total of 30 patients (Talley *et al.*,

TABLE IX. Other Tumors

	Institution	Dose	No. patients evaluated	Response		Reference
				CR	PR	
Renal cell	WSU	90–120 mg/m^2 q 6 wk	9	0	0	Talley *et al.* (1979)
	MSKI	150–175 mg/m^2 5 d	12[a] 23	0 0	0 0	Gralla and Yagoda (1979)
Myeloma	SEG	120 mg/m^2 q 6 wk + prednisone		too early		
	CALGB	120 mg/m^2 q 6 wk	27	0	1	CALGB Minutes (1980)
Brain	WSU	90–120 mg/m^2 q 6 wk	1	0	0	Talley *et al.* (1979)
	BTSG	200 mg/m^2 q 8 wk TZT vs cisplatin vs BCNU + 5-FU		too early		

[a] No prior therapy.

1979; Gralla and Yagoda, 1979) (Table IX). In multiple myeloma, there are two studies: The CALGB observed one PR of 27 patients (CALGB Minutes, 1980); the SECG is studying the combination of chlorozotocin and prednisone. In malignant gliomas, where these nitrosoureas have shown useful activity, the Brain Tumor Study Group is comparing the activity of chlorozotocin at 200 mg/m² versus triazinate (TZT) versus cis-platinum versus BCNU plus 5-FU.

V. CONCLUSION

Chlorozotocin has been adequately tested in phase II trials in melanoma, lung, colon, breast, pancreas, and renal cell cancer. There is evidence of activity only in melanoma, but the activity is no greater than for the other chloroethylnitrosoureas in this disease. However, since the toxicity at effective dose levels is less than that of other chloroethylnitrosoureas, a few trials of combination therapy have been initiated.

Chlorozotocin was originally developed because of evidence in murine experiments that it is a nonmyelosuppressive chloroethylnitrosourea. However, this was not confirmed in human studies. Chlorozotocin has demonstrated all of the major known toxicities of the well-established nitrosoureas. Nevertheless, the potential for improved therapeutic index, that is, lower toxicity at equally effective dose levels, may have been demonstrated in melanoma. Unfortunately, the response rate in melanoma is low, and it is even less in other diseases in which testing is complete. A gain in therapeutic index is important only if the antitumor effectiveness is significant.

REFERENCES

Ahlgren, J., Smith, F., Kerwin, D., Sikic, B., Weiner, J., and Schein, P. S. (1980). Submitted for publication.

Ahmann, D. L., Frytak, S., Kvols, L. K., Hahn, R., Edmonson, J., Bisel, H., and Creagan, E. (1980). *Cancer Treat. Rep. 64*, 721–723.

Anderson, T., McMenamin, M., and Schein, P. (1975). *Cancer Res. 35*, 761–765.

Anderson, T., Fisher, R., Barlock, A., and Young, R. (1976). *Proc. Amer. Assoc. Cancer Res. 19*, 356.

Baker, J., Lokey, J. L., Price, N. A. Winokur, S., Chang, M., and Harman, W. (1979). *New Eng. J. Med. 12*, 662.

Cancer and Leukemia Group B Minutes, April 1980.

Casper, E., and Gralla, R. (1979). *Cancer Treat. Rep. 63*, 459–450.

Comis, R. L. (1976). *Cancer Treat. Rep. 60*, 165–176.

Cornell, C., Hoth, D., Marty, G., Pajak, F., Coleman, M., and Leone, L. (1981). *Cancer Treat. Rep.* In press.

Costanza, M., Nathanson, L., Schoenfeld, D., Wolter, J., Colsky, J., Regelson, W., Cunningham, T., and Sedransk, N. (1977). *Cancer 40*, 1010–1015.

Creagan, E., Eagan, R., Fleming, T., Frytak, S., Kvols, L., and Ingle, J. (1979). *Cancer Treat. Rep. 63*, 11–12.

Eastern Cooperative Oncology Group Minutes, November 1980.

Gastrointestinal Study Group Minutes, February 1980.

Gralla, R. J., Tan, C., and Young, C. (1979a). *Cancer Treat. Rep. 63*, 17–20.

Gralla, R. J., and Yagoda, A. (1979b). *Cancer Treat. Rep. 63*, 1007–1008.

Helman, L., Louie, A., and Slavik, M. (1976). Clinical Brochure: Chlorozotocin (NSC-178248), August.

Hoth, D. F., Butler, T., Winokur, S., Kales, A., Woolley, P., Schein, P. (1978). *Proc. Amer. Assoc. Clin. Oncol. 18*, 309.

Hoth, D. F., Woolley, P. V., Macdonald, J., Green, D., and Schein, P. (1979). *Clin. Pharm. Ther. 23*, 712–722.

Hoth, D. F., Schein, P. S., Winokur, S., Woolley, P., Robichaud, K., Binder, R., and Smith, F. (1980). *Cancer 46*, 1544–1547.

Kovach, J. S., Moertel, C. G., Schutt, A. J., Frytak, S., O'Connell, M. J., Rubin, J., and Ingle, J. N. (1979). *Cancer 43*, 2189–2196.

Magill, G. B., Sordillo, P., Gralla, R., and Golbey, R. (1980). *Proc. Amer. Assoc. Cancer Res. 21*, 362.

New Drug Liaison Minutes, Bethesda, Maryland, September 1978.

New Drug Liaison Minutes, Bethesda, Maryland, September 1979.

Schein, P. S., Bull, J., Doukas, D., and Hoth, D. (1979). *Cancer Res. 38*, 259–260.

Sklaroff, R., Lacher, M., Lee, B., and Young, C. (1979). *Proc. Amer. Assoc. Clin. Oncol.* 182.

Southeastern Cancer Study Group Minutes, April 1980.

Southwestern Oncology Group Minutes, June 1980.

Talley, R. W., Brownlee, R. W., Baker, L., Oberhauser, N., and Pitts, K. (1979). *Proc. Amer. Soc. Clin. Oncol. 20*, 440.

Taylor, S., Belt, R., Haas, C., Stephens, R., and Hoogstraten, B. (1980). *Cancer 46*, 2365–2369.

Van Amburg, A., Ratkin, G., and Presant, C. (1980). *Proc. Amer. Assoc. Cancer Res. 21*, 353.

Wasserman, T., Slavik, M., and Carter, S. (1974). *Cancer Treat. Rep. 1*, 131–151.

Wheeler, G. P., Bowdon, B. J., and Grimsley, J. A. (1974). *Cancer Res. 34*, 194–200.

Woolley, P., Pavlovsky, S., Schein, P. S., and Rosenoff, S. (1980). *Proc. Amer. Assoc. Cancer Res. 21*, 156.

Chapter 34

CURRENT STATUS OF NITROSOUREAS UNDER DEVELOPMENT IN JAPAN[1]

Makoto Ogawa

I. INTRODUCTION

The clinically available nitrosoureas, BCNU, CCNU, and methyl CCNU have demonstrated considerable antitumor activities against brain tumors, lung cancer, lymphomas, malignant melanoma, gastrointestinal tumors, and other tumors (Carter et al., 1977). The major toxicities of these agents are gastrointestinal and hematologic. Leukopenia and thrombocytopenia occur 4 to 6 weeks after a dose, and these are cumulative. In addition, recent literature (Durant et al., 1979; Harmon et al., 1979; Morton, 1979) suggests that irreversible nephrotoxicity and pulmonary toxicity may occur after reaching certain total doses of drug. These hematologic and chronic cumulative toxicities are unfavorable factors for clinical use. Thus, analogs that have a reduction of the toxicities and extension of the antitumor efficacies have been sought.

Since the majority of investigators in Japan have used Karnofsky's criteria (Karnofsky, 1961) for the assessment of responses, the responses were rearranged to correspond to definitions described in the WHO handbook for reporting results of cancer treatment (WHO, 1979) as follows: 1-C: complete response (CR); 1-B: partial response (PR); 1-A: minor response (MR); 0-C and 0-B: no

[1]This study was partially supported by NCI Contract No. NO1-CM-22054 (United States) and by a grant from the Ministry of Health and Welfare (Japan).

change (NC); and 0-A and 0-0: progressive disease (PD). In hematologic malignancies, the definition of response was based on Kimura's criteria (Kimura, 1965; Kimura *et al.*, 1969).

The results reviewed in this chapter are summarized from published literature in which the definition of responses was clearly documented. Most literature reports summarized results obtained during phase I and II studies.

II. ACNU

A. Preclinical Results

Arakawa and co-workers (Arakawa *et al.*, 1974; Shimizu and Arakawa, 1975) synthesized a water-soluble nitrosourea, 1-(4-amino-2-methylpyrimidine-5-yl)methyl-3-(2-chloroethyl)-3-(2-chloroethyl)-3-nitrosourea (ACNU), in 1974. The structure is shown in Fig. 1. A major mode of action of ACNU was the inhibition of DNA synthesis (Nakamura *et al.*, 1978; Kanamaru *et al.*, 1978).

ACNU showed broad distribution in the whole body and was rapidly excreted by the kidney, but significant retention was observed in tumor tissues, liver, and thymus, whereas relatively low levels were observed in brain and cerebrospinal fluid (Shigehara and Tanaka, 1978). The drug demonstrated wide antitumor activity in a variety of experimental tumors including L1210 mouse leukemia, P388 leukemia, Lewis lung carcinoma, and other tumors (Arakawa *et al.*, 1974; Shimizu and Arakawa, 1975; Ogawa and Fujimoto, 1980). In addition, Arakawa and Inoue (1980) reported the prolongation of survival time of mice bearing spontaneous leukemia in AKR mice.

A phase I study was initiated in 1974. The starting dose of 0.2 mg/kg was based on one-tenth of the maximum tolerated dose in dogs, and six to seven steps of dose escalation were necessary to reach a maximum tolerated dose of 4 mg/kg (Cooperative Study Group of Phase I Study of ACNU, 1976). Thereafter, a phase II study employing doses of either 2 mg/kg weekly or 3 mg/kg in 6- to 8-week intervals was performed (Ogawa and Fujimoto, 1980; Saito, 1980).

B. Clinical Efficacy

A phase II study employing either 3 mg/kg alone or the same dosage in combination with an immunopotentiator picibanil was performed by the Tokyo Cancer Chemotherapy Cooperative Study Group (Ishii and Hattori, 1978). They obtained 3 PRs out of 16 patients with gastric cancer, 1 PR of 20 patients with colorectal cancers, 1 PR of 9 patients with gall bladder cancer or hepatoma, and 1 PR of 13 patients with breast cancer.

Recently Saito (1980) summarized the results of a total of 625 patients entered in phase I and II studies and reported response rates of 21.2% in gastric cancer,

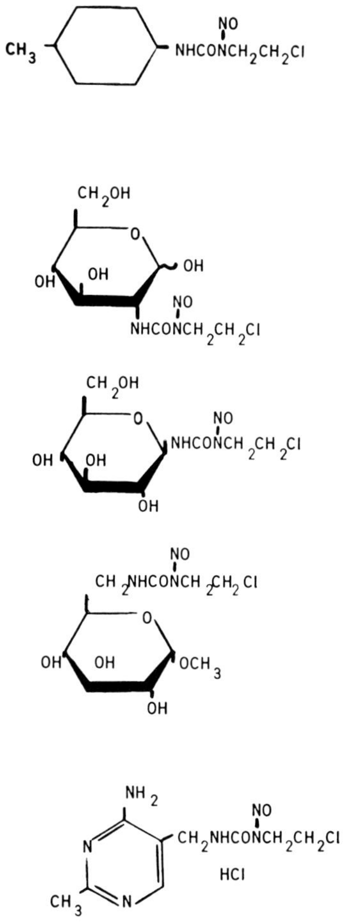

Methyl-CCNU

Chlorozotocin

GANU

MCNU

ACNU

Fig. 1. Structure of methyl CCNU, chlorozotocin, ACNU, GANU, and MCNU.

23.7% in lung cancers, 34.4% in brain tumors, 23.1% in colorectal cancers, and 14.3% in liver cancer.

Saijo and co-workers (1977; 1978) conducted phase II studies in pulmonary metastatic patients with cancers and sarcomas, and reported 2 PRs out of 7 patients with uterine cancer, 2 PRs of 3 patients with head and neck tumors, and 1 CR and PR out of seven patients with sarcomas.

Among 42 patients with brain tumors treated with either ACNU alone or in successive combination with irradiation, there were 7 PRs and 13 PRs defined by CT scan and neurological improvement (Hori *et al.*, 1978; Saito *et al.*, 1978; Kuwana, 1979).

There were several other publications of phase I and II experiences each reporting a few cases of solid tumors (Saito *et al.*, 1977; Kimura *et al.*, 1969;

TABLE I. Antitumor Activity in Phase I and II Studies of ACNU: Solid Tumors

Disease	No. of evaluable patients	CR	PR	MR	SD	PD	CR + PR	(%)
Gastric cancer	63	0	3	4	11	45	3	(4.8)
Brain tumors	42	7	13	10	0	12	20	(47.6)
Colorectal cancer	21	0	1	2	5	13	1	(4.8)
Sarcomas	14	1	0	2	0	11	1	(7.1)
Breast cancer	13	0	1	1	2	10	1	(7.7)
Uterine cancer	12	0	2	1	2	7	2	(16.7)
Gall bladder and hepatoma	11	0	1	2	2	5	1	(9.1)
Ovarian cancer	8	0	0	0	5	3	0	
Pancreatic cancer	8	0	0	0	0	8	0	
Head and neck tumors	6	0	2	0	0	4	2	(33.3)
Renal cell cancer	2	0	0	0	0	2	0	

Ogawa et al., 1978; Kubo et al., 1978; Ishiyama, 1978, Honjo et al., 1979). Summarizing these results, the overall response rate (CR+PR) was 4.8% in gastric cancer, 47.6% in brain tumors, 4.8% in colorectal cancer, 7.1% in sarcomas, 7.7% in breast cancer, 16.7% in uterine–cervical cancer, 9.1% in gall bladder cancer and hepatoma, and 33.3% in head and neck tumors respectively (Table 1). No partial response was reported in patients with ovarian cancer, pancreatic cancer or renal cell carcinoma.

Preliminary experience with ACNU in combination with DTIC (Inoue et al., 1979) suggests that such a combination might be active against malignant melanoma.

Responses were observed in a number of patients with lung cancer (Table II) (Saijo et al., 1977; 1978; Ogawa et al., 1978; Ogawa and Fujimoto, 1980; Kimura et al., 1978; Yokoyama, 1980). There was one PR in epidermoid and large-cell carcinomas. In addition, there was one CR and three PRs out of 13 patients with small cell type. No response was observed in 19 patients with adenocarcinoma and in 22 patients with unclassified histologic type.

The results observed in hematologic malignancies are summarized in Table III

TABLE II. Antitumor Activity in Phase I and II Studies of ACNU: Lung Cancer

Histology	No. of evaluable patients	CR	PR	MR	SD	PD	CR + PR	(%)
Epidermoid	11	0	1	0	0	10	1	(9.1)
Adenocarcinoma	19	0	0	1	1	17	0	
Large cell	3	0	1	0	1	1	1	(33.3)
Small cell	13	3	3	1	1	5	6	(46.2)
Unclassified	22	0	0	4	0	18	0	
Total	68	3 (4.4%)	5 (7.3%)	6 (8.8%)	3 (4.4%)	51 (75%)		

TABLE III. Antitumor Activity in Phase I and II Studies of ACNU: Hematologic Tumors

Disease	No. of evaluable patients	CR	PR	NR	CR + PR	(%)
Non-Hodgkin lymphomas	18	3	2	13	5	(27.8)
Hodgkin's disease	14	2	4	8	6	(42.9)
Acute leukemias	6	0	0	6	0	
CML	10	5	1	4	6	(60.0)
CLL	1	0	0	1	0	

(Cooperative Study, 1976; Kimura *et al.*, 1978; Majima *et al.*, 1978; Ogawa *et al.*, 1978; Saito *et al.*, 1977; Takubo *et al.*, 1978). There were 3 CRs and two PRs out of 18 patients with non-Hodgkin's lymphomas, and 2 CRs and 4 PRs out of 14 patients with Hodgkin's disease. No remissions were observed in acute leukemias. Takubo and co-workers (1978) reported 5 CRs and one PR in ten patients with chronic myelogenous leukemia (CML); furthermore, they observed hematologic improvements in four of five patients with polycythemia vera and in both patients with essential thrombocytopenia.

C. Toxicity

Saito (1980) reported toxicities seen in a total of 526 patients with various solid tumors. Anorexia and vomiting were observed in approximately 15% of patients, and nausea occured in 6.1%. These gastrointestinal toxicities were mild or moderate and well tolerated. Transient and reversible abnormalities of liver function were also observed in a few patients, and renal dysfunction was seen in only one patient. No venous pain or local irritation during intravenous infusion was reported.

The dose limiting factor was hematologic toxicity. When a dose of 3 mg/kg was injected, platelet counts of less that $7 \times 10^4/mm^3$ and white blood cell counts of less than $4000/mm^3$ occurred in about half of the patients (Saijo *et al.*, 1977; 1978; Saito *et al.*, 1977). Thrombocytopenia reached a nadir 3 to 4 weeks after a dose, and it occurred 1 week earlier than leukopenia. Both thrombocytopenia and leukopenia required 2 to 3 weeks for recovery. Furthermore, these hematologic toxicities appeared to be cumulative with repetitive administrations.

D. Discussion

ACNU, a water-soluble nitrosourea that crosses the blood–brain barrier, demonstrated considerable clinical efficacy in patients with brain tumors, small-cell carcinoma in lung, uterine–cervical cancer, malignant lymphomas, and chronic myelogenous leukemia. In addition, the drug appeared to be active against head and neck tumors, sarcoma, and gastrointestinal tumors.

The major dose limiting factor was hematologic toxicity, but it was possible to repeat administrations of 3 mg/kg in every 6 to 8 weeks. Gastrointestinal toxicity was usually milder than that seen in BCNU, CCNU, or methyl CCNU; furthermore, vascular pain occasionally seen by injection of BCNU was not seen in ACNU.

III. GANU

A. Preclinical Studies

A water soluble nitrosourea, 2-(2-chloroethyl)-3-(β-D-glucopyranosyl)-1-nitrosourea (NSC-D254157, GANU), in which the structure resembles that of chlorozotocin, was synthesized by Machinami, Suami, and co-workers (Machinami et al., 1975) (Fig. 1). The major mode of action is the inhibition of DNA synthesis (Yoshida et al., 1976). In experimental tumors, GANU was active against various tumors including L1210 mouse leukemia, sarcoma 180 and others (Hisamatsu and Uchida, 1977). Aoshima and Sakurai (1977) reported that the antitumor activity of GANU was comparable to chlorozotocin but that it was more myelosuppressive than chlorozotocin and less toxic than BCNU. Similar results were reported by Fox and co-workers (1977).

In our comparative study of the antitumor activity of various treatment schedules in L1210 mouse leukemia, GANU showed efficacy similar to that of ACNU, methyl CCNU, chlorozotocin, and MCNU when employed on day 1 only. When the treatment was initiated on day 4, the antitumor efficacies of GANU and chlorozotocin were inferior to those of the other three nitrosoureas; in addition, the results were similar when treatments started on day 5 or 6 (Ogawa and Fujimoto, 1980).

The pharmacokinetics of GANU using radiolabeled compounds showed high concentrations in kidney, liver, urinary bladder, small intestine, bone marrow, and spleen and very low concentrations in brain, suggesting that GANU does not cross the blood–brain barrier (Kanko et al., 1980).

B. Phase I Study Results

The phase I study of GANU was initiated in October 1978, involving 12 major institutions (Ogawa, 1980, Kanko et al., 1980). A total of 67 patients with mostly advanced solid tumors entered the study, and all were included in the analysis. The majority of patients had prior conventional treatments and were judged as being refractory.

Gastrointestinal toxicities such as anorexia, nausea, or vomiting were observed in a few patients in each dosage; however, these were usually mild and well tolerated (Table IV).

Mild elevation of SGOT or SGPT occurred in a few patients who received dosages exceeding 50 mg/m². Furthermore, two patients had mild elevation of BUN and serum creatinine but these were suspected to be related to progressive diseases. Mild diarrhea, fever, and alopecia were seen in one patient each.

Hematologic toxicity was dose dependent. At doses below 50 mg/m² leukopenia was not observed, whereas at doses of 70 mg/m² or more, significant leukopenia occurred (Table V). Median days to the nadir of leukopenia ranged from 34 to 40 days, and 5 to 24 days were required for recovery. Thrombocytopenia occurred less frequently than leukopenia. At doses below 70 mg/m², thrombocytopenia was not observed, but when dosages exceeding 120 mg/m² were administered, thrombocytopenia was severe. Median days to nadir ranged from 9 to 45 days and 2 to 3 weeks were necessary for recovery.

C. Discussion

Gastrointestinal toxicities of GANU were mild and not dose dependent. Hematologic toxicities were dose dependent, and the patterns of leukopenia and thrombocytopenia were similar to those seen in the preceding nitrosoureas, but the recovery pattern appears to be more rapid.

There were several episodes of transient reduction in size of gastric, pancreatic, colorectal, lung, and breast tumors. Metastatic lymph nodes in a patient with gastric cancer were reduced in size more than 50%.

Overall, the results indicated that a maximum tolerated dose of GANU was 160 mg/m², and a recommended dose for phase II study was determined as 120 mg/m² in 6- to 8-week intervals.

TABLE IV. Clinical Toxicity of GANU

Dosage (mg/m²)	Evaluable patients	Nausea	Vomiting	Hepatic		Renal	
				GOT	GPT	BUN	Creatinine
10	3	1	0	0	0	0	0
20	4	1	0	0	0	0	0
33	6	0	0	0	0	0	0
50	13	0	0	2	3	1	1
70	14	1	2	1	1	0	0
90	11	3	1	3	2	1	1
120	5	0	0	1	1	0	0
160	11	1	1	1	1	0	0
Total	67	7	4	8	8	2	2

TABLE V. Hematologic Toxicity of GANU

Dosage (mg/m^2)	Evaluable patients	No. patients with leukopenia ($< 4 \times 10^3$/mm^3)	Leukocyte count [mean (10^3/mm^3)] Pretreatment	Nadir	Time to nadir (mean days)	Time to recovery (mean days)
50	11	2	5.5	3.9	34	5
70	11	2	4.2	2.4	39	—
90	9	3	6.5	3.3	40	7
120	3	2	6.2	2.2	38	24
160	4	3	9.1	2.5	34	14

Dosage (mg/m^2)	Evaluable patients	No. patients with thrombocytopenia ($< 10 \times 10^4$)	Platelet count [mean (10^4/mm^3)] Pretreatment	Nadir	Time to nadir (mean days)	Time to recovery (mean days)
70	10	1	23	7	37	12
90	9	1	19	6.3	9	14
120	3	1	22	3	45	21
160	6	2	39	4.3	30	16

IV. MCNU

A. Preclinical Data

A water soluble nitrosourea, 1-(2-chloroethyl)-3-(methyl-α-D-glucopyra nose-6-yl)-1-nitrosourea (MCNU), was synthesized by Research Laboratory of Tokyo Tanabe Co., Ltd (Sekido et al., 1979). The structure resembles chloro-zotocin and GANU (Fig. 1). MCNU inhibited DNA synthesis of L1210 leu-kemic cells in vitro and did not inhibit RNA synthesis.

The antitumor activity was observed in a variety of tumors including L1210, P388, B16 melanoma, Lewis lung carcinoma, and others. Sekido and co-workers (1979) reported that the antitumor activity of MCNU was superior to that of chlorozotocin and CCNU and similar to that of methyl CCNU.

In our experimental studies (Ogawa and Fujimoto, 1980; Ogawa, 1980) MCNU had activity superior to GANU and chlorozotocin and comparable to ACNU and methyl CCNU. However, when leukopenia was compared, MCNU was more toxic than chlorozotocin or GANU and similar to ACNU or methyl CCNU.

B. Phase I Study Results

A starting dose of 10 mg/m^2 in a single injection was based on one-tenth of LD$_{10}$ in mice. The modified Fibonnaci's method was employed for dose escala-

TABLE VI. Hematologic Toxicity of MCNU: Leukocytes

Dosage (mg/m²)	Evaluable patients	No. patients with leukopenia (<4 × 10³/mm³)	Leukocyte count [mean (10³/mm³)] Pretreatment	Nadir	Time to nadir (mean weeks)	Time to recovery (mean weeks)
20	3	1	5.0	3.1	5	1
33	2	0				
50	5	1	5.8	2.9	2	1
70	5	1	4.0	3.0	1	1
90	5	4	6.7	2.9	4	1–2
120	7	6	7.9	3.2	6	2
160	2	2	3.7	1.2	5–6	1–2
200	1	1	6.9	1.2	5	–

tions. A total of 52 patients entered phase I study, and all were evaluable. The majority of patients with gastrointestinal tumors and non–oat-cell cancers of the lung who had conditions refractory to conventional chemotherapies entered into the study; 11 patients with chronic myelogenous leukemia were also registered (Ogawa, 1980).

Mild gastrointestinal toxicities such as anorexia, nausea, and vomiting occurred in a few patients in each dosage, and it appears to be not related to dose escalations. No hepatic or renal dysfunction has been reported. Mild and transient decreases of leukocytes occurred in one patient from treatment with doses between 20 mg/m² and 70 mg/m² (Table VI).

When a dosage of 90 mg/m² was administered, approximately 50% reduction of leukocytes occurred in four out of five patients, and the time to the nadir was 4 weeks later. This delayed leukopenia was more evident in dosages exceeding 120 mg/m², and 2 weeks were needed for recovery. Similar results were obtained in

TABLE VII. Hematologic Toxicity of MCNU: Platelets

Dosage (mg/m²)	Evaluable patients	No. patients with thrombocytopenia (<10 × 10⁴/mm³)	Platelet count [mean (10⁴/mm³)] Pretreatment	Nadir	Time to nadir (mean weeks)	Time to recovery (mean weeks)
10	7	3	30	7.9	3–4	1
20	2	0	–	–	–	–
33	2	1	11	2.2	2	–
50	4	0	–	–	–	–
70	7	1	18.3	7.7	5	1
90	5	2	11.4	3.4	3.5 (2–5)	1
120	6	6	23.1	5.2	5 (3–6)	2
160	2	2	17.7	5.0	5 (4–6)	2
200	1	1	21.7	1.4	3	3

thrombocytopenia from 90 mg/m^2, and time to nadir was 1 week earlier. Time needed for recovery was also similar (Table VII).

C. Discussion

The phase I study of MCNU has not been completed yet, but it may be compared to those of GANU and ACNU. MCNU seems to be more myelosuppressive. According to our experience, a maximum tolerated dose is 160 mg/m^2, and a recommended dose for Phase II study is between 90 mg/m^2 and 120 mg/m^2. At the dosages of 120 and 160 mg/m^2 leukopenia was more profound than thrombocytopenia.

REFERENCES

Aoshima, M., and Sakurai, Y. (1977). *Gann 68*, 247–250.
Arakawa, M., and Inoue, T. (1980). *Gan to Kagakuryoho 7*, 145–147.
Arakawa, M., Shimizu, F., and Okada, N. (1974). *Gann 65*, 191.
Carter, S. K., Bakowsky, M. T., and Hellman, K. (1977). In "Cancer Chemotherapy," pp. 70–72. Wiley, New York.
Cooperative Study Group of Phase I Study on ACNU (1976). *Jpn. J. Clin. Oncol. 6*, 55–62.
Durant, J. R., Norgard, M. J., Murad, T. M., Bartolucci, A. L., and Langford, K. H. (1979). *Ann. Int. Med. 90*, 191–194.
Fox, P. A., Panasci, L. C., and Schein, P. S. (1977). *Cancer Res. 37*, 783–787.
Harmon, W. H., Cohen, H. J., Schneeberger, E. E., and Grupe, W. E. (1979). *New Eng. J. Med. 300*, 1200–1203.
Hisamatsu, T., and Uchida, T. (1977). *Gann 68*, 819–824.
Hori, M., Nakagawa, H., Hasegawa, H., Mogami, H., Hayakawa, T., and Nakata, Y. (1978). *Cancer Chemother. 5*, 773–778.
Inoue, K., Inagaki, J., Horikoshi, N., Nagura, E., Ueoka, H., Murosaki, S., Kobayashi, T., Fujimoto, S., and Ogawa, M. (1979). *Cancer Chemother. 6*, 81–87.
Ishii, Y., and Hattori, T. (1978). *Cancer Chemother. 5*, 1195–1204.
Ishiyama, K. (1978). *Rinsho to Kenkyu 55*, 309–312.
Kanko, T., Saito, T., and GANU Phase I Study Group (1980). *Gan to Kagakuryoho. Submitted.*
Kanamaru, R., Asamura, M., Saito, H., Hayashi, Y., Saito, T. and Saito, S. (1978). *Sci. Rep. Res. Inst. Tohoku Univ. 30*, 162–170.
Karnofsky, D. A. (1961). *Clin. Pharmacol. Ther. 2*, 709–712.
Kimura, K. (1965). In "Advances in Chemotherapy of Acute Leukemia under the Japan–U.S. Cooperative Science Program," pp. 21–23. September 27–28, Bethesda, Md.
Kimura, K., Sakai, Y., Konda, C., Kashiwada, N., Kitahara, T., Inagaki, J., Mikuni, M., and Sakano, T. (1969). *Saishin Igaku 24*, 816–824.
Kubo, A., Ishisatari, J., and Tobita, Y. (1978). *Rinsho to Kenkyu 55*, 272–280.
Kuwana, N. (1974). *Rinsho to Kenkyu 56*, 276–278.
Machinami, T., Nishiyama, S., Kikuchi, K., and Suami, T. (1975). *Bull. Chem. Soc. Jap. 48*, 3763–3764.
Majima, H., Oguro, M., and Takagi, T. (1978). *Cancer Chemother. 5*, 355–359.
Morton, D. L. (1979). *Cancer Treat. Rep. 63*, 226–227.
Nakamura, T., Sasada, M., Tashima, M., Yamamoto, K., Uchida, M., Sawada, H., and Uchino, H. (1978). *Cancer Chemother. 5*, 991–1000.

Ogawa, M., and Fujimoto, S. (1980). *In* ''Recent Results of Cancer Research,'' in press.

Ogawa, M. (1980). *Cancer Treat. Rev.* In press.

Saijo, N., Kawase, I., Nishiwaki, Y., Suzuki, A., and Niitani, H. (1977). *Cancer Chemother. 4,* 579-584.

Saijo, N., Nishiwaki, Y., Kawase, I., Kobayashi, T., Suzuki, A., and Niitani, H. (1978). *Cancer Treat. Rep. 62,* 139-141.

Saito, T. (1980). *In* ''Recent Results in Cancer Research'' (S. K. Carter and Y. Sakurai, eds.), Vol. 70, pp. 91-106. Springer-Verlag, Berlin, Heiderberg, New York.

Saito, T., Yokoyama, M., Himori, T., Ujiie, S., Sugawara, N., Sugiyama, Z., and Kitada, K. (1977). *Cancer Chemother. 4,* 991-1004.

Saito, Y., Nakaya, Y., Muraoka, K., and Fujiwara, T. (1978). *Cancer Chemother. 5,* 779-794.

Sekido, S., Ninomiya, K., and Iwasaki, M. (1979). *Cancer Treat. Rep. 63,* 961-970.

Shigehara, E., and Tanaka, M. (1978). *Gann 69,* 709-714.

Shimizu, F., and Arakawa, M. (1975). *Gann 66,* 149-154.

Takubo, T., Masaoka, T., Nakamura, H., Ueda, T., Shibata, H., and Yoshitaka, J. (1978). *Cancer Chemother. 5,* 599-603.

Yokoyama, M. (1980). *Gan to Kagakuryoho 7,* 339-349.

Chapter 35

DESIGN OF CLINICAL TRIALS WITH NITROSOUREA

Stephen K. Carter

I. INTRODUCTION

The clinical evaluation of newer nitrosoureas subsequent to BCNU has been a sequential process in which each new compound has entered study against the background of continued active investigation of the earlier nitrosourea drugs (Wasserman *et al.*, 1975). The complexity of interpretation has been increased by the continued progress in the chemotherapy field since the introduction of BCNU in the early 1960s. The phase II populations have become progressively more advanced and pretreated as effective combinations are used as the primary chemotherapeutic approach. For example, BCNU was initially studied in Hodgkin's disease in the era prior to routine use of MOPP therapy. PCNU, the newest nitrosourea, must now be studied in patients exposed to two different combination regimens that might total eight drugs. It is a fair assumption that a phase II study of BCNU today would result in a lower response rate than originally described by Lessner *et al.* (1968) in their original phase II study.

A new analog must struggle for survival in the face of the expectations and prejudices that the parent structure's clinical history has stimulated. BCNU was hailed as a new exciting structure with clinical activity (Carter *et al.*, 1972). The next two compounds to enter clinical trials, CCNU and methyl CCNU, were also met with great enthusiasm. This was due to the newness of the drugs and excitement about nitrosoureas in general and the experimental data indicating a potential superior therapeutic index. After CCNU and methyl CCNU failed to improve dramatically the therapeutic index of nitrosoureas in comparison to BCNU, a sense of disappointment began to set in. This new pessimism was a result of several factors. One factor was a recognition that the rodent transplantable tumor systems were not going to be good analog predictors for this class of drug. A second factor was intrinsic limitations built into a compound with delayed and cumulative myelosuppression. A third factor was the lengthening clinical history of the nitrosoureas; the glamour was now attached to new drug classes such as anthracyclines and bleomycins.

II. STRUCTURE–ACTIVITY/TOXICITY RELATIONSHIPS

The newer generations of nitrosourea analogs have been geared to new experimental rationales and clinical strategies. For a period of time it was hoped that relative alkylation and carbamoylation would offer an approach to rational analog selection. It was hypothesized that alkylation related to clinical efficiency while carbamoylation related to toxicity. A drug that had high alkylating activity and low carbamoylating activity might therefore have a higher therapeutic index. A drug with such characteristics was chlorozotocin (Macdonald *et al.*, 1980). Chlorozotocin was attractive for other reasons as well. It was viewed as an analog of streptozotocin in that both nitrosourea compounds contained a sugar moiety. Streptozotocin, which is methylnitrosourea with a sugar moiety attached, was a drug that did not exhibit myelosuppression. Streptozotocin had clinical activity in pancreatic islet cell tumors and lymphomas. It was limited in its use by renal toxicity and severe nausea and vomiting. Chlorozotocin can be viewed as streptozotocin with a chloroethyl group in place of the methyl group on one end. It is significantly more effective than streptozotocin in L1210 leukemia and other screening models.

The final exciting stimulus for chlorozotocin was experimental data in mice indicating a low potential for myelosuppression. As a result of these studies, chlorozotocin was able to reawaken excitement about the dormant potential for nitrosoureas. Unfortunately, as is so often true, the experimental date was not confirmed by clinical experience. Chlorozotocin is also only weakly active and appears to be distinctly less active than the earlier more established nitrosoureas. The hope that sugar-linked nitrosoureas would not be myelosuppressive has been dashed as well by the experience in Japan with GANU and in France with RFCNU.

III. CLINICAL TRIALS: PAST, PRESENT, AND FUTURE

The six nitrosourea compounds tested in the United States fall into two broad structural categories. The most tested category is the chloroethylnitrosoureas

TABLE I. Comparative Toxicity Spectra of Nitrosoureas

Drug	Organ-specific toxicity			
	Delayed marrow	Renal	Liver	Gastrointestinal
BCNU	Yes	No	No	Yes
CCNU	Yes	No	No	Yes
Methyl CCNU	Yes	No	No	Yes
Streptozotocin	No	Yes	No	Yes
Chlorozotocin	Yes	No	No	Yes
PCNU	Yes	No	No	Yes
ACNU	Yes	No	No	Yes
MNU	Yes	No	No	Yes

without a sugar. Representative of this class are BCNU, CCNU, methyl CCNU, and PCNU. The first three drugs share a common spectrum of antitumor activity and toxicity. There is little of significance in terms of quantitation of antitumor effect or side effects to choose among them.

The second category comprises the sugar-linked nitrosoureas as seen first with streptozotocin and recently with chlorozotocin, which also contains a chloroethyl moiety. Streptozotocin has both an antitumor spectrum and a toxicity spectrum different from the chloroethylnitrosoureas.

The toxicity spectra of the six nitrosoureas from the United States, plus ACNU from Japan and MNU from the USSR, are compared in Table I. Delayed marrow toxicity is seen with all except streptozotocin, whereas only this drug has clinically significant renal toxicity. All show gastrointestinal side effects, and none show significant hepatic toxicity.

The antitumor spectra of the same seven nitrosoureas are compared in Table II. The major roles for this class of compounds include brain tumors and the lymphomas. The activity in gastrointestinal cancer is minimal alone, but combination studies have been extensive.

The testing of a new nitrosourea clinically must take place against the exten-

TABLE II. Comparative Antitumor Spectra of Nitrosoureas

Drug	Brain	Colon	Stomach	Hodgkin's	Lung	Melanoma	Islet cell
BCNU	+	+	±	+	±	+	?
CCNU	+	±	±	+	+	±	?
Methyl-CCNU	±	+	+	+	±	+	?
Streptozotocin	−	−	?	+	−	?	+
Chlorozotocin	?	−	?	?	−	−	?
PCNU	?	?	?	?	?	?	?
MNU	?	?	±	?	+	+	?
ACNU	?	?	+	?	?	?	?

sive background of data that exists for the numerous analogs that have already been placed into clinical trial (Carter, 1980). The existing nitrosoureas have valid toxicologic patterns, although several are quite similar in that regard. In terms of acute toxicity, the most critical problem is delayed hematologic toxicity that becomes cumulative with repeated doses. A nonmyelosuppressive but active nitrosourea would be of great interest particularly because of the potential for more efficacious combination usage. Even a nitrosourea that could be given at full therapeutic doses every 3 weeks would offer significant potential for improvement in the therapeutic index. Unfortunately, something always is dose-limiting, since the magic bullet still remains elusive. With streptozotocin, which is a nonmyelosuppressive nitrosourea, the trade-off has been renal toxicity and severe nausea and vomiting. Significant renal toxicity is a poor trade-off, especially since it limits the combination potential with drugs that are excreted by the kidneys. Large-animal toxicologic studies on other nitrosoureas have given evidence for potential hepatic and pulmonary toxicities with these drugs, which also would not be particularly attractive.

IV. FUTURE CLINICAL TRIALS

What in essence will be required is a nonmyelosuppressive analog with dose-limiting toxicity that is not cumulative and does not preclude repeated dosing. This may be a tall order to fill; and a more reasonable approach might be to aim for a less myelosuppressive analog that will not have other organ-related toxicities and so would lead to an improved therapeutic index. The nitrosoureas still lend themselves to relatively easy phase II study (Carter, 1978; Carter et al., 1972). This is because a low level of nitrosourea activity has been reported for tumors such as large bowel cancer, non–oat-cell lung cancer, malignant melanoma, and renal cell carcinoma. In each of these malignancies it is feasible to contemplate a phase II study of a new nitrosourea in patients previously untreated by drugs. In all of these tumors, drugs such as BCNU, CCNU, and methyl CCNU give objective response rates in the range of 10–20%. There is, therefore, no ethical problem in using a new nitrosourea in place of these older drugs in a phase II approach. The hypothesis that the older nitrosoureas are only weakly active at best in these tumors is supported by the dismal results with nitrosourea-containing combinations. Combinations with 5-FU in large bowel cancer, with dacarbazine in melanoma, with alkylating agents in non–oat-cell lung cancer, and with vinblastine in renal cell cancer have all failed to demonstrate a survival advantage in comparison with single-agent results.

In these tumors an appropriate phase II strategy would be a classic single-arm study of 20 to 30 previously untreated patients. A positive endpoint would be an objective response rate that was at least 30 to 40%. Anything less than this would appear to offer little potential for being effective in combination with other drugs based on prior clinical experience. The only exception to this might be with a drug that had only mild myelosuppression with a 20 to 30% activity range. A reasonable strategic concept could be to demand phase II evidence that a new nitrosourea would be at least twice as effective as earlier "active" nitrosoureas in these tumor types.

The nitrosoureas have definitely established activity in Hodgkin's disease, but their role in the non-Hodgkin's lymphoma is less clearly established. Hodgkin's disease is one of the few cancers where prospective randomized studies comparing some of the nitrosoureas exist. Cancer and Leukemia Group B (CALGB) compared BCNU and CCNU in previously treated patients, and CCNU was shown to be superior in terms of objective response rate. A subsequent study compared CCNU and methyl CCNU in a similar situation, and again CCNU was superior. Based on these studies, CCNU has replaced nitrogen mustard in the primary combination, MOPP. This regimen of CCNU, vinblastine, procarbazine, and prednisone is now the first-line combination of CALGB. Other groups use BCNU in first-line combinations, most particularly BCVPP, in which cytoxan, vinblastine, procarbazine, and prednisone are added. Some groups do not use nitrosoureas at all in either first- or second-line combinations and reserve it for palliative third-line therapy. The major limitation for the nitrosoureas in Hodgkin's disease is delayed and cumulative marrow toxicity. A new analog without marrow toxicity or with marrow damage that is not delayed and cumulative would be welcome. A reasonable place to test a new nitrosourea would be after failure on MOPP and ABVD as a third-line therapy. The strategy will be more difficult if a nitrosourea is utilized as part of a primary combination regimen. A reasonable end point for phase II study of new analogs in Hodgkin's disease would be a 50% response rate with a toxicity spectrum that might give promise for easy inclusion into a new combination regimen.

The malignant gliomas would pose the problem that BCNU has become established as the drug of choice to be used concomitantly with and after postoperative radiation therapy. While patients who develop evidence of progressive disease after such therapy are candidates for phase II studies, this could be with a new nitrosourea unless lack of cross-resistance was being evaluated. For patients who have not been exposed to prior nitrosoureas, a response rate in excess of 50% would be required to make a large-scale phase III combination seem of value.

Oat-cell lung cancer would pose a strategic problem. The nitrosoureas, especially CCNU, are included in many of the highly effective combinations such as CCNU, cytoxan, and methotrexate or the POCC regimen of the Northern California Oncology Group. Still many investigators do use combinations including adriamycin, cytoxan, and vincristine without nitrosoureas, and so phase II study in previously treated patients would be feasible. The therapeutic results with combination chemotherapy and combined modality treatment are so good that previously untreated patients are no longer available for phase II testing.

V. CONCLUSION

The nitrosoureas are an important class of active agents. Analog development and clinical testing have been so extensive that the clinical literature is difficult to interpret. It is critically important that new analogs have a rationale clinical testing strategy.

REFERENCES

Carter, S. K., Schabel, F. M. Jr., Broder, L. E., and Johnston, T. P. (1972). *Adv. Cancer Res. 16,* 273–332.

Carter, S. K. (1977). *Cancer 40,* 5440–557.

Carter, S. K. (1978). *Cancer Chemother. Pharmacol. 1,* 123–129.

Carter, S. K. (1980). *Cancer Res. 70,* 119–123.

Lessner, H. E. (1968). *Cancer 22,* 451–456.

Macdonald, J. S., Hoth, D., and Schein, P. S. (1980). *Cancer Res. 70,* 83–89.

Wasserman, T. H., Slavik, M., and Carter, S. K. (1975). *Cancer 36,* 1258–1268.